Long-Term Coastal Evolution and Morphodynamics: Ecosystem Protection and Coastal Safety

Guest Editors

Fu Wang
Fengling Yu
Yan Li
Xiaohe Lai

MDPI

Basel • Beijing • Wuhan • Barcelona • Belgrade • Novi Sad • Cluj • Manchester

Guest Editors
Fu Wang
Tianjin Center (North China Center of Geoscience Innovation)
China Geological Survey (CGS)
Tianjin
China

Fengling Yu
College of Ocean and Earth Sciences
Xiamen University
Xiamen
China

Yan Li
School of Ocean Sciences
China University of Geosciences
Beijing
China

Xiaohe Lai
College of Civil Engineering
Fuzhou University
Fuzhou
China

Editorial Office
MDPI AG
Grosspeteranlage 5
4052 Basel, Switzerland

This is a reprint of the Special Issue, published open access by the journal *Water* (ISSN 2073-4441), freely accessible at: https://www.mdpi.com/journal/water/special_issues/34S8661K18.

For citation purposes, cite each article independently as indicated on the article page online and as indicated below:

Lastname, A.A.; Lastname, B.B. Article Title. *Journal Name* **Year**, *Volume Number*, Page Range.

ISBN 978-3-7258-8295-3 (Hbk)
ISBN 978-3-7258-8296-0 (PDF)
https://doi.org/10.3390/books978-3-7258-8296-0

Long-Term Coastal Evolution and Morphodynamics: Ecosystem Protection and Coastal Safety

Contents

About the Editors . . . . **vii**

Fu Wang, Fengling Yu, Yan Li and Xiaohe Lai
New Advances in Long-Term Coastal Evolution and Morphodynamics: Ecosystem Protection and Coastal Safety
Reprinted from: *Water* **2026**, *18*, 1113, https://doi.org/10.3390/w18101113 1

Shujun Dong, Jian Cui, Rui Zhou and Qi Zhao
Variation Characteristics of Hydraulic Circulation in Groundwater Circulation Well Under Natural Hydraulic Gradient Influence and Method to Expand Applicability
Reprinted from: *Water* **2026**, *18*, 164, https://doi.org/10.3390/w18020164 7

Eleni Achmakidou, George P. Petropoulos, Anna Karkani, Niki Evelpidou, Spyridon E. Detsikas and Georgios Nektarios Tselos
A Geoinformation-Based Approach for Mapping Coastal Vulnerability in Sweden
Reprinted from: *Water* **2025**, *17*, 3027, https://doi.org/10.3390/w17203027 22

Liyuan Zhao, Linghao Kong, Xuzhen Zhang, Xiangcai Han, Zhigang Zhao, Baofei Li, et al.
Characteristics and Risk Assessment of Soil Salinization in the Yellow River Delta Region, China
Reprinted from: *Water* **2025**, *17*, 2920, https://doi.org/10.3390/w17202920 40

Xiangcai Han, Linghao Kong, Liyuan Zhao, Zhigang Zhao, Yachao Li, Decheng Zhang, et al.
Characterization of Groundwater Chemistry Under the Influence of Seawater Intrusion in Northern Laizhou, Shandong Province, China
Reprinted from: *Water* **2025**, *17*, 2954, https://doi.org/10.3390/w17202954 55

Jiaquan Sun, Wanbing Song, Xiubo Sun, Jian Cui and Hongwei Ma
Analysis of Soil Salinization Characteristics in Coastal Area of Panjin City, Liaoning Province
Reprinted from: *Water* **2025**, *17*, 2666, https://doi.org/10.3390/w17182666 71

Tao Ma, Haolin Li, Yandi She, Yuanyuan Zhao, Xueke Feng and Feng Zhang
Spatial–Temporal Change and Dominant Factors of Coastline in Zhuhai City from 1987 to 2022
Reprinted from: *Water* **2025**, *17*, 2569, https://doi.org/10.3390/w17172569 83

Yong Li, Xuezheng Liu, Junjie Liu and Guangsen Xiong
Numerical Simulations of Extratropical Storm Surge in the Bohai Bay Based on a Coupled Atmosphere–Ocean–Wave Model
Reprinted from: *Water* **2025**, *17*, 2364, https://doi.org/10.3390/w17162364 101

Dejian Wu, Jian Qin and Hao Chen
Monitoring Regional Terrestrial Water Storage Variations Using GNSS Data
Reprinted from: *Water* **2025**, *17*, 2128, https://doi.org/10.3390/w17142128 120

Alirio Sequeira, Carlos Coelho and Márcia Lima
Morphodynamic Interactions Between Sandbar, Beach Profile, and Dune Under Variable Hydrodynamic and Morphological Conditions
Reprinted from: *Water* **2025**, *17*, 2112, https://doi.org/10.3390/w17142112 134

Cheng Chen, Weijia Yang, Zhihui Wang, Kailong Huangfu, Feng Cai, Haoyan Chen and Youlin Chen
Study on Characteristics of the Water Diversion Ratio and Impact of the Diversion Dyke at South and North Ports of the Minjiang River During Wet Season
Reprinted from: *Water* **2025**, *17*, 1183, https://doi.org/10.3390/w17081183 158

Shanshan Zheng, Cheng Chen and Peng Qiu
Spatial and Temporal Shifts and Driving Mechanisms of Embodied Carbon in Water Transport Trade in BRICS Countries
Reprinted from: *Water* **2025**, *17*, 1070, https://doi.org/10.3390/w17071070 . . . . **174**

About the Editors

Fu Wang

Fu Wang is working at the Tianjin Center, China Geological Survey (North China Geological Science and Technology Innovation Center). He has presided over more than ten national and provincial scientific research projects and participated in over thirty projects as a key member. He has won four provincial and ministerial science and technology awards and the Tianjin May 1st Labor Medal, and he has been involved in ministerial-level innovation teams and won municipal popular science honors. He has published over 100 academic papers, including 50 as the first or corresponding author. He holds editorial positions in many academic journals, acts as a journal reviewer, and serves in multiple domestic academic societies.

Fengling Yu

Fengling Yu works at College of Ocean and Earth Sciences, Xiamen University, China. As a Physical Geographer and Marine Geologist, Dr. Yu' s work focuses on coastal geomorphology, climate and sea-level change, coastal hazards, sedimentary organic carbon processes across the land-ocean continuum, and offshore carbon storage solutions. Her interdisciplinary research bridges geological records with contemporary environmental challenges. Her work underscores a commitment to understanding past and present environmental systems to inform sustainable coastal management and climate resilience strategies.

Yan Li

Yan Li is currently working at the School of Ocean Sciences, China University of Geosciences (Beijing). He is a specialist in luminescence dating, coastal geological process, and Quaternary geology. He has held several scientific projects supported by the National Science Foundation of China, Beijing Provincial Science Foundation, and the National Science and Technology Major Project. He has published more than 20 publications in the last five years.

Xiaohe Lai

Xiaohe Lai works at College of Civil Engineering, Fuzhou University, China. His research interests include estuarine and coastal geomorphological evolution, artificial intelligence, and disaster risk. He has led or participated in more than ten research projects supported by the National Natural Science Foundation of China, Fujian Provincial New Think Tanks, and other funding bodies. He has published more than 30 papers over the past five years. His policy advisory reports have been adopted by multiple governmental and political advisory bodies, five of which received positive written comments from leaders of the Fujian Provincial Party Committee and the Fujian Provincial Government.

water

MDPI

Editorial

New Advances in Long-Term Coastal Evolution and Morphodynamics: Ecosystem Protection and Coastal Safety

Fu Wang [1,2,*], Fengling Yu [3], Yan Li [4] and Xiaohe Lai [5]

1 Tianjin Center (North China Center of Geoscience Innovation), China Geological Survey (CGS), CGS Key Laboratory of Coast Geo-Environment, Tianjin 300170, China
2 Tianjin Key Laboratory of Coast Geological Processes and Environmental Safety, Tianjin 300170, China
3 State Key Laboratory of Marine Environmental Science, College of Ocean and Earth Sciences, Xiamen University, Xiamen 361102, China; fengling.yu@xmu.edu.cn
4 School of Ocean Sciences, China University of Geosciences, Beijing 100083, China; yan.li@cugb.edu.cn
5 College of Civil Engineering, Fuzhou University, Fuzhou 350108, China; laixiaohe@fzu.edu.cn
* Correspondence: wfu@mail.cgs.gov.cn

1. Introduction

Coastal regions are characterized by their flourishing economies and dense human settlements. However, their geological and ecological environments are notably fragile. Under the combined effects of global climate change, accelerating sea-level rise, and declining riverine discharge, coastal zones are facing increasing environmental and societal pressures [1–5]. This Special Issue aims to highlight recent advances in long-term coastal evolution, including shoreline migration associated with transgression and regression, coastal and neritic morpho-dynamics, and the adaptation of tidal flat and wetland ecosystems. The goal of this Special Issue is to elucidate the patterns and driving mechanisms of coastal change across multiple spatial and temporal scales, while promoting the development of nature-based solutions for ecosystem protection and coastal safety. Particular emphasis is placed on the integration of monitoring techniques, multi-source data, and interdisciplinary approaches.

2. Review of New Advances

This Special Issue comprises 11 papers focusing primarily on coastal hydrodynamic and hydrological processes, coastal geomorphological evolution and shoreline changes, soil salinization and seawater intrusion, and coastal monitoring technologies and methods. The major findings are summarized as follows.

2.1. Coastal Hydrodynamic and Hydrological Processes

Coastal hydrodynamic and hydrological processes provided a fundamental basis for understanding the evolution of estuarine and coastal systems. In the study of estuarine flow partitioning, a two-dimensional hydrodynamic model based on MIKE 21 revealed significant tidal-scale fluctuations in the water diversion ratio between the South and North Ports of the lower Minjiang River. The results indicate that the water diversion ratio of the North Port increases with the length of the diversion dike. A 450 m long dike is highly sensitive to angular variations, whereas the existing 110 m long dike shows limited regulatory effectiveness. These findings offer quantitative support for channel regulation and ecological management strategies (Contribution 1). In storm surge simulations, a fully coupled atmosphere–ocean–wave model improved the accuracy of peak water level simulations, and ocean–wave coupling played a particularly critical role in predicting wave heights. Northeast winds were identified as the primary driver of storm surges in the Bohai Bay, and the western coast was severely affected (Contribution 2). In the context of groundwater circulation well technology, the natural hydraulic gradient was found to disrupt the flow field symmetry of traditional dual-screen circulation wells. In contrast, a multi-screen structure effectively expands the applicable hydraulic gradient range (from 0–4.3‰ to 0–9.3‰), thereby significantly enhancing circulation efficiency and remediation performance (Contribution 3).

Three studies also identified relevant research gaps, including the limited flow regulation effectiveness of short diversion dikes, the underestimation of storm surges under extreme wind conditions, and the necessity of further investigating the coupling mechanisms between circulation well characteristics and hydraulic gradients. Future research should focus on developing coupled models that integrate diversion dike hydrodynamics, water diversion ratios, riverbed evolution, and field-based monitoring data. In addition, data assimilation and machine learning techniques should be incorporated to enhance the accuracy and robustness of storm surge simulations. Finally, field validation and long-term performance assessments of multi-screen groundwater circulation wells are needed to support their broader application under variable hydrogeological conditions.

2.2. Coastal Geomorphological Evolution and Shoreline Changes

Coastal geomorphological evolution and shoreline changes result from the interplay between natural processes and human activities, directly influencing coastal protection and land use, and ecosystem management. For example, in the study of sandbar–beach–dune interactions, the results showed that a wider initial berm dissipated wave energy more effectively, in addition to having the capability to significantly mitigate dune erosion. This funding validates the effectiveness of artificial sand nourishment as a nature-based solution, with its performance jointly governed by nourishment location, volume, and frequency, thereby highlighting the importance of integrated morphodynamic monitoring for evaluating intervention outcomes (Contribution 4). A quantitative analysis of shoreline evolution in Zhuhai City from 1987 to 2022 was conducted using Landsat and high-resolution satellite imagery, combined with the Digital Shoreline Analysis System (DSAS), to extract spatiotemporal information on shoreline change. The results showed that the total shoreline length of Zhuhai City increased from 761.50 km in 1987 to 798.91 km in 2022, while the natural shoreline decreased by 89.82 km. Furthermore, a shift from the natural-dominated to human-dominated shoreline morphology led to a decline in the Index of Coastline Type Diversity, whereas the Index of Coastline Utilization Degree increased from 146.42 to 216.37. Overall, the study systematically revealed the dual natural–anthropogenic driving mechanisms of shoreline evolution under rapid urbanization through the integration of multi-source remote sensing data and spatial analysis (Contribution 5).

Despite the insight gained from the sandbar–beach–dune interaction and shoreline evolution analyses, several knowledge gaps remain. For artificial sand nourishment projects, the long-term morphological response and optimal maintenance schedules remain poorly understood, as the assessment is limited to short-term physical experiments. Moreover, while the case study in Zhuhai City clearly illustrates the shift from natural- to human-dominated shoreline dynamics, a systematic framework for quantifying the coupled natural–anthropogenic driving mechanisms across different spatiotemporal scales is lacking. To address these gaps, sustained long-term monitoring of nourishment performance, the development of integrated models that explicitly represent both natural processes and human interventions, and comparative studies across diverse coastal settings to unravel the multi-scale controls on shoreline change, are required. Future work should particularly strengthen the integration of filed observations, remote sensing, and

https://doi.org/10.3390/w18101113

process-based modeling to improve the transferability and predictive capacity of shoreline change assessments.

2.3. *Soil Salinization and Seawater Intrusion*

Soil salinization and seawater intrusion are closely interrelated coastal environmental issues that pose serious threats to agricultural sustainability and ecosystem stability. Research on soil salinization revealed that the predominant salinization types in Panjin are the chloride-sulfate (43.06%) and sulfate-chloride (50.35%) types, respectively. Spatially, the degree of salinization gradually decreases landward from the coast, with a zonal succession pattern of salinization types. Groundwater depth was identified as the dominant driving factor for soil salinization (correlation coefficient 0.399), while the evaporation coefficient and groundwater salinity served as secondary factors. The synergistic interaction among these three factors drives the upward accumulation of salts (Contribution 6). In the Yellow River Delta, high-risk zones are primarily distributed along the coastline, accounting for 32.67% of the total area. The main factors influencing soil salinization, ranked in order of importance, are distance from the sea > TDS > DEM > soil type > groundwater depth > evaporation coefficient, with seawater intrusion being the core driving factor in the regional salinization pattern (Contribution 7).

Research on seawater intrusion along the northern coast of Laizhou City, Shandong Province, revealed that groundwater chemistry transitions from freshwater to saline water in the sequence of $Ca\text{-}HCO_3{\cdot}Cl$ type → $Ca{\cdot}Na\text{-}SO_4{\cdot}Cl$ type → Na-Cl type. Seawater intrusion is the primary cause of elevated groundwater salinity. Additionally, cation exchange and water–rock interactions further influence the hydrochemical characteristics. The multi-indicator integrated approach of HFED, ion ratios, GQI, and isotope analysis, can serve as a reference for seawater intrusion studies in other coastal regions and highlight the value of combining hydrochemical and isotopic evidences in coastal groundwater assessment (Contribution 8).

The aforementioned studies still have limitations, primarily in an inadequate understanding of the dynamic mechanisms governing groundwater–soil–ocean interactions and insufficient predictive capacity regarding salinization processes under the coupling of multiple factors. Future research should focus on developing coupled groundwater–soil–salt transport models, as well as integrating remote sensing with in situ ground monitoring so as to enable dynamic early warning and precise regulation of soil salinization. Long-term monitoring and multi-source data assimilation should be strengthened to improve process understanding and predictive accuracy.

2.4. *Coastal Monitoring Technologies and Methods*

Coastal monitoring technologies and methods are critical to advancing coastal scientific research and management capacity. In terrestrial water storage monitoring, a comparative study using GNSS, GRACE and GLDAS data revealed that GNSS based inversion offers distinct advantages in spatial resolution. This approach effectively delineates water storage gradients from the Cascade Mountains to the basin. The results also enabled accurate discrimination between mountainous and basin areas. By contrast, GRACE data exhibit blurred spatial patterns, while GLDAS data show limited applicability in mountainous regions. GNSS enables continuous, and dynamic monitoring, and its results are highly consistent with the seasonal trends observed in GRACE and GLDAS datasets. However, a 1–2-month phase lag relative to precipitation data exists, reflecting the complexity of hydrological processes. This study confirms the unique advantages of GNSS as a hydrogeodetic tool for regional water storage monitoring and highlights the value of integrating ground-based geodetic observations with satellite- and model-derived datasets,

https://doi.org/10.3390/w18101113

in addition to addressing the research gap caused by the insufficient spatial resolution of GRACE data (Contribution 9).

Monitoring carbon emission from water transport underpins the formulation of targeted emission reduction strategies. By employing a coupled framework that combines the Multi-Regional Input–Output Model (MRIO) with the LMDI decomposition method, this study provides an integrated quantification and assessment of the spatiotemporal patterns and driving mechanisms of embodied carbon in water transport trade among BRICS countries. The results showed that China was the largest net exporter, while India and South Africa were net importers; Brazil and Russia transitioned from net importers to net exporters. Spatially, carbon emissions were relatively high in Southeast Asia and the Northern Hemisphere, with changes in China's emissions significantly affecting other BRICS countries (Contribution 10). This coupled approach offers a more systematic and comprehensive analytical framework than that provided by a single model. Temporally, it enables long-term dynamic monitoring over the period from 1995 to 2018, capturing the contribution magnitudes and stage-specific characteristics of the three driving effects-scale, structure, and intensity. Spatially, it establishes a spatial LMDI model, using the average of the five countries as a benchmark, to facilitate comparative visualization and monitoring of cross-border spatial differences and transmission pathways. From an ecological protection perspective, this study reveals the spatiotemporal patterns and key driving factors of embodied carbon in BRICS water-transport trade. Policy recommendations are proposed, including trade structure optimization, energy reform, green technology adoption, regulatory system development, and regional cooperation, providing a scientific basis for formulating targeted emission reduction strategies. This study also illustrates the value of integrating socioeconomic analysis with environmental assessment in coastal monitoring and management.

Coastal vulnerability assessment relies on effective evaluation frameworks. In this study, a Coastal Vulnerability Index (CVI) framework was developed based on GIS and remote sensing technologies, integrating six physical parameters: coastal geomorphology, coastal slope, shoreline change rate, relative sea-level change, mean tidal range, and significant wave height. The framework was applied to the Swedish coastline. The results indicate that 36.5% of the coastline falls into the high or very high vulnerability categories, mainly concentrated in southern Sweden. In contrast, over 41% of the coastline exhibits low or very low vulnerability, particularly in the Gulf of Bothnia. Among the six parameters, mean tidal range and relative sea-level change were identified as the most influential factors, with relative sea-level change being the primary driver of coastal vulnerability in Sweden. In southern Sweden, tectonic uplift can no longer compensate for global sea-level rise, creating a coastal "threshold zone" in the region and underscoring the importance of integrating geomorphological, hydrodynamic, and geospatial evidences in vulnerability assessment (Contribution 11).

These studies have identified several challenges in current coastal monitoring, including insufficient spatial resolution of GRACE data, limited applicability of GLDAS in mountainous areas, significant deficiencies in the monitoring and analytical capabilities of the MRIO and LMDI core methodologies, as well as inadequate integration of socioeconomic and ecological response factors within the CVI framework. To address these issues, future research should focus on the following aspects: Strengthen the interdisciplinary assimilation and fusion of multi-source remote sensing, ground-based observations, and numerical models, and upgrade the integrated monitoring technology system for coastal zones. Deepen the analysis of the nonlinear characteristics and exploration of driving mechanisms of the structural effect from an interdisciplinary perspective. Moreover, expand the vulnerability assessment indicator system integrating multidisciplinary dimensions, and

https://doi.org/10.3390/w18101113

systematically incorporate the interdisciplinary assessment content of dynamic adaptive capacity and ecosystem response.

3. Conclusions

This paper reviews four research directions in coastal zones covered in this Special Issue: hydrodynamic and hydrological processes, geomorphological evolution and shoreline changes, soil salinization and seawater intrusion, and monitoring technologies and methods. Overall, these studies have achieved significant progresses in areas such as estuarine flow diversion ratio regulation, coupled storm surge modeling, groundwater circulation well technology, dune erosion mitigation, artificial shoreline expansion, soil salinization, seawater intrusion, carbon emissions, the integration of multi-source data (including GNSS, GRACE, and GLDAS), and coastal vulnerability assessment frameworks. Although these studies focus on regional evolution, they are scientifically representative and can serve as important references for related research in other areas. Furthermore, the findings obtained from these studies can provide scientific evidence for the development, utilization, and protection of coastal zones in diverse coastal settings.

The studies collected in this Special Issue point toward several promising directions for future coastal zone research. First, the application of coupled atmosphere–ocean–wave models (Contribution 2) and GNSS-based terrestrial water storage monitoring (Contribution 9) highlights the value of integrating multi-source observational and modeling frameworks. Expanding these approaches to incorporate machine learning and data assimilation techniques could further enhance the predictive skill of storm surge simulations and the spatial resolution of water storage estimates. Second, the quantitative analyses of shoreline evolution (Contribution 5) and soil salinization drivers (Contributions 6 and 7) demonstrate the need for a systematic understanding of coupled natural–anthropogenic processes. Developing integrated models that explicitly represent feedback between human interventions (e.g., dike construction, nourishment, land use change) and natural dynamics (e.g., tidal processes, sediment transport, saltwater intrusion) will be essential for predicting long-term coastal change under diverse management scenarios. Third, the nature-based solutions explored in this Special Issue—such as artificial sand nourishment for dune protection (Contribution 4)—require sustained long-term monitoring and adaptive management strategies to validate their effectiveness across different environmental settings. Moving forward, interdisciplinary collaboration across fields such as geology, geography, marine science, biology, environmental science and engineering, urban and rural planning, and even sociology, along with the deep integration of multi-source, multi-scale big observational data from space, air, ground, sea, and borehole platforms, dynamically coupled multi-field simulations, and technologies including satellite remote sensing and unmanned aerial vehicles, will be key to advancing both fundamental understanding and practical coastal management.

Author Contributions: Conceptualization, F.W.; investigation, F.W., F.Y., Y.L. and X.L.; writing—original draft preparation, F.W.; writing—review and editing, F.W., F.Y., Y.L. and X.L. All authors have read and agreed to the published version of the manuscript.

Funding: This work was supported by the major program of NSFC (No. 42293261), China Geological Survey Program (DD20211301).

Conflicts of Interest: The authors declare no conflicts of interest.

https://doi.org/10.3390/w18101113

List of Contributions

1. Chen, C.; Yang, W.; Wang, Z.; Huangfu, K.; Cai, F.; Chen, H.; Chen, Y. Study on Characteristics of The Water Diversion Ratio and Impact of the Diversion Dyke at South and North Ports of the Minjiang River During Wet Season. *Water* **2025**, *17*, 1183. https://doi.org/10.3390/w17081183.
2. Li, Y.; Liu, X.; Liu, J.; Xiong, G. Numerical Simulations of Extratropical Storm Surge in the Bohai Bay Based on a Coupled Atmosphere–Ocean–Wave Model. *Water* **2025**, *17*, 2364. https://doi.org/10.3390/w17162364.
3. Dong, S.; Cui, J.; Zhou, R.; Zhao, Q. Variation Characteristics of Hydraulic Circulation in Groundwater Circulation Well Under Natural Hydraulic Gradient Influence and Method to Expand Applicability. *Water* **2026**, *18*, 164. https://doi.org/10.3390/w18020164.
4. Sequeira, A.; Coelho, C.; Lima, M. Morphodynamic Interactions Between Sandbar, Beach Profile, and Dune Under Variable Hydrodynamic and Morphological Conditions. *Water* **2025**, *17*, 2112. https://doi.org/10.3390/w17142112.
5. Ma, T.; Li, H.; She, Y.; Zhao, Y.; Feng, X.; Zhang, F. Spatial–Temporal Change and Dominant Factors of Coastline in Zhuhai City from 1987 to 2022. *Water* **2025**, *17*, 2569. https://doi.org/10.3390/w17172569.
6. Sun, J.; Song, W.; Sun, X.; Cui, J.; Ma, H. Analysis of Soil Salinization Characteristics in Coastal Area of Panjin City, Liaoning Province. *Water* **2025**, *17*, 2666. https://doi.org/10.3390/w17182666.
7. Zhao, L.; Kong, L.; Zhang, X.; Han, X.; Zhao, Z.; Li, B.; Wang, Y.; Li, Y.; Geng, B.; Zhao, M.; et al. Characteristics and Risk Assessment of Soil Salinization in the Yellow River Delta Region, China. *Water* **2025**, *17*, 2920. https://doi.org/10.3390/w17202920.
8. Han, X.; Kong, L.; Zhao, L.; Zhao, Z.; Li, Y.; Zhang, D.; Zhang, H.; Zhao, Y.; Shan, K. Characterization of Groundwater Chemistry Under the Influence of Seawater Intrusion in Northern Laizhou, Shandong Province, China. *Water* **2025**, *17*, 2954. https://doi.org/10.3390/w17202954
9. Wu, D.; Qin, J.; Chen, H. Monitoring Regional Terrestrial Water Storage Variations Using GNSS Data. *Water* **2025**, *17*, 2128. https://doi.org/10.3390/w17142128.
10. Zheng, S.; Chen, C.; Qiu, P. Spatial and Temporal Shifts and Driving Mechanisms of Embodied Carbon in Water Transport Trade in BRICS Countries. *Water* **2025**, *17*, 1070. https://doi.org/10.3390/w17071070.
11. Achmakidou, E.; Petropoulos, G.P.; Karkani, A.; Evelpidou, N.; Detsikas, S.E.; Tselos, G.N. A Geoinformation-Based Approach for Mapping Coastal Vulnerability in Sweden. *Water* **2025**, *17*, 3027. https://doi.org/10.3390/w17203027.

References

1. McGranahan, G.; Balk, D.; Anderson, B. The rising tide: Assessing the risks of climate change and human settlements in low elevation coastal zones. *Environ. Urban.* **2007**, *19*, 17–37. [CrossRef]
2. Lin, Y.; Li, C.; Wang, Z.; Zhang, X.; Han, M.; Cai, Y. Modern sea-level rise breaks 4000-year stability in southeastern China. *Nature* **2025**, *646*, 856–864. [CrossRef]
3. Syvitski, J.P.; Vörösmarty, C.J.; Kettner, A.J.; Overeem, I. Sinking deltas due to human activities. *Nat. Geosci.* **2009**, *2*, 681–686. [CrossRef]
4. Nicholls, R.J.; Cazenave, A. Sea-level rise and its impact on coastal zones. *Science* **2010**, *328*, 1517–1520. [CrossRef]
5. Vousdoukas, M.I.; Ranasinghe, R.; Mentaschi, L.; Plomaritis, T.A.; Athanasiou, P.; Luijendijk, A.; Feyen, L. Sandy coastlines under threat of erosion. *Nat. Clim. Change* **2020**, *10*, 260–263. [CrossRef]

https://doi.org/10.3390/w18101113

water

MDPI

Article

Variation Characteristics of Hydraulic Circulation in Groundwater Circulation Well Under Natural Hydraulic Gradient Influence and Method to Expand Applicability

Shujun Dong [2,3,4], Jian Cui [1,*], Rui Zhou [2,3,4,*] and Qi Zhao [1]

1 Shenyang Center of China Geological Survey, Shenyang 110034, China
2 Key Laboratory of Groundwater Resources and Environment of Ministry of Education, Jilin University, Changchun 130021, China; dongsj22@mails.jlu.edu.cn
3 Jilin Provincial Key Laboratory of Water Resources and Water Environment, Jilin University, Changchun 130021, China
4 National and Local Joint Engineering Laboratory for Petrochemical Contaminated Site Control and Remediation Technology, Jilin University, Changchun 130021, China
* Correspondence: spidercj@126.com (J.C.); zhour@jlu.edu.cn (R.Z.)

Abstract

Natural hydraulic gradients in aquifers disrupt the symmetry and closure of circulation systems of groundwater circulation wells (GCW). To address the unsuitability of traditional dual-screen GCWs in establishing hydraulic circulation under high hydraulic gradients, this study proposes enhancing the applicability and remediation efficiency via multi-screen GCWs, and assesses the evolution of hydraulic circulation characteristics as hydraulic gradient changes as well as applicable hydraulic gradient ranges for different screen quantities. Results indicate that the influence of hydraulic gradient on the hydraulic circulation characteristics of GCWs with different screen quantities follows similar trends. Influence radius (R) and circulation efficiency (Pr) exhibit an overall decreasing trend with increasing hydraulic gradient. Offset angle of circulation zone is positively correlated with hydraulic gradient, with downstream offset angle (θ_D) exceeding upstream offset angle (θ_U). Triple-screen GCW extends the applicable hydraulic gradient range (0~9.3‰) beyond dual-screen GCW (0~4.3‰), decreases θ_U and θ_D by 3.42° and 4.12°, respectively, and increases Pr by 27.62%. Therefore, these findings demonstrate that a multi-screen GCW can effectively expand the applicable hydraulic gradient range and enhance the remediation efficiency of GCWs. This research deepens the understanding of circulation characteristics under the influence of hydraulic gradient and is expected to expand the engineering applicability of GCW technology.

Keywords: groundwater circulation well; hydraulic gradient; hydraulic circulation characteristic; multi-screen; applicable range

1. Introduction

As a key technology in the field of in situ remediation of contaminated aquifers, groundwater circulation well (GCW) technology, which has been developed since the mid-20th century, has gradually become an important alternative to the traditional pump-and-treat (P&T) technology due to its unique integrated pumping and injection design [1–3]. GCW employs two screens within a single wellbore, one for pumping and the other for injection. Through the coordinated operation of pumping and injection, a closed hydraulic circulation system is established within the aquifer surrounding the wellbore [4–7]. Contaminated groundwater is extracted through one screen and treated through a treatment unit to remove pollutants, and the treated water is reinjected into the aquifer through the other screen [8,9]. This process drives the groundwater to form a circulation zone. Compared to P&T, the closed hydraulic circulation system of GCW effectively reduces excessive groundwater extraction [1,10], avoiding secondary environmental issues such as land subsidence and aquifer depletion caused by long-term pumping. Currently, GCW technology is applied in remediation projects at multiple sites, and it has demonstrated excellent technical adaptability and remediation potential [11–16].

In the aquifer, the formation of naturally occurring hydraulic gradients (I) is closely related to topographical variations, hydrological recharge or discharge conditions, and geological structural heterogeneity. Hydraulic gradients exhibit distinct characteristics across various types of aquifers. In plain aquifers, hydraulic gradients are generally at a lower magnitude [17]. In mountainous fissure aquifers or edge areas of valley terraces, hydraulic gradients significantly increase due to topographical undulations and medium connectivity [18]. As the fundamental driving force for directed groundwater flow, natural hydraulic gradients directly influence contaminant migration that continuously spreads downstream [19–22]. The three-dimensional closed flow field formed by GCWs can encapsulate the contaminant plume within the circulation zone to minimize downstream migration and dispersion [23–25], and can effectively control secondary pollution. Consequently, GCWs can efficiently intercept and remove pollutants migrating downstream under the influence of natural hydraulic gradients within the hydraulic circulation range. Its hydraulic circulation characteristics determine the operational performance of GCWs.

Most current studies on the hydraulic circulation characteristics of GCWs have been limited to ignoring the influence of environmental groundwater flow on the GCW-driven flow field [26–30]. The studies by Vats et al. [23], Tatti et al. [2], and Zhu et al. [27] demonstrate that when a GCW is unaffected by natural hydraulic gradients, the shape of the circulation zone approximates an axisymmetric ellipsoid, with the circulation flow field symmetrically distributed around the central axis of the GCW. The impact of hydraulic gradients on the hydraulic characteristics of GCWs has been partially investigated under low hydraulic gradients [31–33]. The studies by Philip et al. [31] and Xia et al. [33] indicate that groundwater flow dominated by natural hydraulic gradients alters the morphology of the circulation zone to disrupt the symmetry of the circulation system. Some treated and reinjected water is entrained by the background flow toward the downstream and fails to form an effective circulation remediation zone around the well, resulting in a reduction in the circulation range. If the hydraulic circulation zone is smaller than the contamination plume, the background flow field carries contaminated groundwater rapidly downstream, resulting in a reduction in the amount of contaminants effectively captured by GCW. However, existing studies have not quantitatively assessed the variation patterns of three-dimensional hydraulic circulation characteristics of GCWs under different hydraulic gradients. Furthermore, as the hydraulic gradient continues to increase, a GCW may fail to form an effective hydraulic circulation zone, rendering the GCW technology inapplicable. Therefore, it is essential to clarify the evolution of three-dimensional hydraulic circulation characteristics of GCWs with varying hydraulic gradients and to determine the applicable hydraulic gradient range for GCWs. This will provide precise theoretical support for practical engineering applications and design.

To address the issues of limited hydraulic circulation range and reduced remediation efficiency of GCWs under larger hydraulic gradients, a method is needed to enhance the applicability and remediation efficiency of GCWs by considering the effects of hydraulic gradients. During the development of GCW technology, researchers and engineers world-

https://doi.org/10.3390/w18020164

wide have conducted extensive studies focused on optimizing the technology. Currently, optimization approaches are primarily categorized into three types. The first category involves adjusting hydrodynamic parameters [27,34], such as increasing the pumping and injection flow rates, altering the circulation mode, and extending the pumping and injection period. The second category involves improving the well structure [35,36], such as optimizing the length of the screens and adjusting the spacing between screens. The third category involves coupling auxiliary technologies, such as chemical oxidation [37], thermal remediation [38], surfactants [14], and biotechnology [39,40]. However, these methods have been developed and studied under conditions of no natural hydraulic gradient or low hydraulic gradient, failing to overcome the technical limitations imposed by the dominant interference of hydraulic gradient-driven flows on the GCW circulation system. Consequently, their effectiveness in enhancing GCW performance is somewhat limited. Therefore, a new optimization approach is required. Multi-screen GCW incorporates multiple screens at different depths along the wellbore, with each screen equipped with an independent pumping or injection unit. It can transform the single large circulation unit of traditional dual-screen GCWs into multiple smaller circulation units, thereby improving hydraulic circulation efficiency [2,41]. Simultaneously, the interference caused by hydraulic gradients on the stability of circulation units is mitigated by reducing the spatial scale of individual circulation units. To our knowledge, no studies have yet explored the use of multi-screen GCWs to enhance the applicability and remediation efficiency of GCWs under larger hydraulic gradients.

To address the aforementioned issues, this study employs numerical simulation methods and utilizes particle tracking models to visualize the hydraulic circulation of GCWs. The objectives are to quantify the evolution patterns of three-dimensional hydraulic circulation characteristics of GCWs as hydraulic gradients vary, determine the applicable hydraulic gradient ranges for GCWs with different numbers of screens, and clarify the potential for expanding the applicability and enhancing the remediation efficiency of GCW technology under larger hydraulic gradients by constructing multi-screen GCWs. This research provides theoretical support for guiding the practical engineering application of GCW technology in groundwater pollution remediation.

2. Materials and Methods

2.1. Groundwater Flow Model of GCW

This study employed the GMS 10.7 software to establish a numerical model, and the conceptual model of groundwater flow related to the GCW system is shown in Figure 1a,b. The aquifer was conceptualized as a homogeneous, anisotropic, uniformly thick, and confined steady-state flow model. The mathematical model was established based on Darcy's law and the principle of mass conservation. The partial differential equations, initial conditions, and boundary conditions are as follows.

$$\begin{cases} \frac{\partial}{\partial x}\left(K_x\frac{\partial h}{\partial x}\right)+\frac{\partial}{\partial y}\left(K_y\frac{\partial h}{\partial y}\right)+\frac{\partial}{\partial z}\left(K_z\frac{\partial h}{\partial z}\right)+W=0 \ \ (x,y,z)\in\Omega; t\geq 0 \\ h(x,y,z,t)|_{t=0}=h_0(x,y,z) \ \ (x,y,z)\in\Omega; t\geq 0 \\ h(x,y,z,t)|_{\Gamma_1}=h_1(x,y,z) \ \ (x,y,z)\in\Gamma_1; t\geq 0 \\ \left.\frac{\partial h}{\partial N}\right|_{\Gamma_2}=0 \qquad (x,y,z)\in\Gamma_2; t\geq 0 \end{cases} \tag{1}$$

where K_x, K_y, and K_z are the hydraulic conductivities along the x, y, and z coordinate axes, respectively (m/d); W is the source–sink term (L/d); h is the groundwater head (m); h_0 is the initial head (m); h_1 is the known head (m); t is time (d); Γ_1 is the fixed-head boundary; Γ_2 is the confining boundary; Ω is the range of the simulation area; N is the outward normal direction of Γ_2. Table 1 lists the input values used in the model.

https://doi.org/10.3390/w18020164

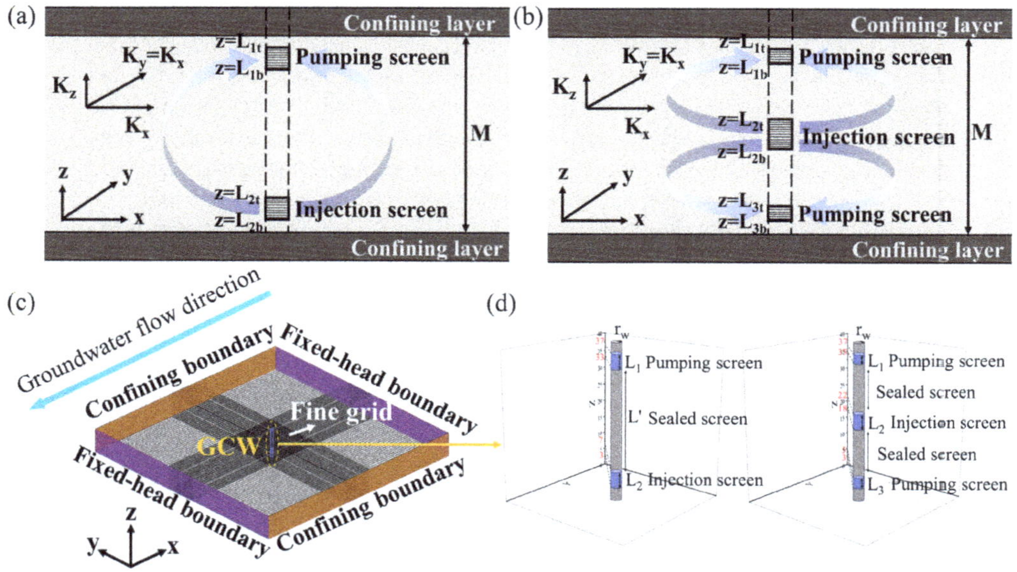

Figure 1. Schematic diagram of the GCW simulation system: (**a**) conceptual model of dual-screen GCW; (**b**) conceptual model of triple-screen GCW; (**c**) grid generation and boundaries; (**d**) source–sink phase.

Table 1. Parameters of the GCW numerical model.

Parameter		Definition	Input Value	Unit
Hydrogeological condition	M	Aquifer thickness	40	m
	K_h	Horizontal hydraulic conductivity (K_x, K_y)	30	m/d
	K_v	Vertical hydraulic conductivity (K_z)	3	m/d
	K_h/K_v	Hydraulic conductivity anisotropy ratio	10	/
	n	Porosity	0.33	/
	μ	Specific yield	0.12	/
	h_0	Initial water head	40	m
Double-screen GCW	r_w	Well radius	1.5	m
	L_{1t}	Vertical coordinate of upper screen top	37	m
	L_{1b}	Vertical coordinate of upper screen bottom	33	m
	L_{2t}	Vertical coordinate of lower screen top	7	m
	L_{2b}	Vertical coordinate of lower screen bottom	3	m
	Q_1	Upper screen pumping flow	−400	m^3/d
	Q_2	Lower screen injection flow	400	m^3/d
Triple-screen GCW	r_w	Well radius	1.5	m
	L_{1t}	Vertical coordinate of upper screen top	37	m
	L_{1b}	Vertical coordinate of upper screen bottom	35	m
	L_{2t}	Vertical coordinate of middle screen top	22	m
	L_{2b}	Vertical coordinate of middle screen bottom	18	m
	L_{3t}	Vertical coordinate of lower screen top	5	m
	L_{3b}	Vertical coordinate of lower screen bottom	3	m
	Q_1	Upper screen pumping flow	−200	m^3/d
	Q_2	Middle screen injection flow	400	m^3/d
	Q_3	Lower screen pumping flow	−200	m^3/d

As shown in Figure 1c, the GCW was positioned at the center of the simulation area. To avoid the interference of the model boundaries on the hydraulic circulation of the GCW, the simulation area was set to 400 m × 400 m, which was significantly larger than the estimated range of hydraulic circulation of GCWs. This area was discretized both horizontally and vertically. In the horizontal direction, the maximum grid size at the model boundaries was 5 m × 5 m. Within a 50 m radius centered on the GCW, the grid was refined to a minimum size of 1 m × 1 m to enhance resolution near the wellbore. The aquifer had a thickness of 40 m, and the model area was vertically divided into 80 layers with a spacing of 0.5 m per layer. The two boundaries parallel to the direction of groundwater flow were confining boundaries. The two boundaries perpendicular to the direction of groundwater flow were fixed-head boundaries, with their head calculated based on the given hydraulic gradient. As shown in Figure 1d, the GCW was conceptualized as screens with identical x and y coordinates but varying z coordinates. For the dual-screen GCW, the total length of the screens and sealed screen was 34 m, with each screen measuring 4 m in length and the sealed screen spanning 26 m. The dual-screen GCW operated in an "upper screen for pumping and lower screen for injection" mode, with pumping and injection flow rates of 400 m^3/d. For the triple-screen GCW, the total length of the screens and sealed screens was 34 m, which was consistent with that of the dual-screen GCW. To ensure uniform flow flux through the screens, the upper and lower screens were each 2 m in length, while the middle screen spanned 4 m. The upper and lower sealed screens were each 13 m in length. The triple-screen GCW operated in a "middle screen injection, upper and lower screen pumping" mode, with pumping flow rates of 200 m^3/d and injection flow rates of 400 m^3/d. The input parameters for the model are provided in Table 1.

2.2. Particle Tracking Model

To visualize the hydraulic circulation characteristics of GCWs, a particle tracking model was constructed using the MODPATH module in GMS software. A total of 100 particles (5 rows × 5 columns × 4 faces) were released in each unit of the injection screens. The migration process of GCW-driven flow was described by the trajectories of these particles.

2.3. Characterization Indexes for the Three-Dimensional Hydraulic Circulation of GCWs

2.3.1. Influence Radius

The characteristics of the circulation zone have always been a focus of GCW research, particularly regarding the study of the influence radius (R), because the successful application of GCW technology requires understanding and predicting its influence area before installation. Currently, in field applications, R is primarily estimated by injecting dyes into the GCW and observing the dye traces in monitoring wells [42]. In laboratory studies, tracers are typically injected to observe their distribution range in the sand tank [35]. In this study, the particle tracking method is employed, and R is determined by measuring the maximum migration distance of particles that constitute the circulation zone. As shown in Figure 2, due to the influence of the natural flow field, the distortion of the circulation zone results in variations in R across different directions. Perpendicular to the hydraulic gradient, the horizontal influence radii (R_T) on both sides of GCW are the same. Parallel to the hydraulic gradient, the longitudinal influence radii (R_L) on both sides of GCW differ and are categorized into upstream longitudinal influence radius (R_{LU}) and downstream longitudinal influence radius (R_{LD}).

https://doi.org/10.3390/w18020164

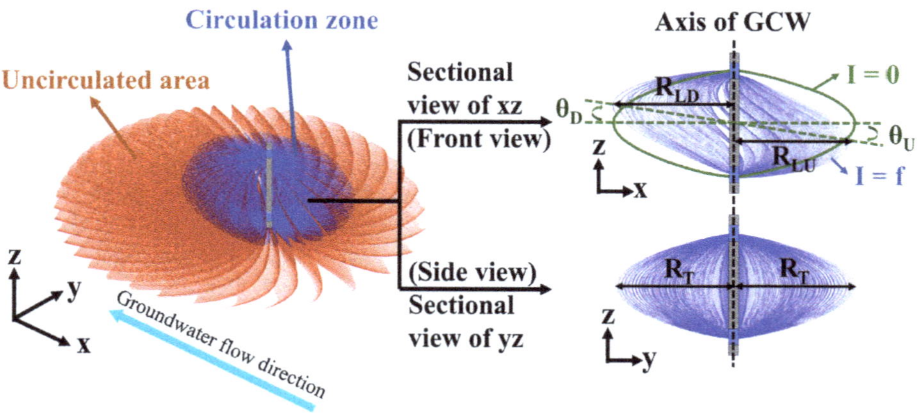

Figure 2. Three-dimensional hydraulic circulation patterns of GCW.

2.3.2. Offset Angle of Circulation Zone

As shown in Figure 2, the offset angle (θ) is defined as the upward (+) or downward (−) angular displacement of the farthest point of the circulation zone formed at a hydraulic gradient of f relative to the farthest point of the circulation zone at a hydraulic gradient of 0. This index can quantify the degree of deformation of the circulation zone caused by the natural flow field.

2.3.3. Circulation Efficiency

The particle recovery rate (Pr) is defined as the ratio of the number of particles recovered from the pumping screen to the number of particles released from the injection screen. This indicator can quantify the circulation efficiency of GCWs, reflecting the closure of the circulation flow field and the ability to establish hydraulic circulation. A higher Pr indicates a stronger capability of GCWs to establish hydraulic circulation, enabling faster capture of more contaminants. It also indicates better closed-loop performance of the circulation flow field, effectively confining contaminants within the circulation unit to prevent their escape into the background flow field, which could cause secondary pollution.

3. Results and Discussion

3.1. Influence of Hydraulic Gradient on the Hydraulic Circulation Characteristics of Dual-Screen GCW

As shown in Figure 3, the regions on the injection screen capable of forming circulation and the release positions of particles constituting the R_L and R_T at one year are marked. The screen is divided into four regions: Area I near the upstream, Area III near the downstream, and Areas II and IV perpendicular to the natural flow field. When the screen is extended with a hydraulic gradient (I) of 0, the region capable of forming circulation is a regular rectangle. As the hydraulic gradient increases, the shape of the region capable of forming circulation changes. In Areas III, II, and IV, particles released near the downstream are carried downstream by the natural flow and fail to form hydraulic circulation, resulting in a reduction in the area of these regions. In contrast, particles in Area I are recovered faster by the pumping screen under the influence of the natural flow field, leading to an increase in the area of the recoverable particle region. When I increases to 1.4‰, all particles released in Area III of the injection screen cannot form hydraulic circulation. When I increases to 2.7‰, only particles in Area I can form hydraulic circulation. However, as the hydraulic gradient continues to increase, some particles in Area I are also carried downstream by the stronger natural flow field, preventing them from forming hydraulic circulation and reducing the area of this region. Consequently, when I reaches 3.6‰, only particles released at the 180° position in Area I can be recovered. This indicates that as the hydraulic gradient increases, the circulation efficiency of GCW becomes very low. Additionally, due to the alteration in the region capable of forming circulation, the release positions of particles constituting the R_L and R_T change, leading to variations in the influence radius of GCW.

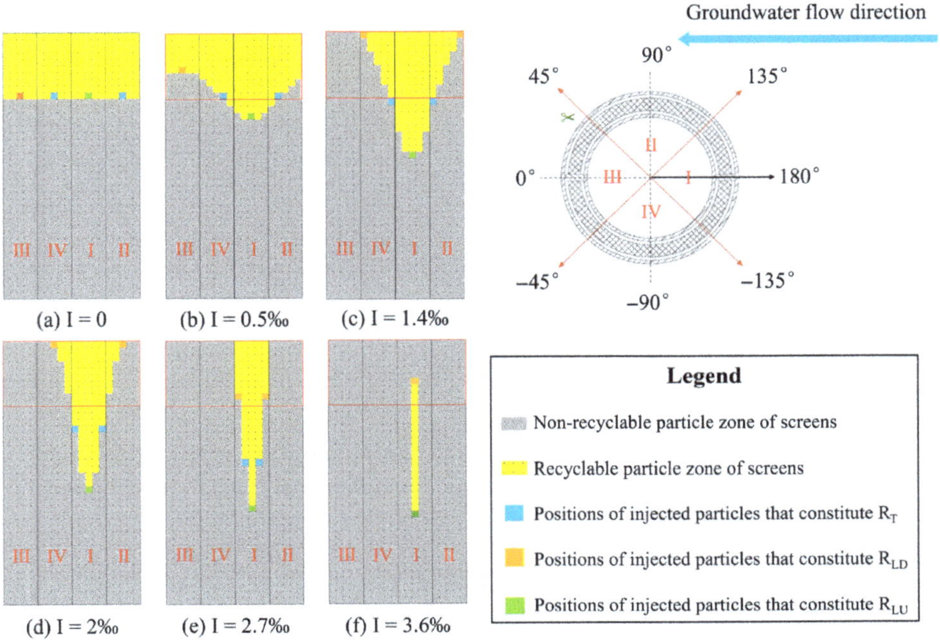

Figure 3. The areas of injection screens that form hydraulic circulation and release positions of particles constituting the longitudinal influence radii (R_L) and the horizontal influence radii (R_T) at one year.

3.1.1. Influence Radius

(1) Longitudinal influence radius

Figure 4a shows the relationship between hydraulic gradient and R_{LU} at one year. As indicated in the figure, when the hydraulic gradient ranges from 0 to 4.3‰, the particle release positions are all at the 180° position in Area I. When the hydraulic gradient ranges from 0 to 2.0‰, R_{LU} fluctuates between 34.26 m and 34.93 m as the hydraulic gradient increases. Within this range, the presence of the hydraulic gradient increases the R_{LU}. When the hydraulic gradient ranges from 2.0‰ to 4.3‰, R_{LU} decreases gradually as the hydraulic gradient increases without any fluctuations. Specifically, R_{LU} decreases from 34.52 m to 27.68 m, a reduction of 6.84 m. When the hydraulic gradient exceeds 4.3‰, R_{LU} is 0. This indicates that the influence of the hydraulic gradient on R_{LU} can be divided into three stages. In the first stage, when the hydraulic gradient is small (less than 2.0‰), it has a weak promoting effect on R_{LU}. In the second stage, when the hydraulic gradient is large (greater than 2.0‰), it exerts a strong inhibitory effect on R_{LU}. In the third stage, when the hydraulic gradient is excessively large (greater than 4.3‰), the driving effect of the hydraulic gradient completely dominates, and the GCW can no longer drive water flow to form a circulation zone.

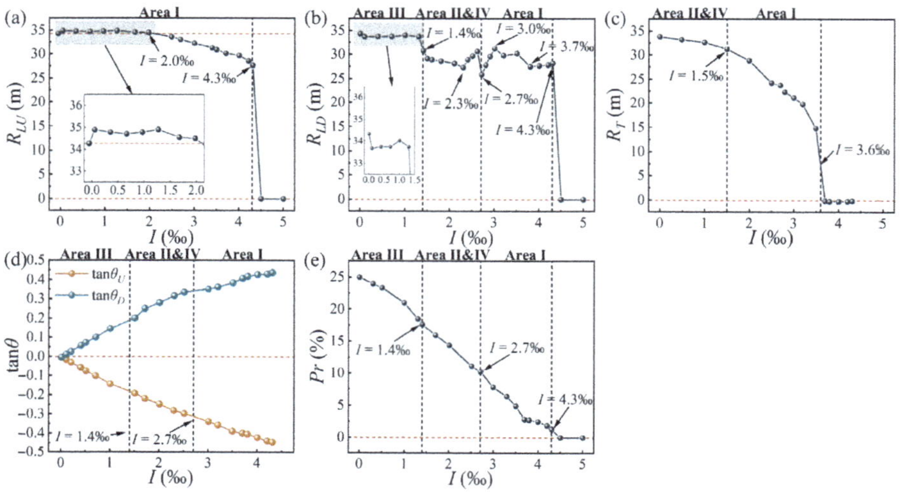

Figure 4. Curves of R_{LU} (**a**), R_{LD} (**b**), R_T (**c**), tanθ (**d**), and Pr (**e**) for dual-screen GCW with hydraulic gradient at one year.

Figure 4b illustrates the relationship between hydraulic gradient and R_{LD} at one year. It can be observed that the variation in R_{LD} with respect to the hydraulic gradient is more complex than that of R_{LU}. When the hydraulic gradient ranges from 0 to 1.3‰, R_{LD} slightly fluctuates between 33.69 m and 34.32 m. At this stage, the particle release positions are all located at the 0° position in Area III and gradually shift toward the position closer to the sealed screen. When the hydraulic gradient increases to 1.4‰, all particles released in Area III are unable to form hydraulic circulation. Consequently, the particle release positions constituting R_{LD} shift from the original 0° position in Area III to the ±68° positions in Areas II and IV, resulting in R_{LD} decreasing from 33.76 m to 30.76 m. Subsequently, as the hydraulic gradient increases to 2.3‰, 2.7‰, 3.0‰, and 3.7‰, the particle release positions constituting R_{LD} change again, causing fluctuations in R_{LD}. In conclusion, when the hydraulic gradient ranges from 1.4‰ to 4.3‰, R_{LD} fluctuates between 25.9 m and 31.25 m. When the hydraulic gradient exceeds 4.3‰, R_{LD} is 0. Therefore, the influence of the hydraulic gradient on R_{LD} can also be divided into three stages. In the first stage, when the hydraulic gradient is small (less than 1.4‰), its inhibitory effect on R_{LD} is minor compared to the water circulation driven by GCW, resulting in negligible changes in R_{LD}. In the second stage, when the hydraulic gradient is large (greater than 1.4‰), its dominant role gradually increases. R_{LD} exhibits strong fluctuations with variations in the release positions of particles constituting R_{LD}, but overall shows a significant inhibitory effect compared to the case of $I = 0$. In the third stage, when the hydraulic gradient is excessively large (greater than 4.3‰), the driving effect of the hydraulic gradient completely dominates, and the GCW can no longer form a circulation zone.

(2) Horizontal influence radius

Figure 4c illustrates the relationship between hydraulic gradient and R_T at one year. When the hydraulic gradient ranges from 0 to 1.5‰, the particles constituting R_T are released from Areas II and IV in the injection screen, with R_T decreasing from 34.32 m to 31.21 m. When the hydraulic gradient ranges from 1.5‰ to 3.5‰, the particles constituting R_T are released from Area I in the injection screen, with R_T decreasing further from 31.21 m to 14.85 m. Although R_T decreases as the hydraulic gradient increases, the rate of decrease gradually becomes more pronounced. When the hydraulic gradient exceeds 3.6‰, the

GCW is unable to form a horizontal influence radius. It can be seen that compared with R_L, the variation in R_T with hydraulic gradient is relatively straightforward, with R_T decreasing monotonically as the hydraulic gradient increases. However, the hydraulic gradient value at which GCW fails to form R_T (3.6‰) is smaller than that for R_T (4.3‰). This is because when $I > 3.6$‰, all particles involved in forming the circulation zone are released at the 180° position in Area I of the injection screen, and the circulation zone appears only as a straight line in the side view.

3.1.2. Offset Angle of Circulation Zone

Figure 4d illustrates the relationship between the hydraulic gradient and the offset angle of the circulation zone at one year. It can be observed that the offset angle of the circulation zone increases as the hydraulic gradient increases, with the θ_D being slightly larger than the θ_U. When $I = 0$, the circulation zone exhibits no deflection, and the farthest point of particle movement lies on the horizontal line passing through the midpoint between the two screens. When $I = 1.4$‰ (where all particles released in Area III of the injection screen cannot form hydraulic circulation), $\tan\theta_U$ is -0.1906, indicating a downward deviation of 10.79°, and $\tan\theta_D$ is 0.2047, indicating an upward deviation of 11.57°. When $I = 2.7$‰ (where all particles released in Areas II and IV of the injection screen cannot form hydraulic circulation), $\tan\theta_U$ is -0.3129, indicating a downward deviation of 17.37°; $\tan\theta_D$ is 0.3861, indicating an upward deviation of 21.11°. When $I = 3.6$‰ (where only particles released in the 180° position of Area I in the injection screen can form hydraulic circulation), $\tan\theta_U$ is -0.3985, indicating a downward deviation of 21.73°; $\tan\theta_D$ is 0.4083, indicating an upward deviation of 22.21°. Therefore, if the hydraulic gradient is excessively high, the circulation zone will undergo severe deformation. This deformation may prevent the circulation zone from encapsulating the contaminant plume, thereby increasing the risk of plume diffusion and secondary contamination.

3.1.3. Circulation Efficiency

Figure 4e shows the relationship between hydraulic gradient and Pr at one year. As shown in the figure, Pr decreases gradually as the hydraulic gradient increases. When the hydraulic gradient is 0, Pr is 25%. When $I = 1.4$‰ (where all particles released in Area III of the injection screen cannot form hydraulic circulation), Pr is 17.63%. When $I = 2.7$‰ (where all particles released in Areas II and IV of the injection screen cannot form hydraulic circulation), Pr is 10.25%. When $I = 3.6$‰ (where only particles released at the 180° position in Area I of the injection screen can form hydraulic circulation), Pr is 3.2%. When $I > 4.3$‰, the GCW is unable to form a circulation zone, resulting in Pr being 0. This clearly shows that, unlike R_L, the hydraulic gradient exhibits a complete inhibitory effect on Pr. The increase in hydraulic gradient will inevitably cause some particles to escape into the background flow field, increasing the risk of contaminant migration and diffusion.

3.2. *Influence of Hydraulic Gradient on the Hydraulic Circulation Characteristics of Triple-Screen GCW*

3.2.1. Influence Radius

(1) Longitudinal influence radius

Figure 5a shows the relationship between hydraulic gradient and R_{LU} at one year. It can be observed that the release positions of particles constituting R_{LU} are all at the 180° position in Area I. The R_{LU} values of the two circulation zones formed by the three screens are essentially identical. When the hydraulic gradient ranges from 0 to 1.2‰, R_{LU} fluctuates between 33.95 m and 35.18 m as the hydraulic gradient increases. Within this range, the presence of the hydraulic gradient increases the R_{LU}. When the hydraulic gradient ranges from 1.2‰ to 9.3‰, R_{LU} decreases gradually with the increase in the hydraulic gradient without any fluctuations. Specifically, R_{LU} decreases from 34.08 m to 17.63 m, a reduction of 16.45 m. When the hydraulic gradient exceeds 9.3‰, R_{LU} is 0. Therefore, the influence pattern of hydraulic gradient on R_{LU} in the triple-screen GCW is similar to that in the dual-screen GCW, which can also be divided into three stages. In the first stage, when the hydraulic gradient is small (less than 1.2‰), it has a weak promoting effect on R_{LU}. In the second stage, when the hydraulic gradient is large (greater than 1.2‰), it exhibits an inhibitory effect on R_{LU}, with R_{LU} decreasing monotonically as the hydraulic gradient increases. In the third stage, when the hydraulic gradient is excessively large (greater than 9.6‰), the GCW is unable to form a circulation zone.

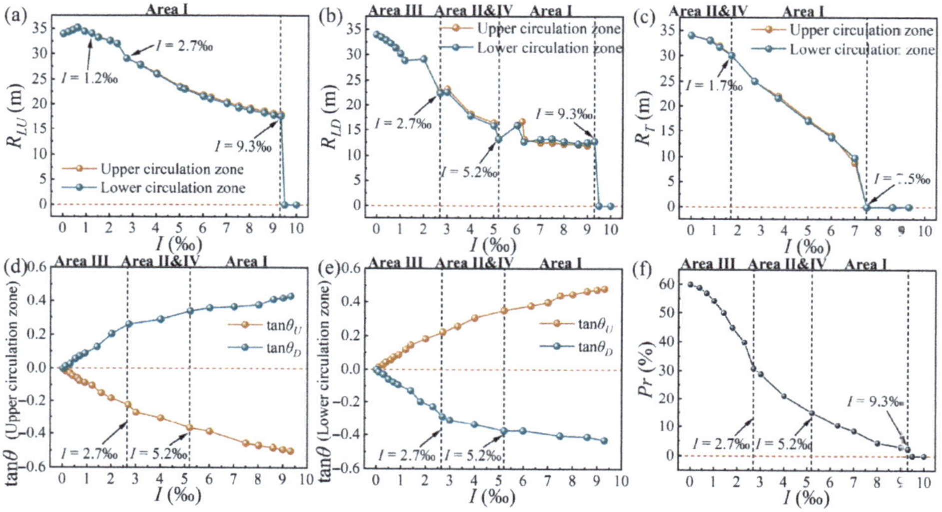

Figure 5. Curves of R_{LU} (**a**), R_{LD} (**b**), R_T (**c**), tanθ (**d**,**e**), and Pr (**f**) for triple-screen GCW with hydraulic gradient at one year.

Figure 5b illustrates the relationship between hydraulic gradient and R_{LD} at one year. The R_{LD} values of the two circulation zones formed by the triple-screen GCW are essentially identical, so only the upper circulation zone is analyzed. When the hydraulic gradient increases to 2.7‰, particles released in Area III cannot form hydraulic circulation, and the particle release positions shift from the original 0° position in Area III to the ±68° positions in Areas II and IV, resulting in R_{LD} decreasing from 33.99 m to 22.66 m. Subsequently, when the hydraulic gradient increases to 5.2‰, particles released in Areas II and IV of the injection screen cannot form a hydraulic circulation, and R_{LD} decreases further from 22.66 m to 13.20 m. Afterward, R_{LD} briefly increases with the increase in hydraulic gradient before stabilizing. When the hydraulic gradient exceeds 9.3‰, the GCW is unable to form a circulation zone, resulting in R_{LD} being 0. It can be seen that compared to the dual-screen GCW, the R_{LD} of the triple-screen GCW is influenced by the hydraulic gradient in a relatively simple manner, generally decreasing with the increase in the hydraulic gradient. Occasional fluctuations still occur due to changes in the release particle positions, but these fluctuations are significantly smaller than that of the dual-screen GCW. This is because the triple-screen GCW reduces the length of the sealed screen, and particles released from positions closer to the sealed screen are less affected by the hydraulic gradient. It also indicates that reducing the spatial scale of individual circulation zones can mitigate the interference of hydraulic gradients in the fluctuation of circulation units.

https://doi.org/10.3390/w18020164

(2) Horizontal influence radius

Figure 5c illustrates the relationship between hydraulic gradient and R_T at one year. It can be observed that the R_T values of the two circulation zones are essentially identical. When the hydraulic gradient ranges from 0 to 1.7‰, the particles constituting R_T are released from Areas II and IV of the injection screen, with R_T decreasing from 34.02 m to 30.08 m. When the hydraulic gradient ranges from 1.7‰ to 7.0‰, the particles constituting R_T are released from Area I, with R_T further decreasing from 30.08 m to 8.83 m. When $I \geq 7.5‰$, R_T is 0. This indicates that R_T decreases monotonically with increasing hydraulic gradient, and the upper limit of hydraulic gradient forming R_T (7.5‰) is lower than that for R_L (9.3‰). This phenomenon and its underlying cause are consistent with those of the dual-screen GCW.

3.2.2. Offset Angle of Circulation Zone

Figure 5d,e illustrate the relationship between the hydraulic gradient and the offset angle of the circulation zone at one year. It can be observed that the upper and lower circulation zones exhibit different offset patterns, which is due to the different circulation modes of the circulation units. If the triple-screen GCW is considered as the superposition of two dual-screen GCWs, the upper GCW involves pumping from the upper screen and injection from the lower screen, representing a reverse circulation mode, while the lower GCW involves injection from the upper screen and pumping from the lower screen, representing a standard circulation mode. The upstream circulation zone always offsets toward the injection screen, while the downstream circulation zone always offsets toward the pumping screen.

Take the upper circulation zone as an example (Figure 5d). When $I = 0$, the circulation zone does not undergo any deviation, and the farthest point of particle movement lies on the horizontal line passing through the midpoint between the two screens. When $I = 2.7‰$ (where all particles released in Area III of the injection screen cannot form hydraulic circulation), $\tan\theta_U$ is -0.2230, indicating a downward deviation of 10.79°, while $\tan\theta_D$ is 0.2652, indicating an upward deviation of 14.85°. When $I = 5.2‰$ (where all particles released in Areas II and IV of the injection screen cannot form hydraulic circulation), $\tan\theta_U$ is -0.3658, indicating a downward deviation of 20.09°, while $\tan\theta_D$ is 0.3396, indicating an upward deviation of 18.75°. When $I = 7.5‰$ (where only particles released in the 180° position of Area I in the injection screen can form hydraulic circulation), $\tan\theta_U$ is -0.4562, indicating a downward deviation of 24.52°, while $\tan\theta_D$ is 0.4390, indicating an upward deviation of 23.70°.

For the upper circulation zone, when the hydraulic gradient is low, the downstream circulation zone exhibits a greater offset angle than the upstream circulation zone. This occurs because particles released at a smaller angle to the groundwater flow direction experience greater displacement due to the hydraulic gradient. Conversely, when the hydraulic gradient is high, the upstream circulation zone shows a greater offset angle than the downstream circulation zone. This is because the proximity of the screen to the aquifer roof limits the deformation of the circulation zone. Similarly, for the lower circulation zone (Figure 5e), the same pattern is observed due to the influence of the confining layer roof.

3.2.3. Circulation Efficiency

Figure 5f illustrates the variation in Pr as the hydraulic gradient ranges from 0 to 9.3‰ at one year. As shown in the figure, Pr gradually decreases with increasing hydraulic gradient. When the hydraulic gradient is 0, Pr is 60%. When $I = 2.7‰$ (where all particles released in Area III of the injection screen cannot form hydraulic circulation), Pr is 31%. When $I = 5.2‰$ (where all particles released in Areas II and IV of the injection screen cannot

https://doi.org/10.3390/w18020164

form hydraulic circulation), Pr is 15%. When $I = 7.5‰$ (where only particles released in the 180° position of Area I in the injection screen can form hydraulic circulation), Pr is 5.5%. It can be observed that the circulation efficiency and the ability to establish hydraulic circulation of the triple-screen GCW are superior to those of the dual-screen GCW, under the same hydraulic gradient.

3.3. Feasibility of Increasing the Number of Screens to Expand GCW Applicability and Enhance Repair Efficiency

Under the conditions of this aquifer within one year, the dual-screen GCW can form a circulation zone within a hydraulic gradient range of 0 to 4.3‰, while the triple-screen GCW can form a circulation zone within a hydraulic gradient range of 0 to 9.3‰. This demonstrates that splitting a large circulation zone into two smaller circulations by increasing the number of screens can increase the upper limit of the hydraulic gradient for the GCW to form circulation zone. Furthermore, comparing the performance indicators of GCW operation with a hydraulic gradient of 2‰ at one year (Table 2) reveals that the influence radii of the dual-screen GCW and triple-screen GCW are not significantly different. However, increasing the number of screens can effectively reduce deformation of the circulation zone and enhance the circulation efficiency. Compared to the dual-screen GCW, the θ_U and θ_D of the triple-screen GCW are reduced by 3.42° and 4.12°, respectively, and the Pr increases by 27.62%. This is attributed to the triple-screen GCW reducing the distance between the screens, thereby minimizing the deformation of released particles near the sealed screen under the influence of the hydraulic gradient. Furthermore, adding screens allows new screens to form new circulation units with the original screens, improving screen utilization and thereby enhancing circulation efficiency. Therefore, under conditions of higher hydraulic gradients, the triple-screen GCW can effectively expand the applicable range of hydraulic gradients for GCWs and significantly improve the circulation efficiency of GCWs.

Table 2. Comparison of performance between dual-screen GCW and triple-screen GCW at one year.

Number of Screens	R_{LU} (m)	R_{LD} (m)	R_T (m)	θ_U	θ_D	Pr
2	34.52	31.96	29.09	−13.83°	15.73°	14.38%
3	32.64	29.24	30.69	−10.41°	11.61°	42%

4. Conclusions

This study visualized the water flow driven by GCWs using a particle tracking model, representing the circulation zone as a three-dimensional graphic composed of countless particle trajectories. Three-dimensional hydraulic circulation characteristics of GCWs were characterized by longitudinal influence radius (R_L), horizontal influence radius (R_T), offset angle of circulation zone (θ), and particle recovery rate (Pr). The research investigates the influence of hydraulic gradient on the hydraulic circulation characteristics of GCWs, elucidates the applicable ranges of hydraulic gradient for GCWs with different numbers of screens, and clarifies the potential to enhance the applicability and operational performance of GCWs under larger hydraulic gradients via multi-screen GCWs. Specifically, the following conclusions were obtained.

(1) The hydraulic circulation characteristics of GCWs with different numbers of screens exhibit similar patterns of influence from hydraulic gradient. The distribution of particle release zones forming hydraulic circulation on the screen continuously varies with the hydraulic gradient. R_L exhibits fluctuating variations with these zone changes, while R_T decreases monotonically as the hydraulic gradient increases. The hydraulic gradient also

https://doi.org/10.3390/w18020164

significantly reduces the circulation efficiency of GCWs and causes deformation of the circulation zone with θ increasing as the hydraulic gradient increases.

(2) The dual-screen GCW can form a circulation zone within a hydraulic gradient range of 0 to 4.3‰ at one year, while the triple-screen GCW can form a circulation zone within a hydraulic gradient range of 0 to 9.3‰. Therefore, increasing the number of screens can effectively increase the upper limit of the hydraulic gradient for GCWs to form circulation zones and expand the applicable range of hydraulic gradients for GCWs.

(3) Compared to the dual-screen GCW at one year, the θ_U and θ_D of the triple-screen GCW are reduced by 3.42° and 4.12°, respectively, and the *Pr* increases by 27.62%. Therefore, under conditions of higher hydraulic gradients, the triple-screen GCW can significantly improve the circulation efficiency of GCWs and reduce the deformation of the circulation zone.

From the perspective of engineering practice requirements, these conclusions can provide a scientific basis for predicting effects and avoiding risks in GCW engineering design. The multi-screen optimization scheme not only expands the technical application scope but also effectively solves the circulation failure problem under high hydraulic gradients, providing a practical optimization path for engineering applications and holding important guiding significance for the practical application of GCWs.

Author Contributions: S.D.: writing—original draft, methodology, investigation, data curation. J.C.: supervision, project administration, funding acquisition. R.Z.: writing—review and editing, project administration, funding acquisition. Q.Z.: formal analysis. All authors have read and agreed to the published version of the manuscript.

Funding: This work was supported by the National Key Research and Development Program of China (Grant No. 2022YFC3703101) and the Joint Program of Liaoning Provincial Science and Technology Plan (Grant No. 2024-MSLH-500).

Data Availability Statement: The original contributions presented in this study are included in the article. Further inquiries can be directed to the corresponding authors.

Acknowledgments: We would like to thank the anonymous reviewers and the editor.

Conflicts of Interest: The authors declare that they have no known competing financial interests or personal relationships that could have appeared to influence the work reported in this paper.

References

1. Ciampi, P.; Esposito, C.; Bartsch, E.; Alesi, E.J.; Petrangeli Papini, M. Pump-and-treat (P&T) vs groundwater circulation wells (GCW): Which approach delivers more sustainable and effective groundwater remediation? *Environ. Res.* **2023**, *234*, 116538. [CrossRef] [PubMed]
2. Tatti, F.; Papini, M.P.; Torretta, V.; Mancini, G.; Boni, M.R.; Viotti, P. Experimental and numerical evaluation of Groundwater Circulation Wells as a remediation technology for persistent, low permeability contaminant source zones. *J. Contam. Hydrol.* **2019**, *222*, 89–100. [CrossRef] [PubMed]
3. Elmore, A.C.; Graff, T. Best Available Treatment Technologies Applied to Groundwater Circulation Wells. *Remediat. J.* **2002**, *12*, 63–80. [CrossRef]
4. Ciampi, P.; Esposito, C.; Bartsch, E.; Alesi, E.J.; Papini, M.P. 3D dynamic model empowering the knowledge of the decontamination mechanisms and controlling the complex remediation strategy of a contaminated industrial site. *Sci. Total Environ.* **2021**, *793*, 148649. [CrossRef]
5. Ciampi, P.; Esposito, C.; Bartsch, E.; Alesi, E.J.; Rehner, G.; Morettin, P.; Pellegrini, M.; Olivieri, S.; Ranaldo, M.; Liali, G.; et al. A data-driven modeling approach for the sustainable remediation of persistent arsenic (As) groundwater contamination in a fractured rock aquifer through a groundwater recirculation well (IEG-GCW®). *Environ. Res.* **2023**, *217*, 114827. [CrossRef]
6. Johnson, R.L.; Simon, M.A. Evaluation of groundwater flow patterns around a dual-screened groundwater circulation well. *J. Contam. Hydrol.* **2007**, *93*, 188–202. [CrossRef]
7. Feng, S.; Cheng, D.; Zhan, H.; Yang, S.; Zhao, X.; Jin, H.; Zhang, L. Evolution characteristics of remediation process of secondary contaminant sources of low-permeability lens driven by circulating well. *J. Hydrol.* **2022**, *613*, 128408. [CrossRef]

8. Wang, P.; Li, J.; An, P.; Yan, Z.; Xu, Y.; Pu, S. Enhanced delivery of remedial reagents in low-permeability aquifers through coupling with groundwater circulation well. *J. Hydrol.* **2023**, *618*, 129260. [CrossRef]
9. Ciampi, P.; Esposito, C.; Petrangeli Papini, M. Review on groundwater circulation wells (GCWs) for aquifer remediation: State of the art, challenges, and future prospects. *Groundw. Sustain. Dev.* **2024**, *24*, 101068. [CrossRef]
10. Kahler, D.M.; Kabala, Z.J. Rapidly Pulsed Pumping Accelerates Remediation in A Vertical Circulation Well Model. *Water* **2018**, *10*, 1423. [CrossRef]
11. Qiu, H.; Xu, J.; Yuan, Y.; Alesi, E.J.; Liang, X.; Cao, B. Low-disturbance land remediation using vertical groundwater circulation well technology: The first commercial deployment in an operational chemical plant. *Sci. Total Environ.* **2024**, *944*, 173804. [CrossRef] [PubMed]
12. Ciampi, P.; Esposito, C.; Papini, M.P. Hydrogeochemical Model Supporting the Remediation Strategy of a Highly Contaminated Industrial Site. *Water* **2019**, *11*, 1371. [CrossRef]
13. Papini, M.P.; Majone, M.; Arjmand, F.; Silvestri, D.; Sagliaschi, M.; Sucato, S.; Alesi, E.; Barstch, E.; Pierro, L. First Pilot Test on the Integration of GCW (Groundwater Circulation Well) with ENA (Enhanced Natural Attenuation) for Chlorinated Solvents Source Remediation. In Proceedings of the 5th International Conference on Industrial Biotechnology (IBIC), Bologna, Italy, 10–13 April 2016; Volume 49, pp. 91–96.
14. Knox, R.; Sabatini, D.; Harwell, J.; Brown, R.; West, C.; Blaha, F.; Griffin, C. Surfactant remediation field demonstration using a vertical circulation well. *Groundwater* **1997**, *35*, 948–953. [CrossRef]
15. Pierre, L.; Matturro, B.; Rossetti, S.; Sagliaschi, M.; Sucato, S.; Alesi, E.; Bartsch, E.; Arjmand, F.; Papini, M.P. Polyhydroxyalkanoate as a slow-release carbon source for in situ bioremediation of contaminated aquifers: From laboratory investigation to pilot-scale testing in the field. *New Biotechnol.* **2017**, *37*, 60–68. [CrossRef]
16. Ciampi, P.; Esposito, C.; Bartsch, E.; Alesi, E.J.; Rehner, G.; Papini, M.P. Remediation of chlorinated aliphatic hydrocarbons (CAHs) contaminated site coupling groundwater recirculation well (IEG-GCW®) with a peripheral injection of soluble nutrient supplement (IEG-C-MIX) via multilevel-injection wells (IEG-MIW). *Heliyon* **2022**, *8*, e11402. [CrossRef]
17. Zhang, Y.; Cao, W.; Wang, W.; Dong, Q. Distribution of groundwater arsenic and hydraulic gradient along the shallow groundwater flow-path in Hetao Plain, Northern China. *J. Geochem. Explor.* **2013**, *135*, 31–39. [CrossRef]
18. Jazayeri, A.; Werner, A.D.; Jenkins, T. Effects of porous media thickness and its hydraulic gradient history on the formation of sand boils: Experimental investigation. *Sci. Total Environ.* **2023**, *880*, 163235. [CrossRef]
19. Wang, X.; Ren, L.; Long, T.; Geng, C.; Tian, X. Migration and remediation of organic liquid pollutants in porous soils and sedimentary rocks: A review. *Environ. Chem. Lett.* **2023**, *21*, 479–496. [CrossRef]
20. Zeng, J.; Tabelin, C.B.; Gao, W.; Tang, L.; Luo, X.; Ke, W.; Jiang, J.; Xue, S. Heterogeneous distributions of heavy metals in the soil-groundwater system empowers the knowledge of the pollution migration at a smelting site. *Chem. Eng. J.* **2023**, *454*, 140307. [CrossRef]
21. Joun, W.-T.; Lee, K.-K. Experimental study for visualizing CO_2-dissolved water plume migration under hydraulic gradient conditions and implication for field monitoring data. *J. Environ. Manag.* **2024**, *357*, 120814. [CrossRef]
22. Li, B.; Zhang, H.; Long, J.; Fan, J.; Wu, P.; Chen, M.; Liu, P.; Li, T. Migration mechanism of pollutants in karst groundwater system of tailings impoundment and management control effect analysis: Gold mine tailing impoundment case. *J. Clean. Prod.* **2022**, *350*, 131434. [CrossRef]
23. Vats, O.P.; Sharma, B.; Stamm, J.; Bhattacharjya, R.K. Groundwater Circulation Well for Controlling Saltwater Intrusion in Coastal aquifers: Numerical study with Experimental Validation. *Water Resour. Manag.* **2020**, *34*, 3551–3563. [CrossRef]
24. Saad, S.; Javadi, A.A.; Abd-Elhamid, H.F.; Farmani, R. Mitigating seawater intrusion in coastal aquifers: Novel approach with treated wastewater injection and groundwater circulation. *J. Hydrol.* **2023**, *626*, 130139. [CrossRef]
25. Ciampi, P.; Felli, G.; Feriaud, D.; Esposito, C.; Papini, M.P. 3D GeoRemediation: A Digital Hydrogeophysical-Chemical Clone and Virtual Hydraulic Barrier with Groundwater Circulation Wells (GCWs) for Groundwater Remediation. *Sustainability* **2024**, *16*, 5216. [CrossRef]
26. Huang, J.Q.; Goltz, M.N. A three-dimensional analytical model to simulate groundwater flow during operation of recirculating wells. *J. Hydrol.* **2005**, *314*, 67–77. [CrossRef]
27. Zhu, Q.; Wen, Z.; Zhan, H.; Yuan, S. Optimization Strategies for in Situ Groundwater Remediation by a Vertical Circulation Well Based on Particle-Tracking and Node-Dependent Finite Difference Methods. *Water Resour. Res.* **2020**, *56*, e2020WR027396. [CrossRef]
28. Chen, J.-S.; Jang, C.-S.; Cheng, C.-T.; Liu, C.-W. Conservative solute approximation to the transport of a remedial reagent in a vertical circulation flow field. *J. Hydrol.* **2010**, *390*, 155–168. [CrossRef]
29. Zhang, F.; Zhu, H.; Zhang, D.; Feng, R.; Zhao, Y.; Liu, H.; Yang, S.; Wang, W. Method of evaluating the radius of influence of a circulating well based on hydraulic head difference. *Hydrogeol. J.* **2024**, *32*, 817–834. [CrossRef]
30. Tu, K.; Wu, Q.; Simunek, J.; Chen, C.; Zhu, K.; Zeng, Y.; Xu, S.; Wang, Y. An Analytical Solution of Groundwater Flow in a Confined Aquifer With a Single-Well Circulation System. *Water Resour. Res.* **2020**, *56*, e2020WR027529. [CrossRef]

https://doi.org/10.3390/w18020164

31. Philip, R.D.; Walter, G.R. Prediction of flow and hydraulic head fields for vertical circulation wells. *Groundwater* **1992**, *30*, 765–773. [CrossRef]
32. Toscani, L.; Stefania, G.A.; Masut, E.; Prieto, M.; Legnani, A.; Gigliuto, A.; Ferioli, L.; Battaglia, A. Groundwater flow numerical model to evaluate the water mass balance and flow patterns in Groundwater Circulation Wells (GCW) with varying aquifer parameters. *Acque Sotter.-Ital. J. Groundw.* **2022**, *11*, 9–19. [CrossRef]
33. Xia, Q.; Zhang, Q.; Xu, M.; Tang, Y.G.; Teng, H.W. Visualizing hydraulic zones of a vertical circulation well in presence of ambient flow. *Desalin. Water Treat.* **2019**, *159*, 151–160. [CrossRef]
34. Jin, Y.; Holzbecher, E.; Sauter, M. Dual-Screened Vertical Circulation Wells for Groundwater Lowering in Unconfined Aquifers. *Groundwater* **2016**, *54*, 15–22. [CrossRef] [PubMed]
35. Chen, S.; Huang, Z.; Zhou, R. Influence of process parameters on the operation effectiveness of groundwater circulation well and multi-objective optimization design. *J. Environ. Manag.* **2025**, *373*, 123589. [CrossRef]
36. Fang, Z.; Ke, H.; Ma, Y.; Zhao, S.; Zhou, R.; Ma, Z.; Liu, Z. Design optimization of groundwater circulation well based on numerical simulation and machine learning. *Sci. Rep.* **2024**, *14*, 11506. [CrossRef]
37. Wang, X.; Zhang, L.; Han, C.; Zhang, Y.; Zhuo, J. Simulation study of oxytetracycline contamination remediation in groundwater circulation wells enhanced by nano-calcium peroxide and ozone. *Sci. Rep.* **2023**, *13*, 9136. [CrossRef]
38. Li, B.; Yuan, Y.; Lu, L.; Liu, S.; Wang, P.; Yao, M.; Dong, J. Synergistic approach of GCW-ISTR for enhanced remediation of semi-volatile organic contaminants in groundwater: Modeling and experimental validation. *Environ. Res.* **2024**, *263*, 119995. [CrossRef]
39. Zhao, Y.; Qu, D.; Zhou, R.; Yang, S.; Ren, H. Efficacy of forming biofilms by *Pseudomonas migulae* AN-1 toward in situ bioremediation of aniline-contaminated aquifer by groundwater circulation wells. *Environ. Sci. Pollut. Res.* **2016**, *23*, 11568–11573. [CrossRef]
40. Xu, L.; Zhu, H.; Wei, J.; Chen, W.; Ma, H.; Pu, S. Integration of immobilized microorganisms with a groundwater circulation well for the remediation of naphthalene-contaminated aquifers. *J. Clean. Prod.* **2025**, *520*, 146127. [CrossRef]
41. Stamm, J.; Eldho, T.; Scholz, M. Flow simulation of a system of groundwater circulation well and pumping well for NAPL site remediation. *WIT Trans. Ecolog. Environ.* **1998**, *23*, 99–106. [CrossRef]
42. Elmore, A.C.; DeAngelis, L. Modeling a ground water circulation well alternative. *Ground Water Monit. Remediat.* **2004**, *24*, 66–73. [CrossRef]

https://doi.org/10.3390/w18020164

water

MDPI

Article

A Geoinformation-Based Approach for Mapping Coastal Vulnerability in Sweden

Eleni Achmakidou [1,2], George P. Petropoulos [3,*], Anna Karkani [2], Niki Evelpidou [2], Spyridon E. Detsikas [3] and Georgios Nektarios Tselos [3]

1 ALTUS Land Sea Air (LSA) Ltd., Thesi Oasi Agias, 73100 Chania, Crete, Greece; sge2000009@uoa.gr
2 Faculty of Geology and Geoenvironment, National and Kapodistrian University of Athens, Panepistimiopolis, Zografou, 15784 Athens, Greece; ekarkani@geol.uoa.gr (A.K.); evelpidou@geol.uoa.gr (N.E.)
3 Department of Geography, Harokopio University of Athens, El. Venizelou St., 70, 17671 Athens, Greece; sdetsikas@hua.gr (S.E.D.); tselosgn@hua.gr (G.N.T.)
* Correspondence: gpetropoulos@hua.gr

Abstract

Coastal areas in Arctic and sub-Arctic regions are getting increasingly vulnerable to climate change, particularly due to sea-level rise and changing wave dynamics. This study assesses the coastal vulnerability of Sweden using the Coastal Vulnerability Index (CVI), based on a GIS-based, semi-quantitative approach integrating Earth Observation (EO) and geoinformatics. Six physical parameters were used: coastal geomorphology, coastal slope, shoreline change rate, relative sea-level change, mean tidal range, and significant wave height. CVI scores were derived from openly available national and international datasets, including satellite imagery and historical records. Mean tidal range and relative sea-level change were the most influential variables. The results indicate significant spatial variability: 36.5% of the coastline is classified as having "High" or "Very High" vulnerability, mainly in the south, while over 41% is "Low" or "Very Low," especially in the Gulf of Bothnia. This assessment offers a harmonized spatial baseline to support sustainable coastal planning and adaptation efforts in Sweden and provides a comparative framework for coastal vulnerability studies across northern Europe.

Keywords: coastal erosion; climate change; GIS; spatial analysis; sea-level rise; glacial isostatic adjustment (GIA); EO-PERSIST

1. Introduction

The shoreline represents the dynamic interface between land and sea, constantly reshaped by waves, tides, currents, and sediment transport [1]. While coastal erosion, defined as the landward retreat of the shoreline due to sediment loss, has always been part of natural coastal evolution [2], it poses a growing risk when it outpaces deposition. This is leading to irreversible land loss that threatens ecosystems, infrastructure, and human settlements. Monitoring coastlines and understanding shoreline behavior are therefore critical for effective coastal risk assessment and long-term planning [3,4]. This is especially urgent in Arctic and sub-Arctic regions, where climate change is accelerating coastal change through rising air and sea temperatures, declining sea ice cover, permafrost thaw, and increasingly frequent and intense storms [5,6]. These combined drivers are causing Arctic coastlines to retreat at rates of several meters per year [7], reshaping landscapes and increasing hazards for local communities [8]. In this context, systematic mapping and monitoring of coastal change are essential for identifying vulnerability hotspots, guiding adaptation efforts, and informing spatial planning in these rapidly evolving environments.

Sweden's extensive and ecologically diverse coastline is also experiencing growing climate-related pressures. These include sea-level rise, intensified storm surges, coastal erosion, and saltwater intrusion [9–11]. Beyond gradual and climate-driven processes, coasts may also be affected by more sudden coastal hazards. Examples include tsunamis generated by submarine landslides [12], or by atmospheric pressure shocks [13] that can amplify wave impacts in semi-enclosed coastlines [14]. Rip currents can intensify under specific wave conditions [15] as well as extreme wave events [16], posing risks to both beach safety and sediment transport. The soft-soil coasts of southern Sweden already face significant erosion, while sub-Arctic and temperate regions show environmental changes comparable to those observed in higher latitudes [17].

In response, several municipalities in southern Sweden have implemented both hard-engineering solutions, like sea walls and groynes, and softer, nature-based measures such as dune stabilization and coastal zoning (Figure 1) [18,19]. The importance of coastal monitoring has grown significantly as local governments and the European Union prioritize climate adaptation and risk mitigation. At the policy level, both the United Nations Sustainable Development Goals (SDGs) and the EU Adaptation Strategy emphasize the need for inclusive, data-driven, and forward-looking planning to enhance climate resilience, particularly in vulnerable coastal zones [10,20].

Figure 1. Examples of coastal management interventions in southern Sweden: (**left**) hard protection using rock revetments in the Malmö region; (**right**) soft, nature-based dune stabilization in Lomma.

To this end, the wide spectrum of geoinformation technologies provide an effective and efficient approach to assessing risks across large regions [21]. In particular, On the one hand, Earth Observation (EO) enables reliable and continuous monitoring of shoreline changes, making it possible to detect rates of coastal change and emerging trends with consistency. When this information is combined with complementary geospatial data, such as elevation, bathymetry, or population density, it becomes possible to conduct more in-depth and accurate assessments of coastal vulnerability. On the other hand, Geographic Information Systems (GIS) serve as an ideal tool for integrating these datasets and producing visual outputs that both enhance analysis and improve communication. By converting complex geospatial information into clear and accessible maps, GIS supports decision-making and stakeholder engagement, especially in remote and data-scarce areas like the Arctic [22]. Overall, the combined use of EO and GIS enables precise and cost-effective vulnerability assessments, presenting important observations into the dynamics of Arctic coastal systems [23].

Among the most widely applied geoinformation-based methods for coastal risk assessment is the Coastal Vulnerability Index (CVI) [24]. First developed by [25] and later refined by [26], the CVI is a semi-quantitative approach that integrates multiple physical parameters. Specifically, coastal slope, geomorphology, relative sea-level change, shoreline change rates, mean tidal range and mean significant wave height are each evaluated and assigned a vulnerability score. These scores are then combined into an overall index classification. The CVI provides a simple numerical basis for evaluating and comparing different coastline segments based on their potential for physical change and can be effectively visualized through maps. The method has been extensively applied in diverse geographical contexts, including the United States, Europe, and Asia [27–30].

Given the growing concerns about coastal vulnerability and the threats associated with climate change, this study contributes to advancing the understanding of shoreline sensitivity in high-latitude regions such as Sweden. The analysis is based on standardized physical parameters within a GIS-based framework and integrates EO (see Section 2.2) and other geospatial datasets to enhance the resolution and accuracy of coastal risk mapping. A further aim of this study is to identify the physical parameters most strongly associated with high vulnerability along Sweden's coastline. By applying an internationally recognized method, the results offer a consistent spatial overview that can inform adaptation planning and support cautious comparisons with other Arctic and sub-Arctic coastal regions.

2. Experimental Setup

2.1. Study Area

Sweden was selected as the study area for three main reasons. First, parts of its coastline face growing environmental pressures, including sea-level rise and erosion, which have already led to measurable shoreline retreat. In certain locations, particularly in low-lying soft-sediment shoreline recession has reached up to 200 m over the past 35 years, posing increasing risks to infrastructure and natural systems [10]. Geographically, Sweden is located in Northern Europe and extends from temperate zones in the south to sub-Arctic and Arctic conditions in the far north, sharing land borders with Norway and Finland. Its coastline spans more than 44,000 km, including islands [31], and borders the Gulf of Bothnia, the Baltic Sea, the Kattegat, and a portion of the Skagerrak, providing an indirect connection to the North Sea (Figure 2). The coastal landscape is highly varied: the southern shores are generally low-lying and erosion-prone, composed of unconsolidated glacial and post-glacial sediments, whereas the northern coasts are more rugged and underlain by resistant crystalline bedrock [31,32].

Combined with a well-documented history of coastal change, these stressors make this area an ideal setting for coastal vulnerability assessment. Second, the country's coastline is highly diverse and geomorphologically complex, stretching from temperate to sub-Arctic zones and encompassing nearly 100,000 islands [33]. This geomorphological variability enables the application of the CVI across different types of coastal environments. Third, the region benefits from long-term, high-quality geospatial and environmental datasets, as well as active climate adaptation policies, all of which support integrated, large-scale assessments. Furthermore, Sweden's inclusion in the ongoing EU-funded EO-PERSIST project (https://eo-persist.eu/, accessed on 25 May 2025) enabled field-based research and facilitated strong institutional collaboration. This combination enhanced both the analytical depth of the study and the broader dissemination and awareness of coastal vulnerability challenges. Given its exposure to the early effects of global change, the Baltic Sea coasts serve as valuable reference point for assessing coastal vulnerability in comparable Arctic and sub-Arctic regions [34].

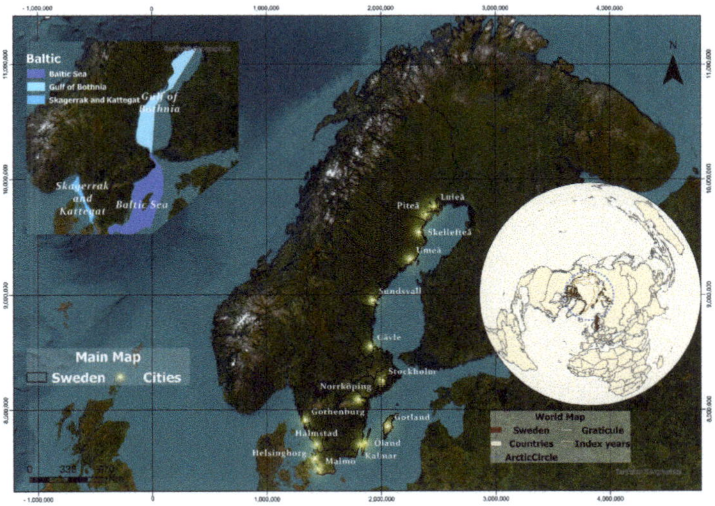

Figure 2. The Baltic Sea area within a broader European or global framework Sweden's three marine planning areas (Baltic Sea, Gulf of Bothnia, Kattegat-Skagerrak), as defined by the Swedish Agency for Marine and Water Management used as a geographic basis for interpreting the results.

2.2. CVI Variables

To assess Sweden's coastal vulnerability, the CVI approach brings together multiple physical variables, including geomorphology, coastal slope, hydrodynamic parameters, and shoreline change rates. In order to generate detailed vulnerability profiles for Sweden's diverse coastal regions, data were sourced from the national institutions Swedish Meteorological and Hydrological Institute (SMHI), the Geological Survey of Sweden (SGU), the Swedish Maritime Administration (SMA), as well as from the European platform, EMODnet. These data sources were selected based on their reliability, recent updates, and the availability of long-term time series, as required for the Coastal Vulnerability Index assessment [35]. Table 1 summarizes the dataset, with the following sections elaborating on each parameter.

Table 1. Data sources used in CVI calculations in this study. All sources are open access.

Variable	Source	Reference Period	Resolution/Accuracy	Format
Geomorphology	EMODnet Geology/coastal type https://emodnet.ec.europa.eu/geoviewer/#!/ (accessed on 25 May 2025)	-	~1:250,000 scale (vector shoreline classification)	Shapefile/WebGIS
Coastal Slope	https://portal.opentopography.org/arcticDem?opentopoID=OTSDEM.112018.3413.3 (accessed on 25 May 2025)	-	~32 m horizontal resolution	Raster (GeoTIFF)
Mean wave height (m)	Wave model data (SGI, LTH, SGU) https://gis.sgi.se/kustdataportal/# (accessed on 25 May 2025)	1979–2018	3-hourly temporal resolution; ~3–5 km spatial resolution	Model output (Time series (CSV))
Relative Sea-level change (mm/y)	SMA 2025 https://www.bshc.pro/wp-content/uploads/Future_sea_levels.pdf (Accessed on 25 May 2025)	IPCC 2100	Tide gauge network (86 stations); vertical datum RH2000	Report (PDF), Tables
Shoreline erosion/accretion (m/y)	EMODnet Geology/coastal migration https://emodnet.ec.europa.eu/geoviewer/#!/ (accessed on 25 May 2025)	-	~1:250,000 scale, regional compilations	Shapefile/WebGIS
Tidal range (m)	Swedish Meteorological and Hydrological Institute (SMHI) https://www.smhi.se/vader/vader-till-havs/vattenstand-och-vagor (accessed on 25 May 2025)	-	-	-

2.2.1. Geomorphology

The geology, geomorphology, and presence of coastal defenses largely determine how resilient or vulnerable a coast is to erosion and flooding integrates information on geological structure, lithological composition, morphodynamics, and the coastal zone's historical evolution. It largely determines the natural resistance of coastlines to erosion processes and long-term geomorphological changes [35]. At the same time, it remains one of the most challenging variables to verify without field measurements or the use of aerial photogrammetry, due to the complexity and variability of coastal landforms [36]. In this study, geomorphological classification was achieved primarily through EMODnet, an initiative involving over 160 organizations that collectively compile, harmonize, and disseminate marine geoscience datasets. Specifically, the "coastal type" layer from the EMODnet geology map was utilized to categorize shoreline features. To validate and enhance the accuracy of this classification, high-resolution satellite imagery from Google Earth Pro was analyzed, allowing for visual confirmation of landform features such as cliffs, beaches, estuaries, and rocky shores. Additionally, geological data from the SGU were consulted to support the identification of underlying lithological formations and assess material erodibility.

2.2.2. Coastal Slope

Coastal slope plays a critical role in shaping nearshore wave behavior by determining how wave energy is dissipated and how sediments are transported and redistributed along the coast [37]. In this study, slope values were calculated using a 32-m resolution Digital Elevation Model (DEM) from OpenTopography, based on ArcticDEM Version 4 mosaics released in July 2023. This source was chosen for its consistent elevation coverage across the region, high spatial resolution, and open access, making it suitable for large-scale coastal analysis. Slope (%) was computed by dividing the vertical rise of 30 m by the horizontal distance between the 0 m shoreline and the inland 30 m contour (Equation (1)):

$$\text{Slope }(\%) = \left(\frac{\text{Rise }(30\text{ m})}{\text{Run (Distance)}}\right) * 100 \quad (1)$$

Variations in the coastal slope strongly influences how much wave energy reaches the base of the coast, especially during storm events. Slopes below 2.5% are generally more vulnerable, as they allow storm surges and wave energy to penetrate further inland. Slopes above 20% contribute to enhanced energy dissipation and resilience against erosion [38].

2.2.3. Hydrodynamic Parameters

To assess the hydrodynamic contribution to coastal vulnerability, three key parameters were incorporated into the CVI framework: Relative Sea-Level Change (RSLC), Mean Significant Wave Height and Tidal Range. Each was derived from national datasets and previous scientific studies, and selected for their direct influence on coastal inundation, erosion, and sediment dynamics [39].

RSLC was calculated using projections from SMA and SMHI, based on the IPCC SSP5-8.5 scenario for 2100 [40]. These projections were already locally adjusted for post-glacial land uplift using the NKG2016LU geodetic model [41] and referenced to Sweden's official vertical datum, RH 2000. RSLC rates (in mm/year) were estimated from 86 tide gauge stations by calculating the difference in projected mean sea level between 2025 and 2100 [41].

Tidal range is defined as the difference in height between the highest and lowest tides [42]. Mean tidal range was derived from SMHI tide gauge records [43], which document Sweden's characteristically low tidal amplitudes. Semi-diurnal tidal ranges vary from approximately 10 cm in the Skagerrak to less than 5 cm in the Baltic Sea. These values were classified uniformly due to minimal spatial variation, recognizing that low tidal range can amplify erosion by focusing wave energy over a narrower vertical profile [39].

Significant wave height, defined as the mean of the highest one-third of waves, is a key control on coastal evolution [43]. Wave energy is directly related to wave height, governing processes such as sediment transport and erosion. Mean significant wave height data were obtained from SGU's numerical wave model outputs [44], covering the western and southern coasts of Sweden, and from long-term Baltic Sea wave records [45], covering the northern coastline. Together, these datasets provide consistent, model-based nearshore wave information along the entire Swedish coastline for the period 1979–2018.

2.2.4. Shoreline Change

Shoreline change is also an important parameter in the CVI, as it shows whether coasts are gradually eroding or gaining land over time, which is a key factor for understanding how stable a coastal area is. These long-term trends can have serious consequences for infrastructure, ecosystems, and coastal communities. As climate change increases pressure on coastlines, shoreline change becomes even more relevant [46,47]. For this study, shoreline change data from the EMODnet Coastal Migration dataset were used, which combines satellite imagery from ESA Sentinel-2 and NASA's Landsat missions (5, 7, and 8). These satellites provide spatial resolutions between 10 and 30 m, with image updates every 1–2 weeks. The EMODnet team processed this imagery (2007–2017) using tools such as Google Earth Engine to extract long-term shoreline trends. Using this dataset allowed for consistent and well-documented shoreline change estimates across Sweden's coast, which helped improve the reliability of our CVI analysis.

3. Methodology

3.1. Pre-Processing

For the CVI estimation, the Swedish coastline was segmented into more than 2000 units in the GIS environment, allowing the index to be calculated separately for each individual stretch of coast. A key focus during this stage was on the geomorphological classification, particularly addressing unique features of the Swedish coastline such as "green beaches", low-lying, vegetated zones with partial marsh characteristics. These areas pose classification challenges due to their hybrid nature, however while vegetation provides stabilizing functions similar to dune systems, their elevation and sediment type often increase vulnerability. To resolve this, updated geomorphological data were integrated from EMODnet and high-resolution imagery (Google Earth Pro), capturing the dual ecological and physical nature of these areas.

Future sea-level projections were derived from SMA and SMHI datasets, based on IPCC SSP5-8.5 median scenarios. These projections were adjusted using the NKG2016LU model to incorporate local glacial isostatic adjustment (GIA). RSLC was calculated for 86 tide gauge stations using the formula (Equation (2)):

$$\mathrm{RSLC}\left(\frac{\mathrm{mm}}{\mathrm{year}}\right) = \frac{\mathrm{MSL2100} - \mathrm{MSL2025}}{75} \tag{2}$$

where MSL2100 and MSL2025 are the calculated mean sea level of each tide gauge for the year 2100 and 2025 relatively [41]. The denominator of the fraction is derived from the time difference 2100–2025.

Mean significant wave height data were sourced from SGU model and Baltic Sea records and interpolated via Thiessen polygons. The analysis was conducted using long-term datasets (3-hourly, 1979–2018), which inherently include seasonal variability and storm events, capturing overall wave exposure rather than short-term extremes. The protective role of archipelagos and offshore islands was integrated into exposure assessments, as these features can significantly attenuate wave energy. Each coastal segment's wave height was then scored from 1 (low exposure) to 5 (high exposure), based on adjusted regional mean significant wave height values.

3.2. CVI Calculation

Once the preprocessing stage was completed, the CVI was calculated using ArcGIS Pro (v. 3.4.3). The methodology requires that each parameter is individually scored on a scale from 1 (very low) to 5 (very high) to reflect increasing levels of vulnerability. Specifically, higher values indicate greater susceptibility to that particular factor. The classification criteria used for scoring are presented in Table 2, while Equation (3) shows the formula used to compute the final index.

$$\mathrm{CVI} = \sqrt{\frac{a*b*c*d*e*f}{6}} \tag{3}$$

where a = geomorphology, b = coastal slope, c = relative sea-level change, d = shoreline displacement rate, e = mean tidal range and f = mean significant wave height. Finally, the resulting CVI map was exported in GeoTIFF format for further visualization and analysis.

Table 2. Classification criteria and numerical ranking used in the Coastal Vulnerability Index (CVI) following the methodology of [26].

Vulnerability	Very Low	Low	Moderate	High	Very High
Variables	1	2	3	4	5
Coastal slope (%)	>20	7.0–20	4.0–7.0	2.5–4	<2.5
Geomorphology	Rocky, cliffed coasts, Fjords, Fiards, Artif. Constructions	Medium Cliffs, Indented coasts	Low cliffs, Glacial drift, Alluvial plains, Beachrocks, Dunes, Mixed material	Cobble beaches, Estuaries, Lagoons	Barrier beaches, Sand beaches, Salt marshes, Mud flats, Deltas, Mangrove, Coral reefs
Shoreline erosion/accretion (m/yr)	>2.0	1.0–2.0	−1.0–+1.0	−1.0–−2.0	<−2.0
Relative sea-level change (mm/yr)	<1.8	1.8–2.5	2.5–3.0	3.0–3.2	>3.2
Mean Wave Height (m)	<0.55	0.55–0.85	0.85–1.05	1.05–1.25	>1.25
Mean Tide range(m)	>6.0	4.1–6.0	2.0–4.0	1.0–1.9	<1.0

4. Results

The results of the CVI assessment reveal substantial spatial variation in the vulnerability of Sweden's coastline, both in the overall index and across the six contributing parameters. Figure 3 shows the percentage distribution of vulnerability classes for each individual CVI variable, highlighting which parameters contribute most to overall coastal vulnerability. Figure 3 maps the spatial distribution of the six variables used in the CVI calculation: geomorphology, coastal slope, shoreline change rate, mean significant wave height, mean tidal range, and relative sea-level change. Figure 4 presents the final CVI classification of the Swedish coastline, divided into five levels of vulnerability ranging from very low to very high, and also includes a pie chart summarizing the proportion of coastline falling into each category. A detailed analysis each parameter and its spatial distribution is presented in the following sections.

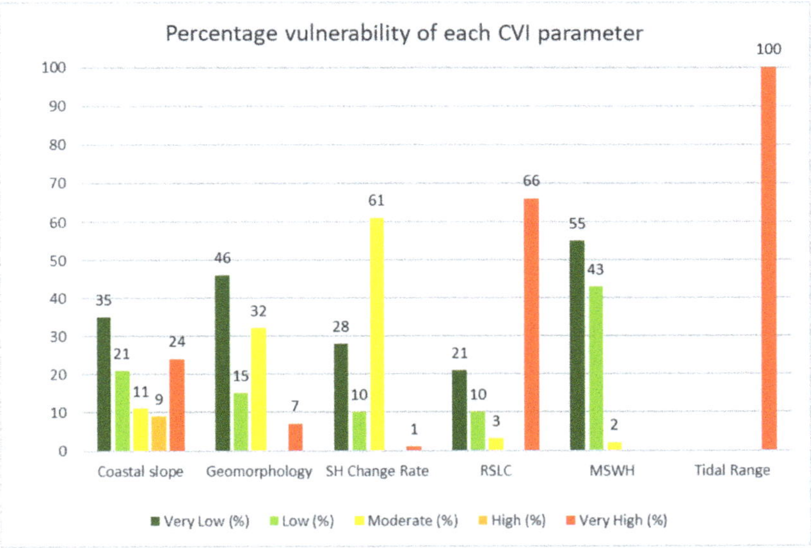

Figure 3. Percentage vulnerability of each CVI variable. Vulnerability is shown using a color scale, where red indicates high vulnerability and green indicates low vulnerability. Parameters include: SH Change Rate = Shoreline Change Rate, RSLC = Relative Sea-Level Change, and MSWH = Mean Significant Wave Height and Tidal Range.

Beginning with coastal slope, approximately 35% of the coastline falls within the steepest slope class (slope > 20%), which is generally associated with very low vulnerability (Figure 3, Table 2). Areas with high slope values (>7%), representing 21% of the coastline, are particularly concentrated along the Bothnian Sea and in the surroundings of major urban centers such as Gothenburg and Stockholm, which lie along the Skagerrak and the Baltic Sea, respectively. In contrast, 24% of the Swedish coastline is classified as highly vulnerable with respect to slope. These segments are primarily located in southern coastal zones exposed to the Baltic Sea, especially in Gotland, Kalmar, Sandviken, and Skanör (Figure 4a). These areas exhibit low slope angles, often below 2.5%.

The geomorphology variable in the CVI analysis shows a distribution comparable to that of coastal slope. Segments classified as having very low vulnerability account for 46% of the total coastline and are primarily located near Gothenburg and Stockholm. The moderate vulnerability class accounts for 32%, while 15% of the coastline falls under the low vulnerability category. Finally, 7% of the coastline is classified as very highly vulnerable based on geomorphological characteristics (Figures 3 and 4b).

When examining shoreline change rates, more than half of the Swedish coastline (61%) falls into the moderate vulnerability class (Figure 3), with the majority of changes ranging between −1 and +1 m per year (Table 2). This is followed by segments classified as exhibiting low vulnerability (10%) with change rates between 1–2 m per year, and very low vulnerability (28%) where rates exceed 2 m per year (Figure 4, Table 2). A small portion of the coastline, less than 1%, exhibits very high vulnerability, while there is no segment characterized as highly vulnerable in terms of shoreline change rate. This corresponds to areas with localized erosion hotspots where shoreline change exceeds these typical ranges.

RSLC emerges as the dominant factor driving coastal vulnerability. Approximately 66% of the coastline is categorized under the very high vulnerability class in this component alone (Figure 3). The spatial distribution of sea-level trends reveals a distinct gradient: southern coastal zones experience the most rapid relative sea-level rise, while northern Baltic regions show reduced or even negative trends. (Figure 4d). This gradient highlights the importance of regional differences in glacio-isostatic adjustment, although the protective effect of land uplift appears insufficient in southern and low-lying areas.

Wave exposure, as measured by MSWH, showed relatively low contributions to overall vulnerability. No regions were classified as having high or very high vulnerability under this factor, with wave heights remaining below 1.05 m along the entire coast (Figure 4). Finally, the tidal range across Sweden is not exceeding 0.55 m. As a result, the entire coastline is classified under very high vulnerability in terms of tidal influence, based on the CVI ranking system, which considers low tidal amplitude a contributing factor to coastal erosion (Table 2, Figure 3).

The spatial distribution of vulnerability reveals distinct regional contrasts (Figure 5). The southern coast of Sweden, particularly the Kattegat region and Gotland, Kalmar, Sandviken, and Skanör, exhibits higher vulnerability, indicated by the red and orange segments on the map. These areas are likely influenced by low-lying terrain and more vulnerable geomorphological features, such as sandy beaches. In contrast, the northern coastline, including the Bay of Bothnia and the Bothnian Sea (Figure 2), shows very low to low vulnerability, which can be attributed to higher coastal elevations, vegetated shorelines, and reduced wave energy exposure.

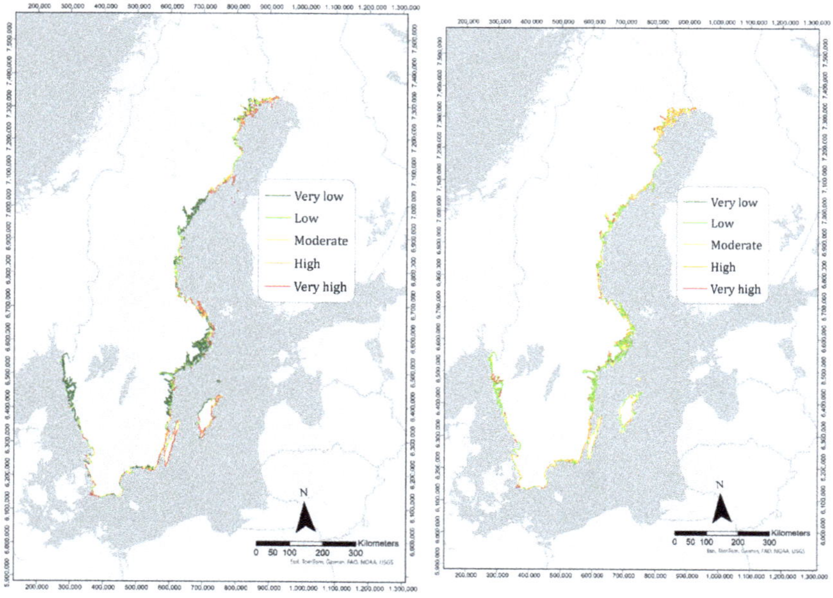

(a) (b)

Figure 4. *Cont.*

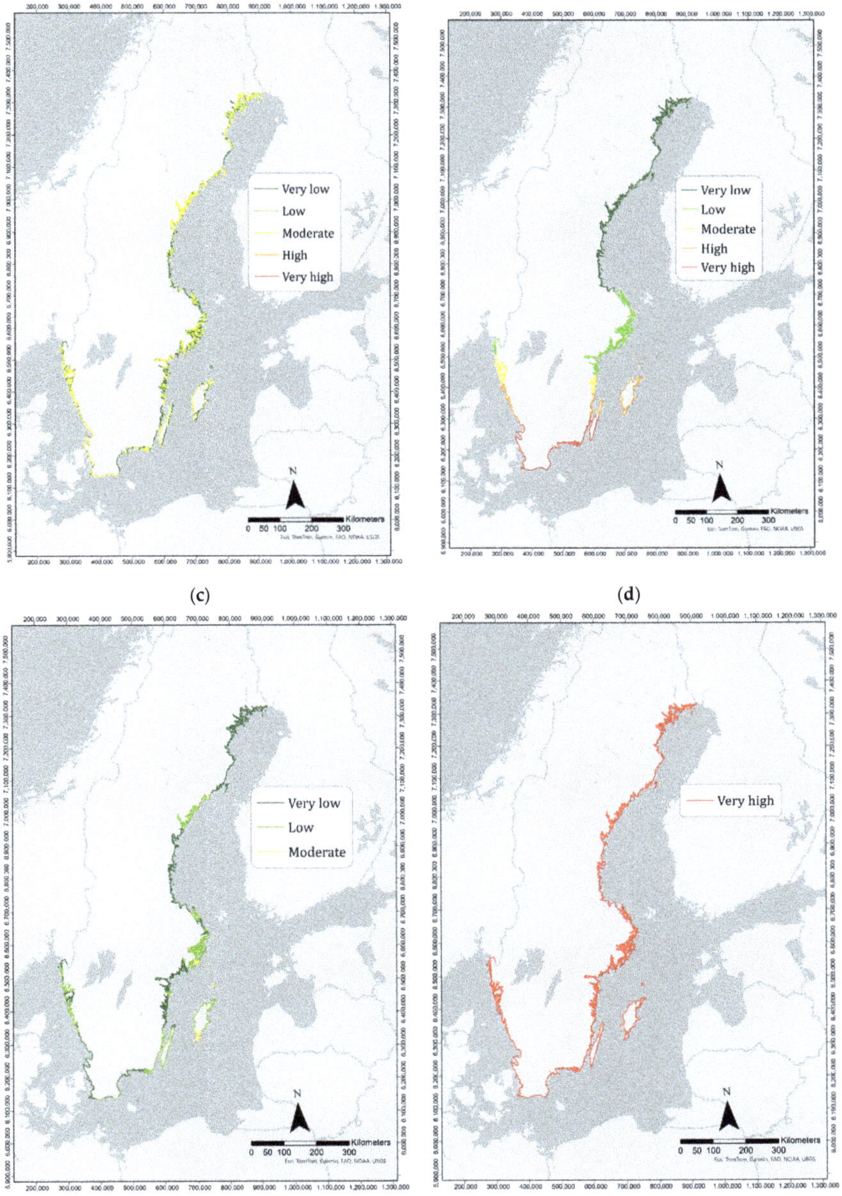

(c) (d)

(e) (f)

Figure 4. Spatial distribution of individual physical variables used in the CVI along the Swedish coastline. Maps (**a**–**f**) displays the relative vulnerability classification for each of the six physical parameters included in the CVI calculation: (**a**) coastal slope, (**b**) geomorphology, (**c**) shoreline change rate, (**d**) relative Sea-level change), (**e**) mean significant wave height, and (**f**) tidal range. The maps use a gradient color scheme, where green hues indicate low vulnerability and red hues indicate high vulnerability. These classifications allow for the identification of spatial trends and high-risk coastal segments based on individual variable contributions.

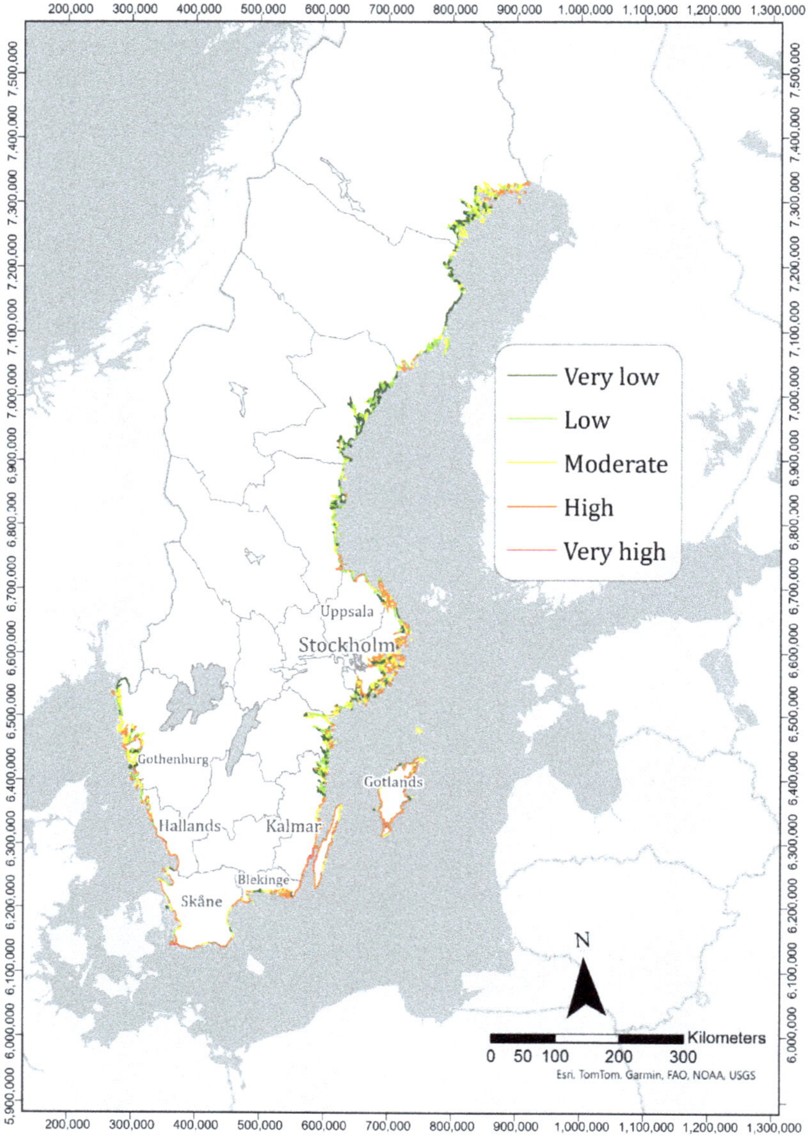

Figure 5. CVI classification of the Swedish coastline.

Overall, CVI results (Figure 6) indicate that the most prevalent vulnerability class is "High", accounting for 34.3% of the Swedish coastline. This is followed by "Moderate" (22.2%) and nearly equal proportions of "Very Low" (20.7%) and "Low" (20.6%). Only a small fraction (2.2%) of the coastline is classified with "Very High" vulnerability. These findings suggest that, while certain regions face considerable risk, a substantial portion of the coastline demonstrates moderate to low susceptibility, likely due to favorable topographic and oceanographic conditions.

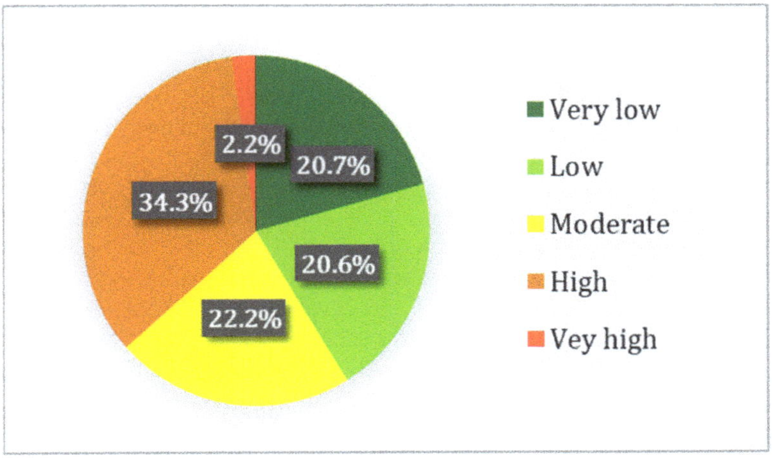

Figure 6. CVI Results Pie chart for coastline vulnerability in percentage (%).

5. Discussion

The present study applied the CVI to evaluate spatial patterns of physical coastal risk along the Swedish shoreline, aiming to identify areas most prone to sea-level rise and erosion. Areas classified as highly vulnerable are predominantly located in the southern regions, particularly along the Baltic Sea, Skagerrak, and Kattegat, while the Gulf of Bothnia shows consistently lower vulnerability. This section discusses these findings in more detail, examining the contribution of each CVI parameter, comparing the results to those of related studies in the Baltic and Arctic regions as well as with other international coastal areas. Finally, it also considers the methodological limitations of the approach.

Regarding coastal slope, about 35% of the Swedish coastline features steep gradients (>20%) and is thus considered less susceptible to wave impact and flooding. These areas are especially concentrated around the Bothnian Sea and urban centers like Stockholm and Gothenburg. On the contrary, 24% of the coast, particularly in Gotland, Kalmar, Sandviken, and Skanör, exhibits gentle slopes and is classified as highly vulnerable (Figure 4a). This spatial pattern is consistent with the distribution of sedimentary and marshy landforms in southern Sweden, indicating areas of reduced natural resilience to coastal hazards.

The geomorphology of the region closely aligns with the spatial patterns observed in coastal slope vulnerability. Areas with low vulnerability in terms of slope are often characterized by erosion-resistant rocky shorelines or artificial coastlines such as harbors and engineered structures, features commonly observed around Gothenburg, Stockholm, and much of the Bothnian Sea coastline (Figure 4b). These regions are typically more resilient to erosion and sea-level rise due to their geomorphological characteristics and the human interventions. Conversely, southern parts of the country, especially around Skåne, show a predominance of sandy, vegetated, or marshy coastal types, associated with softer sediment composition and greater exposure. Approximately 46% of the Swedish coastline falls into the very low vulnerability category based on geomorphology, primarily reflecting the dominance of hard substrates or reinforced shores in the urbanized and glacially uplifted zones. In contrast, about 7% of the coast is classified as very highly vulnerable due to soft sediment environments, emphasizing the necessity for tailored coastal protection in those areas.

Shoreline change rates further illustrate the dynamics of coastal vulnerability. Approximately 61% of the Swedish coastline falls under the moderate vulnerability class, with shoreline changes ranging between −1 to +1 m per year. The remaining segments are characterized as having low (10%) and very low (28%) vulnerability, corresponding to shoreline progradation exceeding 1 m per year, often driven by anthropogenic activities such as land reclamation or coastal development. Although localized erosion hotspots represent less than 1% of the total coastline, they exceed typical erosion thresholds and are of critical importance, as they coincide with areas of minimal natural resistance and pose a significant threat to nearby infrastructure.

Among all factors, RSLC is the most significant contributor to overall vulnerability. Approximately 66% of the coastline is categorized as very highly vulnerable, underscoring the impact of glacio-isostatic differences across the country. While northern regions still benefit from land uplift, southern areas face compounded threats from ongoing subsidence and accelerated sea-level rise. These findings are consistent with previous studies [33,48], which emphasize the diminishing protective effect of land uplift in southern Sweden.

MSWH had only a minor influence on the final vulnerability classification of the Swedish coastline, as values remain consistently below 1.05 m. This limited wave action is largely due to the protective role of the Gulf of Bothnia and the numerous islands, which restrict wave fetch and reduce overall wave energy along the coast.

Tidal range, which remains uniformly low across Sweden (generally below 0.55 m), results in the entire coastline being classified as very highly vulnerable under the CVI framework. This classification reflects the broader understanding that waves remain in continuous contact with the dune foot or coastal base for extended periods, especially during storm surges, thereby increasing the shoreline's exposure to erosion [48].

The spatial patterns of coastal vulnerability identified in this study generally align with previous research on Sweden's southern coastline, particularly in regions such as Gotland, Kalmar, and Skåne. In Gotland, past investigations have emphasized the erosional susceptibility of the island due to its geological composition, especially the presence of marlstone formations, as well as the increasing impact of frost weathering and sea-level rise [49]. These observations are consistent with the present CVI results, which classify much of Gotland's coastline as highly vulnerable. Similarly, Kalmar and the adjacent island of Öland have been repeatedly cited in national assessments [50,51] as erosion-prone areas, largely due to their low topographic relief and sensitivity to storm surges. The CVI mapping confirms these findings (Figure 6), with high vulnerability scores attributed to their gentle slopes and ongoing shoreline retreat. In the Skåne region, coastal sites such as Ystad, Bjärred, Lomma, and Ängelholm also display high erosion potential, driven by easily erodible coastal geomorphology, sea-level rise, and intense wave activity during storm events [52,53]. These patterns mirror the current study's classification of these localities within the high to very high vulnerability categories, reinforcing the role of RSLC, shoreline change rates, slope, and coastal geomorphology as critical drivers of coastal risk. Despite regional variations, the results of this study broadly validate the trends reported in previous research while offering updated spatial insights into areas of emerging concern.

To place these findings in a broader international context, post-glacial North American coasts offer a useful comparison. The uplift rate there exceeds contemporary sea-level rise, leading to long-term negative relative sea-level trends [54]. Such uplift rates are comparable to, or in some cases greater than, those typically reported along the Swedish coast. This suggests that certain coastal margins can be classified as uplift-dominated systems, where vertical land motion is the primary determinant of shoreline evolution rather than eustatic forcing [55].

A similar comparison applies to tidal influence. In macrotidal environments such as the Bay of Fundy in Canada, tidal oscillations can exceed several meters. This, results in buffering storm-surge impacts by distributing water levels across a wider vertical range [56]. By contrast, microtidal coasts, where tidal amplitudes are only a few centimeters, result in persistent wave impact at a fixed shoreline level. The exposure becomes a threat when it is consistent across shallow lagoonal and insular coasts, where tidal ranges typically below one meter [57]. Given that tidal amplitudes along the Swedish coast remain below 0.5 m, the study area aligns more closely with other microtidal margins than with macrotidal ones.

The influence of small islands on coastal exposure has been documented across several archipelagic regions. In the Scottish Hebrides, wave models indicate that islands reduce and redirect energy [58]. A similar contrast has been reported in southern Finland, where wave energy dissipates before it reaches the coast [59]. A comparable mechanism is likely present along the Swedish archipelago, where island formations may effectively act as natural buffers. This study provides a spatial overview of coastal vulnerability in Sweden using the CVI framework; however, several limitations should be acknowledged. The use of medium-resolution datasets, such as 32-m elevation model and EMODnet-derived geomorphology, reduces spatial precision, especially in areas with abrupt terrain changes [29]. Moreover, the CVI assigns equal weights to all parameters and uses fixed classification thresholds. While this promotes comparability, many researchers have noted, this lack of weighting may oversimplify the influence of dominant factors and overlook localized or site-specific vulnerability factors [60]. In the Swedish context, this may underrepresent key processes such as post-glacial rebound and low wave energy. Furthermore, the CVI framework inherently classifies microtidal coasts as highly vulnerable, which in this study resulted in 100% of the Swedish coastline being scored as very high risk with respect to tidal range. This outcome reflects a methodological limitation of the CVI and should be considered with caution, while also indicating the potential value of future refinements tailored to microtidal settings. Lastly, the static nature of CVI does not reflect dynamic processes like storm surges, ice melt, or short-term erosional cycles [61].

Although this study focuses on physical drivers, the approach can complement broader assessments that include socio-economic factors. Future research could integrate data on population density, infrastructure, land use, and adaptive capacity to support more locally tailored planning strategies [62]. Additionally, conducting a sensitivity analysis would clarify how variability in input parameters influences CVI results, improving methodological robustness [63]. Finally, using higher-resolution datasets and performing targeted ground-truthing would enhance spatial precision [36] and increase the decision-making value of coastal vulnerability assessments in Sweden. Given the uneven global distribution of monitoring capacity, a pragmatic next step is to account explicitly for data availability and quality in CVI applications so that outcomes remain comparable between data-rich regions and island-dominated coasts where reanalysis and satellite proxies are often the only viable inputs.

Beyond the spatial assessment itself, this study also offers a useful dataset for the wider coastal research community, especially for those working on high-latitude regions. The full CVI output is openly available and can be reused in other studies that investigate shoreline change or sea-level rise impacts. The method applied is fully reproducible and can be adjusted depending on the needs of each study area, for example, by changing the spatial resolution, the weighting of parameters or the classification thresholds. This makes it possible to apply the same approach to other high-latitude coasts in order to study spatial or temporal variations in vulnerability. Another advantage is that the results are easy to understand and can therefore be used not only by researchers but also by planners and stakeholders. This helps connect the physical processes with their social and economic consequences, and supports more informed decision making.

6. Closing Remarks

This study aimed to assess the physical vulnerability of Sweden's coastline through the first nationwide application of the CVI. To our knowledge, this is the first study of its kind in Sweden, offering a spatially consistent and scalable method for identifying areas most at risk from sea-level rise, erosion, and land subsidence. By integrating six standardized geophysical parameters into a GIS-based framework, it provides new insights into Sweden's exposure patterns and contributes a robust decision-support tool for national and regional coastal management.

The results show that relative sea-level rise is the most influential driver of coastal vulnerability in Sweden. Approximately 66% of the coastline is classified as highly vulnerable under this parameter, particularly in the southern regions, where ongoing land uplift is no longer sufficient to offset rising sea levels. Additionally, the uniformly low tidal range across the country amplifies vulnerability by allowing persistent wave energy to act directly on the coastline, especially during storm surges, increasing erosion risks. In contrast, about 41% of the coastline is classified as having low to very low vulnerability, primarily along the northern coast, including the Bay of Bothnia and the Bothnian Sea. In these areas, wave energy is limited, and geomorphological characteristics, such as rocky or glacially uplifted shorelines, contribute significantly to reducing exposure and maintaining natural resilience. These patterns highlight the pressing need for site-specific monitoring and the integration of socio-economic and cultural indicators in future assessments to support targeted adaptation and management strategies.

Importantly, this study demonstrates that positive glacial isostatic adjustment (GIA) does not automatically guarantee long-term coastal stability. While uplift acts as a natural buffer in northern Sweden, the CVI results reveal 'threshold zones' in southern and central regions where uplift and eustatic sea-level rise are approaching equilibrium. These transition areas, represent emerging hotspots of future submergence and could serve as early-warning indicators for other Arctic and sub-Arctic coastlines.

Furthermore, by disentangling the relative influence of uplift, tidal amplitude, and wave energy within the CVI framework, this study offers a methodological guidance for adapting coastal indices to high-latitude environments dominated by vertical land motion rather than hydrodynamic forcing. This approach may be transferable to uplift-dominated regions.

Beyond the physical assessment, the openly available CVI dataset and reproducible workflow provide an operational foundation for integrating socio-economic indicators—such as population density, infrastructure exposure, and adaptive capacity—into future risk-oriented assessments. Rather than treating physical and social vulnerability separately, this framework allows planners in uplift-dominated regions to distinguish between coastlines where GIA still acts as a protective buffer and those where human assets are already exposed despite ongoing uplift. Future research should build on this work by integrating social vulnerability indicators, such as population density, infrastructure exposure, and adaptive capacity, for a more comprehensive understanding of coastal risk. Incorporating sensitivity analyses, higher-resolution EO data, and localized field observations would further enhance the reliability and policy relevance of CVI-based assessments in Sweden. Together, these steps will strengthen the foundation for climate adaptation planning and ensure more targeted, locally informed responses to coastal hazards.

Author Contributions: Conceptualization, E.A., N.E. and G.P.P.; methodology, E.A., N.E. and A.K.; formal analysis, E.A.; writing—original draft preparation, E.A.; writing—review and editing, E.A., G.P.P., A.K., N.E., G.N.T. and S.E.D.; visualization, E.A.; supervision, G.P.P. and N.E. All authors have read and agreed to the published version of the manuscript.

Funding: This research was financially supported by the project "EO-PERSIST", funded through the European Union's Horizon Europe Research and Innovation Programme (HORIZON-MSCA-2021-SE-01-01) under grant agreement No. 101086386.

Data Availability Statement: The data presented in this study are derived from the following resources available in the public domain The data used in this study are derived from publicly available resources: EMODnet: https://emodnet.ec.europa.eu/geoviewer/#!/ (accessed on 25 May 2025); ArcticDEM (OpenTopography): https://portal.opentopography.org/arcticDem?opentopoID=OTSDEM.112018.3413.3 (accessed on 25 May 2025); Swedish Coastline Data Portal: https://gis.sgi.se/kustdataportal/# (accessed on 25 May 2025); Future Sea Levels Report (BSHC): https://www.bshc.pro/wp-content/uploads/Future_sea_levels.pdf (accessed on 25 May 2025); SMHI: https://www.smhi.se/vader/vader-till-havs/vattenstand-och-vagor (accessed on 25 May 2025).

Conflicts of Interest: Author Eleni Achmakidou was employed by the company ALTUS LSA. The remaining authors declare that the research was conducted in the absence of any commercial or financial relationships that could be construed as a potential conflict of interest.

References

1. Short, A.D. Coastal Processes and Beaches. *Nat. Educ. Knowl.* **2012**, *3*, 15.
2. Mörner, N.A.; Finkl, C.W. Coastal Erosion. In *Encyclopedia of Coastal Science*; Encyclopedia of Earth Sciences Series; Finkl, C.W., Makowski, C., Eds.; Springer: Cham, Switzerland, 2019. [CrossRef]
3. Alessa, L.; Valentine, J.; Moon, S.; McComb, C.; Hicks, S.; Romanovsky, V.; Xiao, M.; Kliskey, A. Toward a Permafrost Vulnerability Index for Critical Infrastructure, Community Resilience and National Security. *Geographies* **2023**, *3*, 522–542. [CrossRef]
4. Toumasi, P.; Petropoulos, G.P.; Detsikas, S.E.; Kalogeropoulos, K.; Tselos, N.G. Coastal Vulnerability Impact Assessment under Climate Change in the Arctic Coasts of Tromsø, Norway. *Earth* **2024**, *5*, 640–653. [CrossRef]
5. Casas-Prat, M.; Wang, X.L. Projections of Extreme Ocean Waves in the Arctic and Potential Implications for Coastal Inundation and Erosion. *J. Geophys. Res. Ocean.* **2020**, *125*, e2019JC015745. [CrossRef]
6. Nielsen, D.M.; Pieper, P.; Barkhordarian, A.; Overduin, P.; Ilyina, T.; Brovkin, V.; Dobrynin, M. Increase in Arctic Coastal Erosion and Its Sensitivity to Warming in the Twenty-First Century. *Nat. Clim. Chang.* **2022**, *12*, 263–270. [CrossRef]
7. Tanguy, R.; Bartsch, A.; Nitze, I.; Irrgang, A.; Petzold, P.; Widhalm, B.; von Baeckmann, C.; Boike, J.; Martin, J.; Efimova, A.; et al. Pan-Arctic assessment of coastal settlements and infrastructure vulnerable to coastal erosion, sea-level rise, and permafrost thaw. *Earth's Future* **2024**, *12*, e2024EF005013. [CrossRef]
8. Whalen, D.; Forbes, D.L.; Kostylev, V.; Lim, M.; Fraser, P.; Nedimović, M.R.; Stuckey, S. Mechanisms, Volumetric Assessment, and Prognosis for Rapid Coastal Erosion of Tuktoyaktuk Island, an Important Natural Barrier for the Harbour and Community. *Can. J. Earth Sci.* **2022**, *59*, 945–960. [CrossRef]
9. Ebert, K.; Ekstedt, K.; Jarsjö, J. GIS Analysis of Effects of Future Baltic Sea Level Rise on the Island of Gotland, Sweden. *Nat. Hazards Earth Syst. Sci.* **2016**, *16*, 1571–1582. [CrossRef]
10. Swedish Ministry of the Environment. *Sweden's Adaptation Communication—A Report to the United Nations Framework Convention*; Government Offices of Sweden: Stockholm, Sweden, 2022. Available online: https://unfccc.int/sites/default/files/ACR/2022-11/ADCOM_Sweden_November_221114.pdf (accessed on 1 October 2025).
11. Leszczyńska, K.; Alexanderson, H.; Clemmensen, L.B.; Giza, A.; Lorenz, S.; Moskalewicz, D.; Oliński, P.; Paprotny, D.; Rosentau, A.; Rutgersson, A.; et al. A Review of Storms and Marine Coastal Flooding in the Baltic Sea—Insights from Instrumental, Historical and Sedimentary Record. *Earth-Sci. Rev.* **2025**, *266*, 105137. [CrossRef]
12. Kopp, H.; Chiocci, F.L.; Berndt, C.; Çağatay, M.N.; Ferreira, T.; Fortes, C.J.E.M.; Gràcia, E.; González Vega, A.; Kopf, A.J.; Yeo, I.A. *Marine Geohazards: Safeguarding Society and the Blue Economy*; Position Paper 26 of the European Marine Board; European Marine Board: Ostend, Belgium, 2021; Volume 26, 100p. [CrossRef]
13. Ren, Z.; Higuera, P.; Liu, P.L.-F. On Tsunami Waves Induced by Atmospheric Pressure Shock Waves after the 2022 Hunga Tonga-Hunga Ha'apai Volcano Eruption. *J. Geophys. Res. Ocean.* **2023**, *128*, e2022JC019166. [CrossRef]
14. Wang, Y.; Zamora, N.; Quiroz, M.; Satake, K.; Cienfuegos, R. Tsunami Resonance Characterization in Japan due to Trans-Pacific Sources: Response on the Bay and Continental Shelf. *J. Geophys. Res. Ocean.* **2021**, *126*, e2020JC017037. [CrossRef]
15. Xu, I.; Xu, C.; Ren, Z.; Wang, D. Numerical Investigation of Rip Currents near a Barred Beach Induced by Irregular Waves. *Proc. Inst. Mech. Eng. Part M J. Eng. Marit. Environ.* **2024**, *239*, 418–434. [CrossRef]
16. Ji, X.; Wang, H.; Lu, X.; Wang, D.; Zou, L.; Ren, Z. Hydrodynamic Performance of Riprap Structures in Reef Lagoon under Extreme Wave Loading: Insights from Porous Media Simulations. *Phys. Fluids* **2025**, *37*, 051702. [CrossRef]

17. Sugiyama, S.; Yamaguchi, A.; Watanabe, T.; Tojo, Y.; Hayashi, N.; Thiebot, J.-B.; Tomiyasu, M.; Hasegawa, K.; Mitani, Y.; Otsuki, M.; et al. Changes in the Coastal Environments and Their Impact on Society in the Qaanaaq Region, Northwestern Greenland. *Polar Sci.* **2025**, 101206. [CrossRef]
18. LIFE Coast Adapt. Protection of Coastal Values. 2024. Available online: https://utveckling.skane.se/siteassets/publikationer/life-coast-adapt-engelska.pdf (accessed on 25 May 2025).
19. Donatus, B.; Angnuureng, B.; Charuka, B.; Almar, R.; Dada, O.A.; Asumadu, R. Challenges and Lessons Learned from Global Coastal Erosion Protection Strategies. *iScience* **2025**, *28*, 112055. [CrossRef]
20. Fuldauer, L.I.; Adshead, D.; Thacker, S.; Gall, S.; Hall, J.W. Evaluating the Benefits of National Adaptation to Reduce Climate Risks and Contribute to the Sustainable Development Goals. Glob. *Environ. Change* **2022**, *76*, 102575. [CrossRef]
21. Tsatsaris, A.; Kalogeropoulos, K.; Stathopoulos, N.; Louka, P.; Tsanakas, K.; Tsesmelis, D.E.; Krassanakis, V.; Petropoulos, G.P.; Pappas, V.; Chalkias, C. Geoinformation Technologies in Support of Environmental Hazards Monitoring under Climate Change: An Extensive Review. *ISPRS Int. J. Geo-Inf.* **2021**, *10*, 94. [CrossRef]
22. Kim, J.-I.; Hyun, C.-U.; Han, H.; Kim, H.-C. Digital Surface Model Generation for Drifting Arctic Sea Ice with Low-Textured Surfaces Based on Drone Images. *ISPRS J. Photogramm. Remote Sens.* **2021**, *172*, 147–159. [CrossRef]
23. Petropoulos, G.P.; Petsini, T.; Detsikas, S.E. Geoinformation Technology in Support of Arctic Coastal Properties Characterization: State of the Art, Challenges, and Future Outlook. *Land* **2024**, *13*, 776. [CrossRef]
24. Azlan, A.; Sahat, S.; Daud, M.E.; Kaamin, M.; Sultan, Z. Assessing Coastal Erosion Risks: A Comparative Study of Coastal Vulnerability Index and Coastal Erosion Risk Assessment Methods. *J. Adv. Res. Des.* **2025**, *132*, 91–102. [CrossRef]
25. Gornitz, V.; White, T.; Cushman, R. *Vulnerability of the U.S. to Future Sea Level Rise*; Oak Ridge National Laboratory: Oak Ridge, TN, USA, 1991.
26. Thieler, E.R.; Hammar-Klose, E.S. National Assessment of Coastal Vulnerability to Future Sea-Level Rise: Preliminary Results for the U.S. Pacific Coast. In *U.S. Geological Survey Open-File Report*; U.S. Geological Survey: Woods Hole, MA, USA, 2000. Available online: http://pubs.usgs.gov/of/2000/of00-178/ (accessed on 1 October 2025).
27. Pendleton, E.A.; Hammer-Klose, E.S.; Thieler, E.R.; Williams, S.J. *Coastal Vulnerability Assessment of Gulf Islands National Seashore (GUIS) to Sea Level Rise; U.S. Geological Survey Open-File Report 03-108*; U.S. Geological Survey: Woods Hole, MA, USA, 2004; 18p. Available online: http://pubs.usgs.gov/of/2003/of03-108/ (accessed on 1 October 2025).
28. Noor, N.M.; Abdul Maulud, K.N. Coastal Vulnerability: A Brief Review on Integrated Assessment in Southeast Asia. *J. Mar. Sci. Eng.* **2022**, *10*, 595. [CrossRef]
29. Rocha, C.; Antunes, C.; Catita, C. Coastal Indices to Assess Sea-Level Rise Impacts—A Brief Review of the Last Decade. *Ocean Coast. Manag.* **2023**, *237*, 106536. [CrossRef]
30. Theocharidis, C.; Doukanari, M.; Kalogirou, E.; Christofi, D.; Mettas, C.; Kontoes, C.; Hadjimitsis, D.; Argyriou, A.V.; Eliades, M. Coastal Vulnerability Index (CVI) Assessment: Evaluating Risks Associated with Human-Made Activities along the Limassol Coastline, Cyprus. *Remote Sens.* **2024**, *16*, 3688. [CrossRef]
31. Statistics Sweden (SCB). 2024. Available online: https://www.statistikdatabasen.scb.se/ (accessed on 1 October 2025).
32. Łabuz, T.A. Environmental Impacts—Coastal Erosion and Coastline Changes. In *Second Assessment of Climate Change for the Baltic Sea Basin*; The BACC II Author Team, Ed.; Regional Climate Studies; Springer: Cham, Switzerland, 2015. [CrossRef]
33. SGU Geology of Sweden. 2020. Available online: https://www.sgu.se/en/geology-of-sweden/ (accessed on 1 October 2025).
34. Hieronymus, M.; Kalén, O. Sea-Level Rise Projections for Sweden Based on the New IPCC Special Report: *The Ocean and Cryosphere in a Changing Climate*. *Ambio* **2020**, *49*, 1587–1600. [CrossRef] [PubMed]
35. SCB Islands in Sweden. Number, Area and Perimeter by Region. 2013. Available online: https://www.statistikdatabasen.scb.se/pxweb/en/ssd/START__MI__MI0812__MI0812T/OarArealOmk/ (accessed on 1 October 2025).
36. Reusch, T.B.H.; Dierking, J.; Andersson, H.C.; Bonsdorff, E.; Carstensen, J.; Casini, M.; Zandersen, M. The Baltic Sea as a Time Machine for the Future Coastal Ocean. *Sci. Adv.* **2018**, *4*, eaar8195. [CrossRef]
37. Komi, A.; Petropoulos, A.; Evelpidou, N.; Poulos, S.; Kapsimalis, V. Coastal Vulnerability Assessment for Future Sea Level Rise and a Comparative Study of Two Pocket Beaches in Seasonal Scale, Ios Island, Cyclades, Greece. *J. Mar. Sci. Eng.* **2022**, *10*, 1673. [CrossRef]
38. Pantusa, D.; D'Alessandro, F.; Riefolo, L.; Principato, F.; Tomasicchio, G.R. Application of a Coastal Vulnerability Index. A Case Study along the Apulian Coastline, Italy. *Water* **2018**, *10*, 1218. [CrossRef]
39. Xu, H.; Xu, N.; Zhang, C.; Chi, S.; Li, Y.; Li, W.; Lu, H. A Machine Learning Based Estimation Method of Beach Slopes at a National Scale: A Case Study of New Zealand. *Geo-Spat. Inf. Sci.* **2025**, 1–20. [CrossRef]
40. Manno, G.; Azzara, G.; Lo Re, C.; Martinello, C.; Basile, M.; Rotigliano, E.; Ciraolo, G. An Approach for the Validation of a Coastal Erosion Vulnerability Index: An Application in Sicily. *J. Mar. Sci. Eng.* **2023**, *11*, 23. [CrossRef]
41. Intergovernmental Panel on Climate Change (IPCC). Climate Change 2021—The Physical Science Basis: Working Group I Contribution to the Sixth Assessment Report of the Intergovernmental Panel on Climate Change. 2021. Available online: https://www.ipcc.ch/report/ar6/wg1/ (accessed on 1 October 2025).

42. SMA Future Sea Levels in Sweden. 2025. Thomas Hammarklint Published 2019-11-30. Available online: https://www.bshc.pro/wp-content/uploads/Future_sea_levels.pdf (accessed on 1 October 2025).
43. Beluru, J.; Akshaya, H.; Hegde, A.V. GIS Based Approach for Vulnerability Assessment of the Karnataka Coast, India. *Adv. Civ. Eng.* **2016**, *2016*, 5642523. [CrossRef]
44. SMHI. Ocean Waves 2016. 2024. Available online: https://www.smhi.se/klimat/klimatet-da-och-nu/arets-vatten/vattenaret/2017-03-31-havsvagor-2016?utm (accessed on 1 October 2025).
45. Adell, A.; Nunes De Brito Junior, A.; Almström, B.; Goodfellow, B.; Bokhari Irminger, S.; Hallin, C.; Nyberg, J. Open-Access Portal with Hindcast Wave Data for Skåne and Halland. *Vatten Tidskr. Vattenvård/J. Water Manag. Res.* **2021**, *77*, 81–90. Available online: https://www.tidskriftenvatten.se/wp-content/uploads/2021/06/VATTEN_21_2_81-90.pdf (accessed on 1 October 2025).
46. Sokolov, A.; Chubarenko, B. Baltic Sea Wave Climate in 1979–2018: Numerical Modelling Results. *Ocean Eng.* **2024**, *320*, 117088. [CrossRef]
47. Luijendijk, A.; Hagenaars, G.; Ranasinghe, R.; Baart, F.; Donchyts, G.; Aarninkhof, S. The State of the World's Beaches. *Sci. Rep.* **2018**, *8*, 6641. [CrossRef]
48. Vousdoukas, M.I.; Ranasinghe, R.; Mentaschi, L.; Plomaritis, T.A.; Athanasiou, P.; Luijendijk, A.; Feyen, L. Sandy Coastlines under Threat of Erosion. *Nat. Clim. Change* **2020**, *10*, 260–263. [CrossRef]
49. Różyński, G. Coastal Protection Challenges after Heavy Storms on the Polish Coast. *Cont. Shelf Res.* **2023**, *266*, 104997. [CrossRef]
50. Glansholm, H. Coastal Erosion on Högklint, Gotland: Processes, Rate, and Future Implications. Master's Thesis, Uppsala University, Department of Earth Sciences, Uppsala, Sweden, 2015. Available online: https://uu.diva-portal.org/smash/record.jsf?pid=diva2%3A856411&dswid=6354 (accessed on 1 October 2025).
51. Þórisdóttir, V. Future Challenges of Kalmar City: Coastal Adaptation Strategies Design in Relation to Sea Level Rise. Master's Thesis, Swedish University of Agricultural Sciences (SLU), Department of Urban and Rural Development, Uppsala, Sweden, 2023. Available online: http://urn.kb.se/resolve?urn=urn:nbn:se:slu:epsilon-s-500682 (accessed on 1 October 2025).
52. SGU Kartvisaren: Erosionsförhållanden på Stranden. 2023. Available online: https://apps.sgu.se/kartvisare/kartvisare-skanestrand.html (accessed on 1 October 2025).
53. Hamilton, A. Coastal Erosion in Ängelholm: A Case Study Tracking Recent and Historical Changes to Vejbystrand. Bachelor's Thesis, Lund University, Lund, Sweden, 2019. [CrossRef]
54. Engström, S.; Olsson, E. Coastal Response to Sea Level Rise in Ystad Municipality: Development, Comparison and Application of Bruun Rule-Based Methods. Master's Thesis, Lund University, Division of Water Resources Engineering, Department of Building & Environmental Technology, Lund, Sweden, 2021. Available online: https://lup.lub.lu.se/luur/download?func=downloadFile&recordOId=9060526&fileOId=9060529 (accessed on 1 October 2025).
55. Shugar, D.H.; Walker, I.J.; Lian, O.B.; Eamer, J.B.R.; Neudorf, C.; McLaren, D.; Fedje, D. Post-Glacial Sea-Level Change along the Pacific Coast of North America. *Quat. Sci. Rev.* **2014**, *97*, 170–192. [CrossRef]
56. Greenberg, D.A.; Blanchard, W.; Smith, B.; Barrow, E. Climate Change, Mean Sea Level and High Tides in the Bay of Fundy. *Atmos.-Ocean.* **2012**, *50*, 261–276. [CrossRef]
57. Tosi, L.; Da Lio, C.; Cosma, M.; Donnici, S. Vulnerability of Tidal Morphologies to Relative Sea-Level Rise in the Venice Lagoon. *Sci. Total Environ.* **2024**, *905*, 173006. [CrossRef]
58. Gleizon, P.; Woolf, D. Wave Energy Assessment in Scottish Seas. In Proceedings of the 10th European Wave and Tidal Energy Conference (EWTEC), Aalborg, Denmark, 2–5 September 2013. Available online: https://www.researchgate.net/publication/263279982_Wave_Energy_Assessment_in_Scottish_Seas (accessed on 1 October 2025).
59. Björkqvist, J.-V.; Tuomi, L.; Fortelius, C.; Pettersson, H.; Tikka, K.; Kahma, K.K. Improved Estimates of Nearshore Wave Conditions in the Gulf of Finland. *J. Mar. Syst.* **2017**, *171*, 43–53. [CrossRef]
60. Sekovski, I.; Del Río, L.; Armaroli, C. Development of a Coastal Vulnerability Index Using Analytical Hierarchy Process and Application to Ravenna Province (Italy). *Ocean Coast. Manag.* **2020**, *183*, 104982. [CrossRef]
61. Barzehkar, M.K.; Parnell, K.; Soomere, T. Extending Multi-Criteria Coastal Vulnerability Assessment to Low-Lying Inland Areas: Examples from Estonia, Eastern Baltic Sea. *Estuar. Coast. Shelf Sci.* **2024**, *311*, 109014. [CrossRef]
62. Tita, E.P.; Petropoulos, G.P.; Varvatsouli, D. Exploring the Relationships between Coastal Vulnerability and Socio-Economic Variables in the Context of Climate Change Using Geoinformatics for the Region of Longyearbyen, Svalbard. *Remote Sens. Appl. Soc. Environ.* **2025**, *30*, 101642. [CrossRef]
63. Boumpoulis, V.D.; Depountis, N.; Dimas, A.; Grivas, E.; Karymbalis, E. Presentation and Analysis of the Geotechnical Coastal Vulnerability Index and Validation of Its Application to Coastal Erosion Problems. *Sci. Rep.* **2025**, *15*, 1424. [CrossRef] [PubMed]

water

MDPI

Article

Characteristics and Risk Assessment of Soil Salinization in the Yellow River Delta Region, China

Liyuan Zhao [1,2,3], Linghao Kong [1,2,3,*], Xuzhen Zhang [1,2,3], Xiangcai Han [1,2,3], Zhigang Zhao [1,3], Baofei Li [1,3,*], Yongfeng Wang [1,3], Yuyan Li [1,3], Baili Geng [1,3], Mingjie Zhao [1,3], Haiyu Wang [1,3], Shigao Liu [1,3], Qingzhuang Miao [1,3], Kai Shan [4] and Yajie Zhao [4]

[1] Yantai Center of Coastal Zone Geological Survey, China Geological Survey, Yantai 264000, China; lyzhao0705@163.com (L.Z.); zxz0381@163.com (X.Z.); hanxcgeology@163.com (X.H.); zhaozhigangcug2025@163.com (Z.Z.); wangyongfeng1204@163.com (Y.W.); 18553555788@163.com (Y.L.); 18243899585@163.com (B.G.); 13789616043@163.com (M.Z.); 2042218435@163.com (H.W.); 15053511507@163.com (S.L.); miaoqz@foxmail.com (Q.M.)

[2] Observation and Research Station of Seawater Intrusion and Soil Salinization, Laizhou Bay, Ministry of Natural Resources, Qingdao 266061, China

[3] Observation and Research Station of Land-Sea Interaction Field in the Yellow River Estuary, Yantai 264000, China

[4] Shandong Yellow River Delta National Nature Reserve Management Committee, Dongying 257091, China; bbsbird@163.com (K.S.); zhaoyajie-2005@163.com (Y.Z.)

* Correspondence: konglinghao9115@163.com (L.K.); 18563868811@163.com (B.L.)

Abstract

Analyzing the characteristics of soil salinization and conducting risk assessments are crucial for ensuring the sustainable development of agriculture and ecosystems. In order to analyze the characteristics of soil salinization and conduct a risk assessment in the Yellow River Delta region, 63 surface soil samples and 37 groundwater samples were collected from this area in August 2023. Based on the test results of the samples and using soil salt content as the criterion, the types, degrees, and risks of soil salinization in the Yellow River Delta region were analyzed separately. The results revealed a relatively high average soil salt content of 4.59 g/kg, with Na^+ and Cl^- as the dominant ions. The primary salinization types were chloride and sulfate-chloride, covering 46.69% and 51.54% of the area, respectively. Moderate salinization was the most widespread, accounting for 45.35% of the region. Severe salinization, extremely severe salinization classes were mainly found in the coastal lowlands of the north and east, constituting 19.73% and 16.25% of the area, respectively. Groundwater exhibited transitional freshwater-saltwater characteristics, indicating widespread seawater intrusion across the region, which significantly contributed to soil salinity. Proximity to the Bohai Sea was the most critical factor influencing salinization, with areas closer to the sea showing a higher risk. High-risk zones, primarily along the coastline, covered 32.67% of the total area. The research findings can serve as valuable references for local wetland management and protection, the scientific enhancement of saline soils, rational soil utilization, effective prevention and control of soil salinization, and the sustainable development of water and soil resources.

Keywords: soil salinization; The Yellow River Delta; soil salt content; groundwater; seawater intrusion; risk assessment

1. Introduction

Soil is an extremely important natural resource [1,2], an integral component of the ecosystem [3,4], and the foundation upon which human survival depends [5]. Agricultural production, industrial development, and daily life all rely on soil [6]. Soil salinization is a common form and a key driver of land degradation [7]. The accumulation of salt in the soil adversely affects plant growth and development [8], leading to a decline in land productivity [9,10] and triggering a series of ecological deterioration processes [11], which severely restrict sustainable agricultural development and regional ecological security [12]. Global warming not only intensifies evapotranspiration in coastal areas but also causes sea level rise. Both factors have contributed to the further aggravation of soil salinization in these regions [13]. Currently, soil salinization is a widespread environmental issue affecting many regions worldwide [14]. Saline soil is widely distributed in over 100 countries and regions around the world, affecting more than 3% of the world's soil resources [15], and it is growing at a rate of 10^4 to 10^5 hectares per year [16]. The coastal areas of China also have extensive areas of saline-alkali land [17]. Soil salinization significantly reduces soil productivity potential, posing a serious threat to the stability of agricultural production worldwide [18].

Soil salinization is a multifaceted process driven by a complex interplay of various factors [19]. The accumulation of salt ions in the soil is a natural cause of salinization [20]. Additionally, climate change, fluctuations in groundwater levels, improper farming practices, and seawater intrusion are significant drivers of salinization [21]. In salinized soil, a high concentration of Na^+ can cause soil particles to disintegrate, expand, and disperse, leading to the destruction of soil aggregate structure and a reduction in the content of large aggregates. Consequently, the existing carbon and nitrogen are no longer protected by aggregates, making them more susceptible to decomposition. This significantly disrupts the mineralization process of soil carbon and nitrogen, thereby adversely affecting nutrient absorption by crops [22]. High concentrations of soluble salts in saline-alkali soils increase osmotic pressure, which affects the contraction of plant stomata. This leads to physiological and biochemical metabolic disorders, as well as nutritional imbalances in plants, ultimately hindering their normal growth [23]. Additionally, soil salinity has a negative effect on soil microorganisms, such as decreasing soil microbial activities, microbial biomass, and enzymatic activity through ionic toxicity and osmotic stress [24]. Overall, soil salinization has a serious impact on agricultural productivity and the ecological environment [25]. Therefore, a comprehensive understanding of the spatial distribution and the fundamental driving mechanisms of soil salinization is essential for safeguarding agricultural productivity and mitigating environmental issues [26]. Moreover, assessing the risk of soil salinization not only provides a quantitative measure of its severity but also offers a critical foundation for guiding prevention and control efforts [27].

The Yellow River Delta region represents the most extensive, well-preserved, and youngest wetland ecosystem in the warm temperate zone worldwide and is rich in saline soil resources [28]. The groundwater in this area is relatively shallow and is also affected by seawater intrusion, resulting in an uneven spatial distribution of soil salinization and complex variations in water and salt content. This has seriously hindered ecological protection and the sustainable use of land resources in the region [29,30]. Due to the long-term impact of seawater intrusion in the Yellow River Delta region, factors closely associated with seawater intrusion are likely the primary influences affecting the type, extent, and risk of soil salinization in this area. Understanding the extent and risk of salinization in this area is essential for scientifically managing resources, increasing land productivity, and enhancing the environment—all crucial steps toward sustainable development. Accordingly, this paper focuses on the Yellow River Delta region, analyzing the characteristics and influencing factors of soil salinization based on soil salinity distribution and groundwater chemical properties, and conducting a risk assessment. The research findings can serve as valuable references for local wetland management and protection, the scientific enhancement of saline soils, rational soil utilization, effective prevention and control of soil salinization, and the sustainable development of water and soil resources.

2. Materials and Methods

2.1. Study Area

The study area is located in the Yellow River Delta region, China (Figure 1), primarily encompassing five counties: Hekou, Kenli, Lijin, Dongying, and Guangrao, all under the jurisdiction of Dongying City, Shandong Province. It borders Bohai Bay and covers an area of approximately 7019.25 km^2. Characterized by four distinct seasons, the region experiences a warm temperate, semi-humid continental monsoon climate, featuring dry and windy conditions in spring, high temperatures with substantial rainfall in summer, clear and crisp weather in autumn, and cold, dry conditions in winter. The climate experiences a mean yearly temperature of 12.3 °C and an average precipitation of 537.3 mm, with 70% occurring during the summer months. The annual average evaporation rate is 1962.1 mm, which is 3.6 times greater than the precipitation. The terrain mainly consists of alluvial plains, with elevations ranging from 0 to 20 m. According to the results of the second soil survey in Shandong Province, the soil in the Yellow River Delta region is primarily classified into three categories: fluvo-aquic soils, coastal solonchaks, and alluvial soils. Fluvo-aquic soils and coastal solonchaks are the predominant soil types in the Yellow River Delta, accounting for 47% and 44% of the soil area, respectively. Fluvo-aquic soils are the main cultivated soil and, after proper cultivation and improvement, are suitable for growing crops such as wheat, corn, and cotton. Coastal solonchaks can be found in coastal areas and have a relatively high salt content, making them unsuitable for agricultural cultivation. Alluvial soils mainly form from cultivation on alluvial deposits, fluvial deposits, or flood diversion silt of the Yellow River. It represents a transitional soil type between fluvo-aquic soils and coastal solonchaks, accounting for 9% of the area. The largest river in the region is the Yellow River, which extends approximately 128 km within the study area.

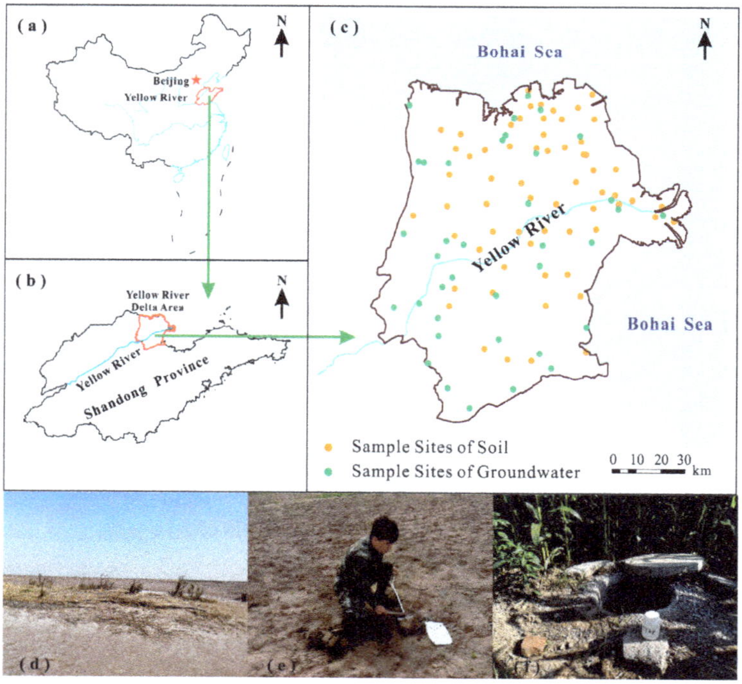

Figure 1. Map of the Yellow River Delta region. (**a**) Map of China; (**b**) Location of the Yellow River Delta region; (**c**) Distribution map of the sample sites; (**d**) The saline-alkali land of the Yellow River Delta; (**e**) Collection of soil samples; (**f**) Collection of groundwater samples.

2.2. *Sample Collection and Test Method*

2.2.1. Sample Collection

To conduct this research, 63 surface soil samples and 37 groundwater samples were collected from the Yellow River Delta region in August 2023 (Figure 1).

The sampling points for the soil samples were determined by comprehensively considering factors such as natural geography, hydrogeology, the history of seawater intrusion, soil salinization, and transportation accessibility in the Yellow River Delta region. This was performed after consulting relevant materials and conducting research and evidence collection. The sampling points were approximately evenly distributed throughout the study area and covered various soil types and land use categories within it. Groundwater samples were collected from local civilian wells or hydrological monitoring wells.

Surface soil samples were taken from a depth of 0 to 20 cm. When collecting soil samples, recently accumulated soil and areas affected by human pollution were avoided. Plant roots, stones, and other debris were removed from the samples. Each sample weighed no less than 1000 g. Natural air drying, grinding, and sieving were applied to the collected soil samples for further analysis. Groundwater samples were collected using 500 mL polyethylene bottles, which were rinsed three times with the groundwater prior to sampling. Care was taken to avoid bubbles in the samples during collection. After sampling, the bottles were sealed and labeled. The samples were promptly transported to the laboratory for testing.

2.2.2. Test Method

All the surface soil samples were prepared with an extraction solution of a 1:5 soil-to-water ratio. The HCO_3^- was determined by the standard H_2SO_4 double neutralization titration method. Ca^{2+}, Mg^{2+}, and SO_4^{2-} were measured by the EDTA complexometric titration method. Cl^- was measured by the standard $AgNO_3$ titration method. K^+ and Na^+ were determined by the flame photometry method. The soil salt content (SSC) was determined by the method of drying and weighing.

The K^+, Na^+, Ca^{2+}, and Mg^{2+} in the groundwater samples were determined by inductively coupled plasma atomic emission spectrometry. Cl^-, SO_4^{2-}, and total dissolved solids (TDS) were measured by ion chromatography. HCO_3^- was tested by potentiometric titration.

2.3. *Methods of Assessment*

2.3.1. Types of Soil Salinization

According to the "Code of Practice for Coastal Soil Salinization Monitoring and Evaluation" [31] (HY/T 0320-2021) issued by the Ministry of Natural Resources of the People's Republic of China, soil salinization types in the Yellow River Delta region can be classified into four categories based on the ratio of Cl^- to SO_4^{2-} content in the soil. The classification details are presented in Table 1.

Table 1. Classification of Soil Salinization Types.

Types of Soil Salinization	Cl^-/SO_4^{2-}
Sulfate type (SO_4^{2-})	<0.5
Chloride-sulfate type (Cl^--SO_4^{2-})	0.5–1.0
Sulfate-chloride type (SO_4^{2-}-Cl^-)	1.0–4.0
Chloride type (Cl^-)	>4.0

2.3.2. Degree of Soil Salinization

According to the "Technical Guidelines for Environmental Impact Assessment-Soil Environment (Trial)" [32] (HJ964-2018) issued by the Ministry of Ecology and Environment of the People's Republic of China, the degree of soil salinization in the Yellow River Delta region is classified based on the level of salt content in the soil. The classification details are presented in Table 2.

Table 2. Classification of Soil Salinization Degree.

Soil Salinization Degree	Soil Salt Content (SSC)/(g/kg)
No salinization	SSC < 1
Mild salinization	1 ≤ SSC < 2
Moderate salinization	2 ≤ SSC < 4
Severe salinization	4 ≤ SSC < 6
Extremely severe salinization	SSC ≥ 6

2.3.3. Risk Assessment of Soil Salinization

Soil salinization risk assessment is a comprehensive evaluation system that considers multiple driving factors contributing to soil salinization [33]. The steps are as follows:

Calculate the comprehensive evaluation index CR for soil salinization risk:

$$CR = \sum(K_i \times Q_i)$$

where K_i represents each influencing factor, and Q_i represents the value of each influencing factor.

To ensure the reliability of the CR value, standardization processing is carried out:

$$CR' = \frac{(CR - CR_{min})}{(CR_{max} - CR_{min})}$$

where CR' represents the normalized comprehensive evaluation index of soil salinization, CR_{max} and CR_{min} represent the maximum and minimum values of the comprehensive evaluation index of soil salinization, respectively.

In this study, by integrating the advantages of the correlation coefficient method and the factor detection method, the comprehensive weights of the risk factors were determined. The correlation coefficient method primarily considers the linear relationship between the influencing factors of soil salinization and soil salt content. In contrast, the factor detection method measured the correlation between the independent and dependent variables, capturing both linear and nonlinear components. This integration provided a theoretical basis for accurately determining the weights of influencing factors [34]. Integrating these two methods can effectively minimize the influence of human subjectivity, accurately identify the relationships among multiple factors, and overcome the limitations of traditional analytical approaches.

A weighted average method was used in the assessments. The risk index values of soil salinization were calculated using the Raster Calculator. According to the previous research and considering the characteristics of the study area, the risk index values were classified into five risk levels [35], as detailed in Table 3. The higher the risk index is, the higher the potential risks brought about by soil salinization are.

Table 3. Classification of Soil Salinization Risk Levels.

Risk of Soil Salinization	Comprehensive Evaluation Index
High risk	$CR' \geq 0.7$
Medium–high risk	$0.5 \leq CR' < 0.7$
Medium risk	$0.3 \leq CR' < 0.5$
Medium–low risk	$0.1 \leq CR' < 0.3$
Low risk	$CR' < 0.1$

2.4. Statistical Analysis

The data were processed and statistically analyzed using Microsoft Excel 365 and SPSS 26. The graphs were created with CorelDRAW X8, Origin 2022, and ArcGIS 10.8.

3. Results

3.1. Distribution Characteristics of Soil Salinity

3.1.1. Descriptive Statistical Analysis

Table 4 presents the content of salt ions in the surface soil. In the surface soil layer, the concentration of cations followed the order $Na^+ > Ca^{2+} > Mg^{2+} > K^+$, with Na^+ accounting for 71.03% of the total cation concentration and an average content of 1.18 g/kg, making it the dominant ion. The concentration of anions was $Cl^- > SO4^{2-} > HCO_3^-$, with Cl^- comprising 72.97% of the total anion concentration and an average content of 2.05 g/kg, also making it the dominant ion. The total salt content of the soil ranged from 0.497 to 14.97 g/kg, with a significant variation in the content. The average value was 4.59 g/kg, indicating that the salt content was generally at a high level, but the distribution was uneven.

Table 4. Descriptive Statistics of Salt Content in Surface Soil.

					Unit: g/kg
	Min	**Max**	**Average**	**Stdev**	**CV**
Ca^{2+}	0.001	1.55	0.23	0.32	137.53%
HCO_3^-	0.118	0.56	0.26	0.10	40.00%
K^+	0.005	0.34	0.06	0.07	124.88%
Mg^{2+}	0.001	1.74	0.19	0.32	168.71%
Na^+	0.010	5.11	1.18	1.35	114.53%
SO_4^{2-}	0.021	4.40	0.50	0.63	125.68%
Cl^-	0.128	6.97	2.05	1.95	94.79%
SSC	0.497	14.97	4.59	3.80	82.89%

The coefficient of variation (CV) is the ratio of the standard deviation to the mean. A higher CV indicates a greater degree of spatial heterogeneity for a given ion [36]. According to the statistical results, except for HCO_3^-, which exhibited a relatively low CV of 40.00%, the CVs for the other ions ranged from 94.79% to 168.71%. This suggests that the concentration of HCO_3^- varied little within the soil of the study area, whereas the concentrations of other ions were unevenly distributed and exhibited greater variability, often showing elevated levels in certain localized areas. The CV of the total soil salt content was 82.89%, which was also at a relatively high level. These findings indicated that the overall soil salinity in the region was relatively high but unevenly distributed, with localized areas of concentrated salt content.

3.1.2. Zoning of Soil Salinization Types

Table 5 and Figure 2 provided a detailed classification of soil salinization types in the Yellow River Delta region, identifying three categories: chloride type, chloride-sulfate type, and sulfate-chloride type. No sulfate-type soils were observed. Among these, chloride-type and sulfate-chloride-type soils were predominant, accounting for 46.69% and 51.54% of the area, respectively. The chloride-type soils were primarily distributed along the coastline, whereas sulfate-chloride-type soils were mostly found farther inland. The chloride-sulfate type soils comprised only 1.77% of the area and appeared in scattered, localized patches. These findings indicated that chloride was the dominant salt component in the soils of the study area.

Table 5. Classification of Soil Salinization Types in the Yellow River Delta Region.

Soil Salinization Types	Area (km^2)	Proportion (%)
Chloride type	3277.30	46.69%
Sulfate-chloride type (SO_4^{2-}-Cl^-)	3617.72	51.54%
Chloride-sulfate type (Cl^--SO_4^{2-})	124.23	1.77%

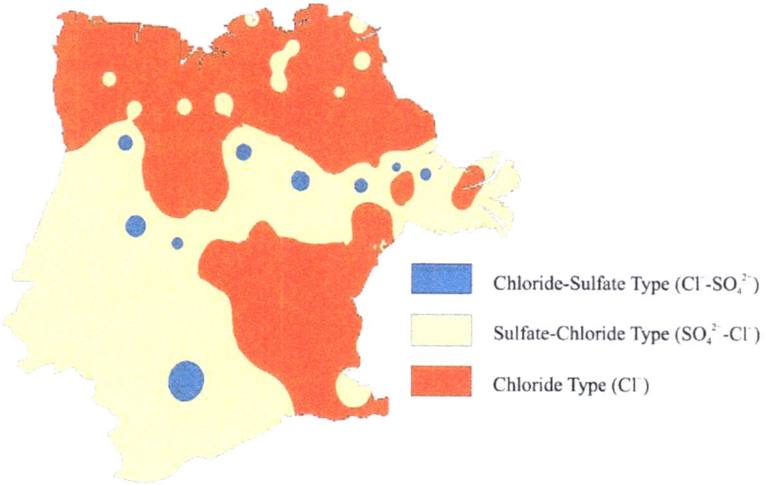

Figure 2. Zoning of soil salinization types.

3.1.3. Zoning of Soil Salinization Degree

The detailed distribution of soil salinization degrees is presented in Figure 3 and Table 6. The spatial distribution revealed that the most severely salinized soils were concentrated in the coastal lowlands of the north and east, accounting for 19.73% (extremely severe) and 16.25% (severe) of the total area, respectively. Moderately salinization soils represented the largest proportion, at 45.35%, and were mainly distributed in the inland regions. An area comprising 18.06% was characterized by mild salinization, with these soils predominantly located in the southern and central regions. Areas with no-salinization soils were scattered throughout the study area, comprising only 0.61% of the total.

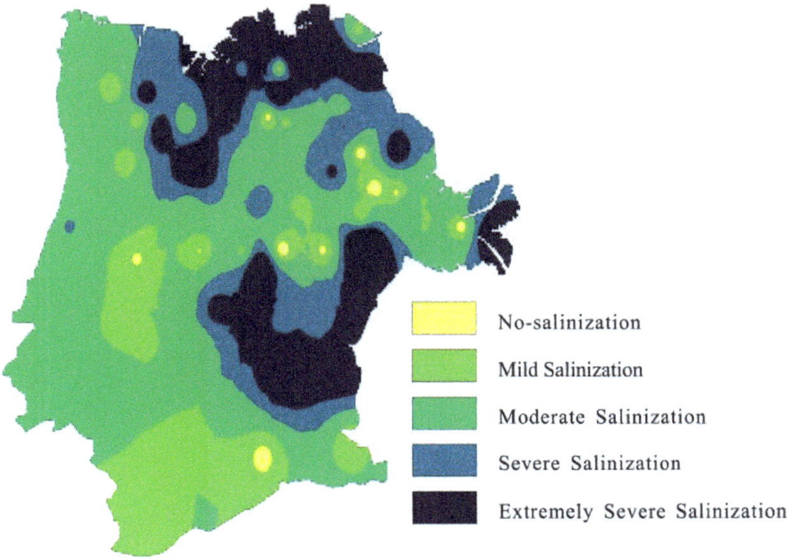

Figure 3. Zoning of soil salinization degree.

Table 6. Area and proportion of soil salinization types.

Degree of Soil Salinization	Area (km^2)	Proportion (%)
No salinization	43.10	0.61%
Mild salinization	1268.33	18.06%
Moderate salinization	3183.88	45.35%
Severe salinization	1141.02	16.25%
Extremely severe salinization	1385.40	19.73%

3.2. Hydrochemical Characteristics

3.2.1. Piper Diagram

The Piper diagram illustrates the hydrochemical composition of water samples by displaying ion proportions [37] and can illustrate the primary ionic composition of a water body [38]. The Piper diagram for groundwater in the Yellow River Delta region is shown in Figure 4. Cations were predominantly Na^+, while anions were mainly Cl^-. The water chemistry type was relatively stable. According to the classification of water chemistry types, the Cl-Ca type was the dominant water chemistry type, representing a transition from freshwater to saltwater, followed by the Cl-Na type, which was characteristic of seawater. Therefore, the groundwater chemical types in most areas were in the transitional stage from freshwater to saltwater, with a few areas already classified as saltwater. This indicated that the entire study area was affected by seawater intrusion, with some areas experiencing more severe impacts. When groundwater was influenced by seawater, large amounts of Na^+ and Cl^- ions were introduced into the groundwater and subsequently into the soil, increasing the risk of soil salinization.

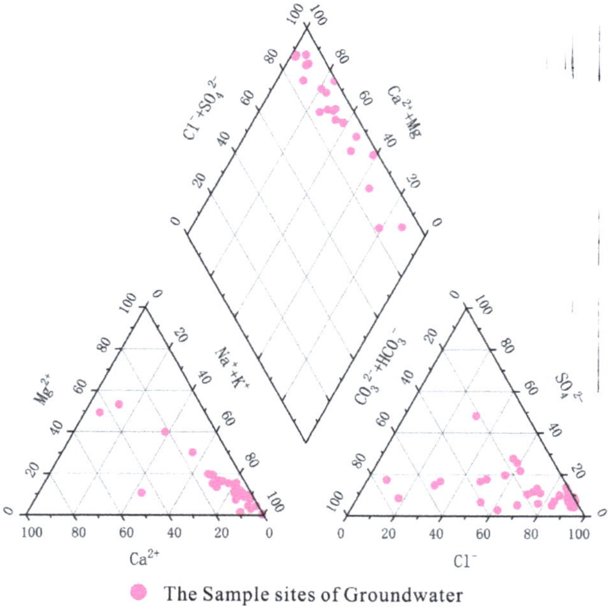

Figure 4. Piper diagram of groundwater in the Yellow River Delta region.

3.2.2. Gibbs Diagram

The Gibbs diagram utilizes the relationship between TDS and the ratio of cation mass concentration ($Na^+/(Na^+ + Ca^{2+})$) and the relationship between TDS and the ratio of anion mass concentration ($Cl^-/(Cl^- + HCO_3^-)$) [39] to macroscopically reflect the main ion control factors in water: atmospheric precipitation, rock weathering, and evaporation–crystallization [40,41].

Figure 5 presents the Gibbs diagram of groundwater in the Yellow River Delta region. At most sampling points, the ratio of cation to anion mass concentration exceeded 0.5, with total dissolved solids (TDS) ranging from 166 to 9604 mg/L. Approximately half of the sampling points fell within the Gibbs model, located in the upper right section of the diagram. The chemistry of shallow groundwater in the Yellow River Delta was primarily influenced by oceanic factors, indicating evaporation and concentration following seawater intrusion. As the groundwater evaporated and became more concentrated, it brought salts to the surface, thereby playing a significant role in soil salinization. However, about half of the sampling points fell outside the Gibbs model, exhibiting a relatively high cation-to-anion mass concentration ratio (>0.8) but relatively low TDS (<1000 mg/L). This suggested that these groundwater samples were influenced not only by seawater intrusion but also by other factors, such as recharged from the Yellow River or atmospheric precipitation.

3.3. *Influencing Factors Analysis and Risk Assessment of Soil Salinization*

The selection of soil salinization risk assessment indicators was based on the specific environmental attributes of the Yellow River Delta region. Six factors influencing soil salinization risk were chosen as independent variables: digital elevation model (DEM), soil type (ST), distance from the Bohai Sea (Dis), groundwater depth (Dep), evaporation coefficient (ET), and total dissolved solids (TDS). Soil salt content (SSC) was selected as the dependent variable to construct a salinization risk assessment index system. The specific conditions of each evaluation factor are illustrated in Figure 6.

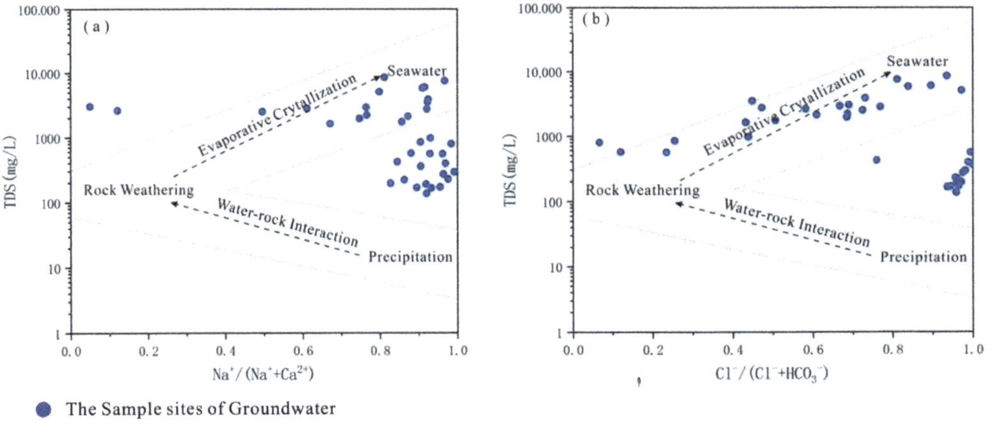

Figure 5. Gibbs Diagram of Groundwater in the Yellow River Delta region. (**a**) The relationship between TDS and $Na^+/(Na^+ + Ca^{2+})$. (**b**) The relationship between TDS and $Cl^-/(Cl^- + HCO_3^-)$.

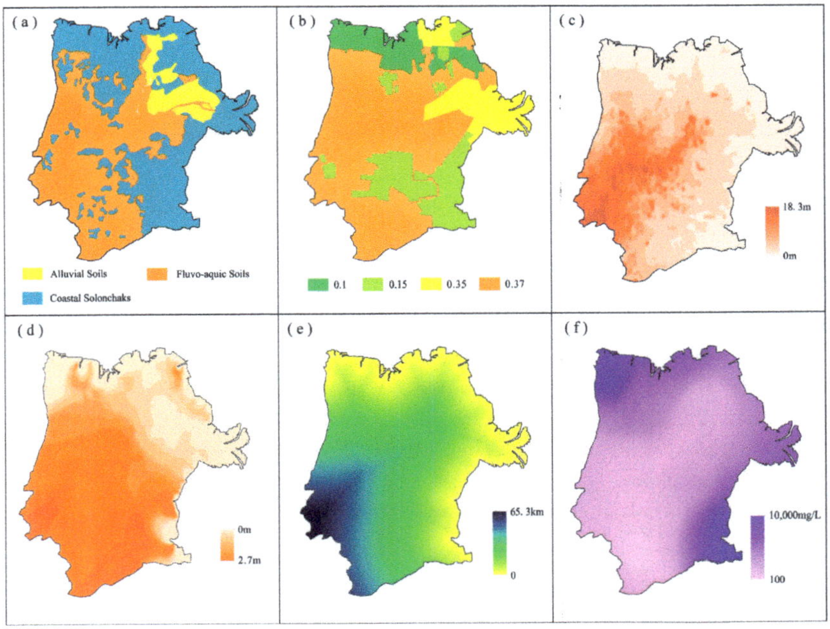

Figure 6. Influencing factors of soil salinization. (**a**) Soil type; (**b**) Evaporation coefficient; (**c**) Digital elevation; (**d**) Depth of groundwater; (**e**) Distance from the Bohai Sea; (**f**) Total dissolved solids.

3.3.1. Analysis of Influencing Factors of Soil Salt Content

By combining the advantages of the correlation coefficient method and the factor detection method, the comprehensive weights of each influencing factor were determined.

Correlation analysis was conducted on SSC, ST, Dis, DEM, Dep, TDS, and ET to examine the relationships between each influencing factor and the current status of soil salinization. The results (Table 7) revealed generally strong correlations, both positive and negative, among the various indicators. Specifically, SSC was positively correlated with TDS, ST, DEM, Dep, Dis, and ET. When considering the absolute values of the correlations with SSC, the influencing factors ranked from strongest to weakest were Dis > ST > DEM > TDS > ET > Dep.

Table 7. Weights of each soil salinization influencing factor.

Influencing Factors	Correlation Index	Weights	q	Weights	Comprehensive Weights	Rank
Dis	−0.551	0.22	0.929	0.29	0.26	1
TDS	0.366	0.15	0.863	0.27	0.21	2
Dem	−0.445	0.18	0.646	0.20	0.19	3
ST	0.467	0.19	0.218	0.07	0.13	4
Dep	−0.273	0.11	0.421	0.13	0.12	5
ET	−0.349	0.14	0.139	0.04	0.09	6

Then the factor detection method was carried out. Using SSC as the dependent variable and DEM, Dep, Dis, ET, ST, and TDS as independent variables, the influence of each independent variable on SSC was determined (denoted as q). According to the analysis results (Table 7), Dis and TDS had the greatest influence on SSC, with q values of 0.929 and 0.863, respectively. The influence values, ranked from largest to smallest, were Dis > TDS > DEM > Dep > ST > ET.

Based on the results of both correlation and influence analyses, the comprehensive weights of each dependent variable were calculated. According to the results (Table 7), the influence of each factor on soil salinity, ranked from greatest to least, was Dis > TDS > DEM > ST > Dep > ET.

3.3.2. Risk Assessment of Soil Salinization in the Yellow River Delta Region

Based on the results of the above analysis, a risk assessment of soil salinization was conducted in the Yellow River Delta region. As shown in Figure 7 and Table 8, the high-risk soil salinization zone was the largest area, distributed along the shoreline and accounting for 32.67% of the total region. The medium-high risk, medium risk, and medium-low risk zones each covered approximately 20% of the area. The low-risk zone was the smallest, comprising only 4.69%. Distance from the Bohai Sea was identified as the most significant influencing factor. It was clearly demonstrated that the closer the area is to the Bohai Sea, the higher the risk of soil salinization; conversely, the farther from the Bohai Sea, the lower the risk. This indicated that seawater intrusion was the primary driver of soil salinization in the Yellow River Delta region. The combined area of high-risk and medium–high risk zones accounted for more than 52% of the total. Therefore, it is recommended to strengthen monitoring of soil salt content in the Yellow River Delta and implement appropriate measures to control soil salinization.

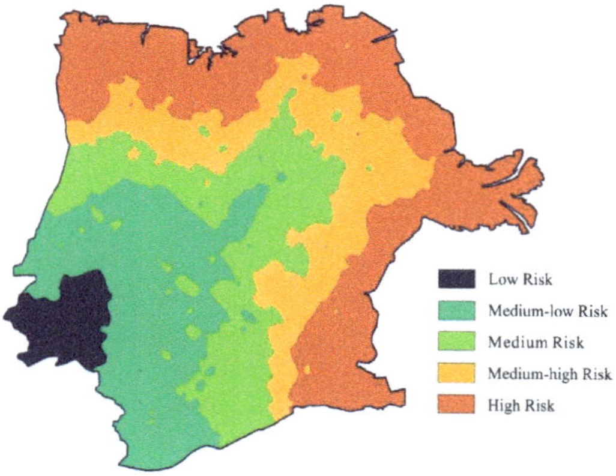

Figure 7. Risk Assessment of soil salinization in the Yellow River Delta Region.

Table 8. Results of soil salinization risk assessment.

Risk of Soil Salinization	Area (km^2)	Proportion (%)
High risk	2293.10	32.67%
Medium–high risk	1362.83	19.42%
Medium risk	1433.33	20.42%
Medium–low risk	1600.58	22.80%
Low risk	329.41	4.69%

4. Discussion

The findings of this study confirm that seawater intrusion is the primary driver of soil salinization in the research area. This process begins within the groundwater system: seawater intrusion initially increases groundwater salinity, shifting it toward brackish water [42]. Subsequently, through evaporative concentration, highly mineralized groundwater moves upward via capillary action, transporting dissolved salts to the soil surface, where they accumulate [43]. This mechanism effectively explains the predominance of sodium and chloride ions and the generally high salt content observed in the area's soils. Typically, regions experiencing more severe seawater intrusion exhibit higher groundwater salinity, which in turn leads to elevated levels of sodium and chloride ions in the soil.

Zonation results of soil salinization types and degrees further support this conclusion. In the Yellow River Delta, the spatial distribution of extremely and severely salinized soils closely corresponds with that of chloride-type saline soils, which are predominantly found in coastal zones. This spatial consistency indicates that seawater intrusion is the primary factor controlling both the type and severity of soil salinization in the study area, fully aligning with the initial research hypothesis.

It is important to note that the risk of soil salinization is influenced by multiple interacting factors [44]. This study identified key variables, including soil type, distance from the Bohai Sea, surface elevation, groundwater depth, groundwater mineralization, and surface evapotranspiration coefficient. Among these, distance from the Bohai Sea and groundwater mineralization (TDS) emerged as the most significant factors, both directly linked to the intensity of seawater intrusion. Coastal areas, characterized by low-lying topography (near sea level) and shallow groundwater tables, are more susceptible to seawater influence. The continuous upward movement of salts through evaporation from mineralized groundwater leads to the formation of typical coastal saline soils, significantly increasing the risk of soil salinization.

Risk assessment results indicate that long-term seawater intrusion has impacted the Yellow River Delta, with high-risk and moderately high-risk zones together comprising more than 52% of the area. It is recommended to strengthen monitoring of soil salinity dynamics and implement targeted measures—such as regulating groundwater levels and optimizing vegetation cover—to inhibit salt migration to the surface and effectively control soil salinization.

5. Conclusions

(1) The average salt content in the soil of the Yellow River Delta region was 4.59 g/kg, indicating a relatively high salinity overall. The dominant cation in the soil was Na^+, comprising 71.03% of the total cation concentration, while the dominant anion was Cl^-, accounting for 72.97% of the total anion concentration. The primary types of soil salinization were chloride type and sulfate-chloride type in the study area, representing 46.69% and 51.54% of the total area, respectively. The degree of soil salinization was predominantly moderate, accounting for 45.35% of the area. Extremely severe salinization and severe salinization were mainly distributed in the coastal lowlands

in the northern and eastern parts of this region, covering 19.73% and 16.25% of the area, respectively.

(2) The results of the Piper and Gibbs diagrams indicated that the groundwater chemical types in most areas of the Yellow River Delta were in a transitional stage between freshwater and saltwater, while a few areas had already become saline. This suggested that the entire study area had been affected by seawater intrusion, with some areas experiencing more severe impacts. Therefore, seawater intrusion in this region was a significant factor contributing to the increased salt content in the soil and played a crucial role in soil salinization.

(3) Based on the results of both correlation and influence analyses, the factors affecting soil salt content in the Yellow River Delta region, ranked from greatest to least impact, were: Dis > TDS > DEM > ST > Dep > ET. These findings indicated that the primary cause of soil salinization in the Yellow River Delta was seawater intrusion.

(4) The results of the risk assessment for soil salinization in the Yellow River Delta region indicated that areas with high risk and medium–high risk accounted for a significant proportion, exceeding 52%. The distance from the Bohai Sea was the most influential factor, demonstrating a clear trend: the closer the area was to the Bohai Sea, the higher the risk of soil salinization. It is recommended to enhance the monitoring of soil salt content in the Yellow River Delta region and implement appropriate measures to control soil salinization.

(5) All these findings above can provide valuable references for local wetland management and protection, the scientific enhancement of saline soils, rational soil utilization, effective prevention and control of soil salinization, and the sustainable development of water and soil resources.

Author Contributions: Conceptualization, L.Z., L.K. and X.Z.; Formal analysis, L.Z., X.H., Z.Z., Y.L., B.G., M.Z. and H.W.; Funding acquisition, L.K., B.L., Y.W., S.L. and Q.M.; Investigation, L.Z., X.H., Z.Z., Y.L., B.G., M.Z., H.W., K.S. and Y.Z.; Methodology, L.Z., L.K. and X.Z.; Project administration, L.K., X.Z. and Y.W.; Supervision, L.K. and Y.W.; Visualization, L.K., X.Z. and Y.W.; Writing—original draft, L.Z., L.K., X.Z. and Z.Z.; Writing—review and editing, L.Z., L.K., X.Z. and B.L. All authors have read and agreed to the published version of the manuscript.

Funding: This work was funded by the Open Research Fund of Observation and Research Station of Seawater Intrusion and Soil Salinization, Laizhou Bay, Ministry of Natural Resources (2024LZORS003) and the Geological Survey Programs of the People's Republic of China (DD20230511, DD20251300107, DD20230701308, and DD20242687).

Data Availability Statement: According to the regulations of the authors' institution, the data used in the paper must be shared upon application. Requests to access the datasets should be directed to the corresponding author, Linghao Kong.

Conflicts of Interest: The authors declare no conflicts of interest.

References

1. Khadija, Z.; Latifa, M.; Abdelaziz, M.; Khattabi, R.S.; Ilham, N.; Salah, S. Soil metal pollution in freeway traffic (Mohammedia-Bouznika section, Morocco). *Int. J. Environ. Sci. Technol.* **2025**, *22*, 1385–1402. [CrossRef]
2. Wu, J.; Li, P.; Qian, H.; Fang, Y. Assessment of soil salinization based on a low-cost method and its influencing factors in a semiarid agricultural area, northwest China. *Environ. Earth Sci.* **2014**, *71*, 3465–3475. [CrossRef]
3. Li, D.; Wu, W.; Zhu, H.; Guo, W.; Liu, Y.; Li, N. Complex Resistivity Dispersion for Monitoring Soil Contaminated by Selected Organic Pollutants. *Environ. Res.* **2025**, *279*, 121756. [CrossRef] [PubMed]
4. Fang, S.; Hua, C.; Yang, J.; Liu, F.; Wang, L.; Wu, D.; Ren, L. Combined pollution of soil by heavy metals, microplastics, and pesticides: Mechanisms and anthropogenic drivers. *J. Hazard. Mater.* **2025**, *485*, 136812. [CrossRef]
5. Hou, D.; Jia, X.; Wang, L.; McGrath, S.P.; Zhu, Y.; Hu, Q.; Zhao, F.; Bank, M.S.; O'Connor, D.; Nriagu, J. Global soil pollution by toxic metals threatens agriculture and human health. *Science* **2025**, *388*, 316–321. [CrossRef]

6. Yu, Y.; Liu, Y.; Zhao, Z.; Cui, H. An Auxiliary Method for Evaluating Soil Erosion and Pollution by Magnetic Susceptibility on Black Soil Farmland. *Water Air Soil Pollut.* **2025**, *236*, 1–14. [CrossRef]
7. Shrivastava, P.; Kumar, R. Soil salinity: A serious environmental issue and plant growth promoting bacteria as one of the tools for its alleviation. *Saudi J. Biol. Sci.* **2015**, *22*, 123–131. [CrossRef] [PubMed]
8. Song, Y.; Sun, J.; Cai, M.; Li, J.; Bi, M.; Gao, M. Effects of management of plastic and straw mulching management on crop yield and soil salinity in saline-alkaline soils of China: A meta-analysis. *Agric. Water Manag.* **2025**, *308*, 109309. [CrossRef]
9. Xiao, Y.; Zhao, G.; Li, T.; Zhou, X.; Li, J. Soil salinization of cultivated land in Shandong Province, China-Dynamics during the past 40 years. *Land Degrad. Dev.* **2019**, *30*, 426–436. [CrossRef]
10. Anna, O.C.; Jin, L. Chemical and hydrological controls on salt accumulation in irrigated soils of southwestern US. *Geoderma* **2021**, *391*, 114976. [CrossRef]
11. Mao, J.; Wu, J.; Liu, Y.; Guo, C.; Xiao, C.; Lu, Y.; Zhao, L.; Zhang, R.; Zhang, H. A model for simulating evaporation from seasonally frozen saline soil. *J. Hydrol.* **2025**, *659*, 133259. [CrossRef]
12. Feng, H.; Xing, X.; Xing, J.; Du, J.; Li, D. Performance and modeling of infiltration flow in cracked saline soils. *J. Hydrol.* **2025**, *656*, 133054. [CrossRef]
13. Xie, X.; Pu, L.; Zhu, M.; Xu, Y.; Wang, X. Linkage between soil salinization indicators and physicochemical properties in a long-term intensive agricultural coastal reclamation area, Eastern China. *J. Soils Sediments* **2019**, *19*, 3699–3707. [CrossRef]
14. Wang, Y.; Deng, C.; Liu, Y.; Niu, Z.; Li, Y. Identifying change in spatial accumulation of soil salinity in an inland river watershed, China. *Sci. Total Environ.* **2018**, *621*, 177–185. [CrossRef]
15. Ajay, S. Soil salinization management for sustainable development: A review. *J. Environ. Manag.* **2021**, *277*, 111383. [CrossRef]
16. Xia, J.; Ren, J.; Zhang, S.; Wang, Y.; Fang, Y. Forest and grass composite patterns improve the soil quality in the coastal saline-alkali land of the Yellow River Delta, China. *Geoderma* **2019**, *349*, 25–35. [CrossRef]
17. Miao, Q.; Wu, S.; Li, J.; Zhang, J.; Wang, Y.; Wu, G.; Sun, Y.; Ying, H.; Cui, Z. Productivity of saline soils can be improved by integrating soil–crop system management. *Eur. J. Agron.* **2025**, *168*, 127618. [CrossRef]
18. Song, Y.; Gao, M.; Li, Z. Impacts of straw return methods on crop yield, soil organic matter, and salinity in saline-alkali land in North China. *Field Crops Res.* **2025**, *322*, 109752. [CrossRef]
19. Zhang, Y.; Hou, K.; Qian, H.; Gao, Y.; Fang, Y.; Xiao, S.; Tang, S.; Zhang, Q.; Qu, W.; Ren, W. Characterization of soil salinization and its driving factors in a typical irrigation area of northwest China. *Sci. Total Environ.* **2022**, *837*, 155808. [CrossRef]
20. Yang, J.; Zhao, J.; Mao, X.; Zhang, Y.; Hu, F. Applicability analysis of weakly supervised semantic segmentation for identifying salinized soil boundaries. *J. Arid. Environ.* **2025**, *229*, 105372. [CrossRef]
21. Wu, C.; Du, X.; Meng, B.; Guo, H. Characteristics and mechanisms of soil salinization in humid climate areas. *J. Hydrol. Reg. Stud.* **2025**, *60*, 102607. [CrossRef]
22. Thangarasu, T.; Mengash, H.A.; Allafi, R.; Mahgoub, H. Spatial prediction of soil salinity: Remote sensing and machine learning approach. *J. South Am. Earth Sci.* **2025**, *156*, 105440. [CrossRef]
23. Rath, K.M.; Fierer, N.; Murphy, D.V.; Rousk, J. Linking bacterial community composition to soil salinity along environmental gradients. *ISME J.* **2018**, *13*, 836–846. [CrossRef] [PubMed]
24. Zhu, H.; Yang, J.; Li, Y.; Liu, X.; Jin, H.; Li, J.; Yao, R. Effects of Soil Salinity on Nitrification and Ammonia-Oxidizing Microorganisms in Coastal Reclaimed Farmland Soil. *J. Soil Sci. Plant Nutr.* **2022**, *22*, 2743–2754. [CrossRef]
25. Sliti, A.; Singh, V.; Pande, A.; Shin, J.-H. Soil holobiont interplay and its role in protecting plants against salinity stress. *Pedosphere* **2025**, *35*, 97–115. [CrossRef]
26. Wang, J.; Ding, J.; Wang, Y.; Ge, X.; Lizaga, I.; Chen, X. Soil salinization in drylands: Measure, monitor, and manage. *Ecol. Indic.* **2025**, *175*, 113608. [CrossRef]
27. Gharemahmudli, S.; Sadeghi, S.H. Inhibiting soil and water loss in a saline soil through cyanobacterization. *Soil Tillage Res.* **2025**, *248*, 106446. [CrossRef]
28. Mai, Z.; Niu, Z.; Zhao, Y.; Li, P.; Wang, Y.; Lv, Y.; Wang, B.; Zhang, M. Land use and cover change significantly enhanced evapotranspiration in the Yellow River Delta from 2000 to 2023. *J. Hydrol. Reg. Stud.* **2025**, *58*, 102220. [CrossRef]
29. Bian, L.; Wang, J.; Liu, J.; Han, B. Spatiotemporal Changes of Soil Salinization in the Yellow River Delta of China from 2015 to 2019. *Sustainability* **2021**, *13*, 822. [CrossRef]
30. Fang, X.; Liu, Z.; Li, J.; Lai, J.; Gong, H.; Sun, Z.; Ouyang, Z.; Dou, W.; Fa, K. Non-Synergistic Changes in Migration Processes between Soil Salt and Water in the Salt Patch of the Coastal Saline Soil. *Agronomy* **2023**, *13*, 2403. [CrossRef]
31. *HY/T 0320-2021*; National Technical Committee 283 on Ocean of Standardization Administration of China. Code of Practice for Coastal Soil Salinization Monitoring and Evaluation. Ministry of Natural Resources of the People's Republic of China: Beijing, China, 2021.
32. *HJ 964-2018*; Ministry of Ecological Environment. Technical Guidelines for Environmental Impact Assessment-Soil Environment (Trial). China Environmental Science Press: Beijing, China, 2018.

33. Zhou, D.; Lin, Z.; Liu, L.; Zimmermann, D. Assessing secondary soil salinization risk based on the PSR sustainability framework. *J. Environ. Manag.* **2013**, *128*, 642–654. [CrossRef]
34. Wang, Z.; Zhang, F.; Zhang, X.; Chan, N.W.; Kung, H.-T.; Zhou, X.; Wang, Y. Quantitative Evaluation of Spatial and Temporal Variation of Soil Salinization Risk Using GIS-Based Geostatistical Method. *Remote Sens.* **2020**, *12*, 2405. [CrossRef]
35. Hou, J.; Rusuli, Y. Assessment of Soil Salinization Risk by Remote Sensing-Based Ecological Index (RSEI) in the Bosten Lake Watershed, Xinjiang in Northwest China. *Sustainability* **2022**, *14*, 7118. [CrossRef]
36. Zhao, L.; Wang, Z.; Kong, L.; Zhao, Z.; Geng, B.; Gu, S.; Li, Y.; Han, X.; Liu, S.; Li, B.; et al. Risk assessment of soil heavy metals in mining activity areas: A case study in Eastern Shandong Province, China. *Environ. Earth Sci.* **2023**, *82*, 513. [CrossRef]
37. Piper, A.M. A graphic procedure in the geochemical interpretation of water-analyses. *Eos Trans. Am. Geophys. Union* **1944**, *25*, 914–928.
38. Liu, J.; Gao, Z.; Wang, Z.; Xu, X.; Su, Q.; Wang, S.; Qu, W.; Xing, T. Hydrogeochemical processes and suitability assessment of groundwater in the Jiaodong Peninsula, China. *Environ. Monit. Assess.* **2020**, *192*, 1–17. [CrossRef]
39. Gibbs, R.J. Mechanisms controlling world water chemistry. *Science* **1970**, *170*, 1088–1090. [CrossRef] [PubMed]
40. Kong, L.; Chu, H.; Hou, G.; Zeng, J.; Zhao, L.; Han, X.; Li, Y.; Zhou, F.; Yin, M.; Zhang, X. Characteristics and mechanisms of sea (salt) water intrusion in Laizhou City, Shandong Province, China. *Front. Mar. Sci.* **2025**, *12*, 1578175. [CrossRef]
41. Zhao, L.; Geng, B.; Zhao, M.; Li, B.; Miao, Q.; Liu, S.; Zhao, Z.; Wang, H.; Li, Y.; Jin, W.; et al. Evaluation of Groundwater Quality and Health Risk Assessment During the Dry Season in the Xin'an River Basin, China. *Water* **2025**, *17*, 2412. [CrossRef]
42. Arslan, H.; Demir, Y. Impacts of seawater intrusion on soil salinity and alkalinity in Bafra Plain, Turkey. *Environ. Monit. Assess.* **2013**, *185*, 1027–1040. [CrossRef] [PubMed]
43. Cui, G.; Liu, Y.; Li, X.; Wang, S.; Qu, X.; Wang, L.; Tong, S.; Zhang, M.; Li, X.; Zhang, W. Impacts of groundwater storage variability on soil salinization in a semi-arid agricultural plain. *Geoderma* **2025**, *454*, 117162. [CrossRef]
44. Wang, H.; Wang, Z.; Bai, Y.; Wan, W.; Wang, W.; Wang, K. Risk assessment method for soil secondary salinization based on hydrological frequency analysis. *Ecol. Indic.* **2025**, *176*, 113688. [CrossRef]

water

MDPI

Article

Characterization of Groundwater Chemistry Under the Influence of Seawater Intrusion in Northern Laizhou, Shandong Province, China

Xiangcai Han [1,2,3], Linghao Kong [1,2,3,*], Liyuan Zhao [1,3], Zhigang Zhao [1,3], Yachao Li [1,3], Decheng Zhang [1,3], Huankai Zhang [1,3], Yajie Zhao [4] and Kai Shan [4]

1 Yantai Center of Coastal Zone Geological Survey, China Geological Survey, Yantai 264004, China; hanxcgeology@163.com (X.H.); lyzhao0705@163.com (L.Z.); zhaozhigangcug2025@163.com (Z.Z.); liyachao1116@163.com (Y.L.); zhangdecheng0618@163.com (D.Z.); 13188773259@163.com (H.Z.)
2 Observation and Research Station of Seawater Intrusion and Soil Salinization, Laizhou Bay, Ministry of Natural Resources, Qingdao 266061, China
3 Ministry of Natural Resources Observation and Research Station of Land-Sea Interaction Field in the Yellow River Estuary, Yantai 264004, China
4 The Yellow River Delta National Nature Reserve Management Committee Shandong, Dongying 257000, China; zhaoyajie-2005@163.com (Y.Z.); bbsbird@163.com (K.S.)
* Correspondence: konglinghao9115@163.com; Tel.: +86-18510980395

Abstract

The rise in sea levels due to global warming and the excessive extraction of groundwater in coastal regions significantly encourages seawater intrusion, resulting in a cascade of ecological and environmental issues, including water quality degradation and soil salinization. The northern sector of Laizhou City, situated on the eastern coast of Laizhou Bay, exemplifies a typical location of seawater intrusion in China, where the rising salinity of groundwater has adversely affected local economic development and public health. This investigation involved the collection of 115 groundwater samples and 13 isotope samples from the northern region of Laizhou City. Statistical analysis, Piper's trilinear diagrams, and various analytical techniques were employed to examine the chemical properties of the groundwater in the study area; characteristic ion ratios, Gibbs diagram, and hydrogen–oxygen isotope methods were utilized to analyze the sources of salinity and groundwater recharge; and a seawater intrusion groundwater quality index, which was applied to the present condition of seawater intrusion, was assessed utilizing the seawater intrusion groundwater quality index (GQI_{SWI}). The findings indicate that the chemical composition of groundwater in the research area is notably intricate. From freshwater to saline water, the groundwater chemistry transitions from Ca-HCO_3·Cl-type water to Ca·Na-SO_4·Cl-type water, and finally to Na-Cl-type water. Seawater intrusion in the research area is the primary cause of elevated groundwater salinity, alongside cation exchange and water–rock interactions that affect water chemistry. Seawater intrusion is predominantly focused in the northern region of the research area. The primary source of groundwater recharge is atmospheric precipitation.

Keywords: north of Laizhou City; chemical characteristics of groundwater; seawater intrusion; cation exchange

1. Introduction

Coastal regions are favored for human settlement due to their distinctive geography and aesthetic appeal. Nonetheless, the rise in population, economic expansion, and engineering infrastructure in coastal regions has heightened the demand for fresh groundwater, resulting in diminishing water levels and subsequent seawater intrusion into coastal aquifers. The literature indicates that seawater intrusion has affected around 500 coastal towns globally [1], encompassing regions such as East Asia [2–5], South Asia [6,7], the Mediterranean [8,9], North America [10], and Europe [11,12].

Seawater intrusion events have resulted in extensive contamination of freshwater aquifers and ecological deterioration along the Chinese coastline. Seawater intrusion in China predominantly occurs in the Bohai Sea region (Shandong, Hebei, Tianjin, and Liaoning) [4,13], the Yangtze River Delta, the Pearl River Delta, the Beibu Gulf coastal area, Hainan, and the central and western plains of Taiwan [3,14].

Current research on seawater intrusion encompasses inquiry and assessment, modeling of seawater intrusion [15], and strategies for its prevention and control [16]. Methods for investigating and evaluating seawater intrusion encompass water chemistry techniques [17,18], isotope analysis [19], and geophysical approaches [20], among others. The water chemistry approach serves as the most direct and complete indicator; however, its drawbacks include an excessive number of indicators, susceptibility to human impact, interference from hydrogeological circumstances, and inconsistencies among indicators. The extensive application of various approaches is the prevailing trend in the investigation of seawater intrusion [13,21–23].

Laizhou Bay, recognized as a critical region experiencing seawater intrusion in China, has been adversely impacted in terms of residential water consumption and industrial and agricultural output [24–29]. The south coast of Laizhou Bay is a hot research area for seawater intrusion and saline (brine) water intrusion, and the methods used to study it are mostly hydrogeochemical and isotope geochemical methods, etc. In the case of saline (brine) water, there are debates on the evaporation and concentration of ancient seawater and the dissolution of evaporite rocks in sediments [4,30–35]. Since the discovery of seawater intrusion in the study area located on the east coast of Laizhou Bay in the 1970s, scholars have also carried out a great deal of research. Currently, it is generally agreed that groundwater salinization along the east coast of Laizhou Bay is mainly caused by seawater intrusion, and the presence of brine has not been found [25,36]. In the 1990s, Xue Yuqun and Yin Zesheng examined the current state of seawater intrusion, the transport dynamics at the saline-freshwater interface, as well as the causes, mechanisms, and classifications of seawater intrusion [25,29]. They employed various methodologies, including water chemistry analysis, groundwater isotope research, statistical groundwater assessments, monitoring section construction, numerical simulations, remote sensing interpretation, ecological investigations, and modeling, resulting in significant insights and accomplishments. In recent years, certain scholars conducted predictive modeling in the study area [37], performed a disaster risk assessment of seawater intrusion in the coastal zone of Laizhou City [38], and proposed control measures for seawater intrusion [39]; however, they neglected to analyze the water chemistry affected by seawater intrusion.

Groundwater serves as the primary water source in the study area, a significant grain-producing region and mining development zone in Laizhou City. The 2017–2020 Yantai City Water Resources Bulletin indicates that groundwater constitutes over 50% of the total water supply in the study region. Seawater intrusion has significantly affected the water supply for agricultural, industrial, and residential use in the studied region. This research project aims to assess the existing condition of the chemical properties of groundwater in the study area due to persistent seawater intrusion. To accomplish these objectives, a comparative schematic map illustrating the progression of seawater intrusion over the years was created, and the origins of groundwater salinity and recharge were examined utilizing hydrochemical and isotopic techniques. The study area is crucial for water resource management, prevention and control of saltwater intrusion disasters, and conservation of the coastal ecological environment.

2. Study Area

The study area is situated in the northern region of Laizhou City, Yantai City, Shandong Province, along the eastern coast of Laizhou Bay (Figure 1). The study area exhibits a continental climate within the northern temperate East Asian monsoon zone, characterized by an average annual precipitation of 616.05 mm and an average temperature of 12.6 °C from 1970 to 2021.

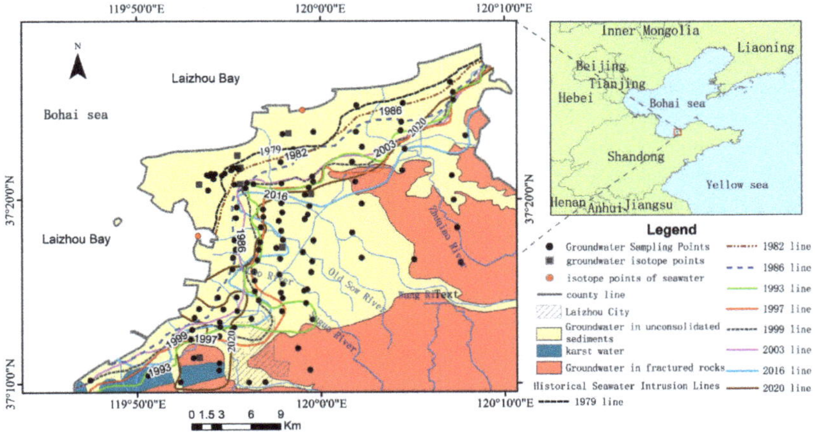

Figure 1. Location map of the study area.

The research area's topography is predominantly elevated in the southeast and depressed in the northwest, characterized primarily by low hills and plains, with an elevation range of 0–478 m. The primary stratigraphy of the study area comprises the Neoproterozoic Jiao Dong Rock Group, the Paleoproterozoic Jingshan Group, and the Powder Mountain Group, alongside the Cenozoic Paleocene and Quaternary. Additionally, a significant expanse of magmatic intrusion affects the exposed strata, with the Quaternary predominantly constituting the principal aquifer, which includes Quaternary alluvial deposits, fine sand, coarse sand, and pebble gravels. The primary aquifers consist of Quaternary alluvial and floodplain medium and fine sands, coarse sands, and pebble gravel. Typically, there exists a singular aquifer layer, though there may be two to four levels, with a close hydraulic relationship between them, and thickness varying from 2 to 30 m. In the southern region of the city, the ancient Yuan dynasty rock formation is visible, and the aquifer consists of carbonate rock, with the water level depth varying from 14.50 to 19.00 m. In the eastern and southeastern regions of the city, granite and metamorphic rock are exposed, with weathering fissure water as the primary source. The area is characterized by limited tectonic water management, and the thickness of the weathering zone typically ranges from 30 to 40 m. With the exception of a portion of the karst fissure water, the groundwater is predominantly submerged. Groundwater discharge routes mostly consist of runoff into the ocean and artificial extraction [40,41].

Groundwater recharge primarily originates from the infiltration of atmospheric precipitation, supplemented by surface water recharge during flood and irrigation events. The flow direction of groundwater runoff aligns with that of surface water systems, predominantly towards the north and northwest. The watershed for surface water coincides with that of groundwater, with the exception of the Xiaogu River, which discharges into Jiaozhou Bay to the south; all other rivers flow northward and northwestward into the Bohai Sea. The coastal region of the study area features extensive mariculture bases, a significant gold mining zone in the north, and a marble mining area in the south. The predominant land use type is arable land, primarily cultivated with winter wheat and maize.

3. Materials and Methods

3.1. Data Sources

Informed by the historical seawater intrusion line (1979–2016) [39,41], the depositional environment of the aquifer, and the groundwater storage types in the study area, a total of 115 groundwater samples were systematically collected at the end of August and the beginning of September 2020 (Figure 1). There are eight samples of fissure water, three samples of karst water, and 104 samples of pore water. The sampling depth is concentrated in 3–17 m, and a few of them can reach more than 25 m. Additionally, 13 hydrogen and oxygen isotope samples were obtained in mid-November, comprising two seawater samples and 11 groundwater samples. Water was taken from the well using a Baylor tube, the sampling bottle was rinsed three times, and two bottles of water samples were collected: 1000 mL and 500 mL. A foam box was used to hold the water samples, with ice packs inside the box. The water samples were filtered through a 0.45 μm filter membrane, and one of the samples was added to a protective reagent, nitric acid, to make the pH less than 2.

Testing of the samples was done by the University of Science and Technology Beijing. Groundwater samples were tested for: pH, Na^+, K^+, Ca^{2+}, Mg^{2+}, Cl^-, $SO_4{}^{2-}$, HCO_{3^-}, and total dissolved solids (TDS). pH was determined by the glass electrode method using a multi-parameter meter (S220, Shanghai Yidian Analytical Instrument Co., Ltd. Shanghai, China); potassium and sodium ions were measured by flame atomic absorption spectrophotometry using a flame photometer (FP640, Shanghai Yidian Analytical Instrument Co., Ltd. Shanghai, China); calcium and magnesium ions were determined by titration with disodium EDTA; chloride ion was determined by volumetric method with silver nitrate; potassium chromate was used as the indicator; sulphate was determined by the turbidimetric method using a double-beam ultraviolet-visible spectrophotometer (UV-1800, Shimadzu, Kyoto, Japan); bicarbonate and carbonate were determined by titrimetric method; and the total dissolved solids were determined by gravimetric method. Determination of hydroxide isotope content by isotope mass spectrometry using a stable isotope mass spectrometer (ISOPRIM 100, Elementar, Hamburg, Germany). The limit of quantification is shown in Table 1.

Table 1. The limit of quantification (LOQ).

Test Items	LOQ
Na^+, K^+	0.5 mg/L
Ca^{2+}	4 mg/L
Mg^{2+}	3 mg/L
Cl^-	3 mg/L
$SO_4{}^{2-}$	1 mg/L
HCO_{3^-}	5 mg/L
TDS	4 mg/L

3.2. Method

3.2.1. Hydrochemical Facies Evolution Diagram

The hydrochemical facies evolution diagram is based on the classification method proposed by Stuyfzand [42–44], presented in 2010. It has been refined since then and is now widely used [34,45,46]. The hydrochemical facies evolution diagram provides a simple method of identifying the state of the intrusion/freshening processes occurring over time within a seawater intrusion zone and is determined using the distribution of anions and cations in the diagram.

The horizontal axis of HFE-D denotes the percentage of Ca^{2+} and Na^{+} + K^{+} concentrations within the total cation concentration, while the vertical axis indicates the percentage of HCO_{3^-} and Cl^{-} concentrations within the total anion concentration (Figure 2). HFE-D categorizes the hydrochemical evolution of seawater intrusion into two phases: the invasion phase, which denotes intrusion, and the recovery phase, which signifies freshening. Furthermore, HFE-D excludes the segment of the axis from 0 to 33.3% to elucidate the hydrochemical progression more distinctly [42,43].

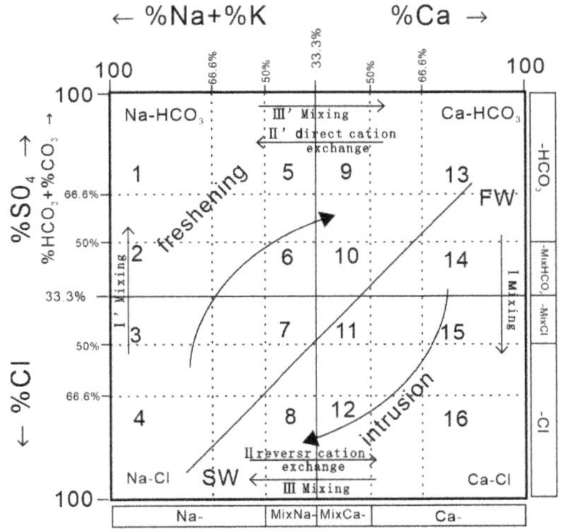

Figure 2. Hydrochemical facies evolution diagram. The hydrochemical constituents of the entire HFE-D are as follows: 1: Na-HCO_3/SO_4, 2: Na-MixHCO_3/MixSO_4, 3: Na-MixCl, 4: Na-Cl, 5: MixNa-HCO_3/SO_4, 6: MixNa-MixHCO_3/MixSO_4, 7: MixNa-MixCl, 8: MixNa-Cl, 9: MixCa-HCO_3/SO_4, 10: MixCa-MixHCO_3/MixSO_4, 11: MixCa-MixCl, 12: MixCa-Cl, 13: Ca-HCO_3/SO_4, 14: Ca-MixHCO_3/MixSO_4, 15: Ca-MixCl, 16: Ca-Cl.

3.2.2. Ion Ratios

The principal ion concentrations were transformed from mg/L to meq/L, a normalizing process to exclude the influence of charges.

This study employed the seawater ratio, defined as the concentration of each standard seawater ion relative to Cl^{-} concentration, to investigate the extent of groundwater and saltwater mixing, as well as the dissolution of rocks and minerals in the study area. Figures 6–8 illustrate the relationship between the ratios of the primary ions Ca^{2+}, Mg^{2+}, Na^{+}, K^{+}, SO_4^{2-}, and HCO_{3^-} to Cl^{-}, as well as Cl^{-} concentration, using logarithmic coordinates. This comparison with seawater ratios elucidates the mechanism of groundwater salinization. The horizontal line in the illustration represents the standard seawater ratio line for the respective ions. The ratio of the primary ions to Cl^{-} along the line corresponds to the seawater ratio, so it is referred to as the seawater ratio line [17].

In the chemical composition of water, certain proportionality coefficients of groundwater formed by different causes or under different conditions have relatively significant differences in values, so such coefficients can be used to determine the cause of groundwater formation.

3.2.3. Gibbs Diagram

The Gibbs diagram is a widely employed technique for qualitatively analyzing the sources of compounds in water, primarily ascribed to natural phenomena such as evaporation-dominated, rock-dominated, and precipitation-dominated water [47]. A Gibbs plot was constructed with $Cl^-/(Cl^- + HCO_{3^-})$ as the abscissa, and the logarithm of TDS was used as the ordinate.

3.2.4. GQI_{SWI}

The assessment of seawater intrusion employs various methodologies, each possessing distinct advantages and disadvantages. The seawater intrusion groundwater quality index (GQI_{SWI}), introduced in 2014, is a multi-index synthesis method that integrates the proportion of seawater contribution (*fsea*) and Piper diagrams, while considering the intricate hydrogeochemical reactions involved in seawater intrusion. The GQI_{SWI} may assess the extent of seawater intrusion by utilizing the scores directly, offering a more comprehensive approach than the single-indicator evaluation method and a simpler alternative to numerical simulation [48].

The diamond-shaped area of the Piper diagram can be segmented into six distinct domains: I, II, III, IV, V, and VI, denoting Ca-HCO_3, Na-Cl, mixed Ca·Na-HCO_3, mixed Ca·Mg-Cl, Ca-Cl, and Na-HCO_3 types of water, respectively. Freshwater is typically depicted in domain I, while saltwater (seawater) is portrayed in domain II. The basic amalgamation of freshwater and seawater is illustrated by a horizontal line at the center of the picture, designated as $GQI_{Piper\,(mix)}$ (Figure 3):

$$GQI_{Piper(mix)} = \left[\frac{(Ca^+ + Mg^+)}{\text{Total cations}} + \frac{(HCO_3^-)}{\text{Total anions}}\right] \times 50\ (\text{meq/L}) \quad (1)$$

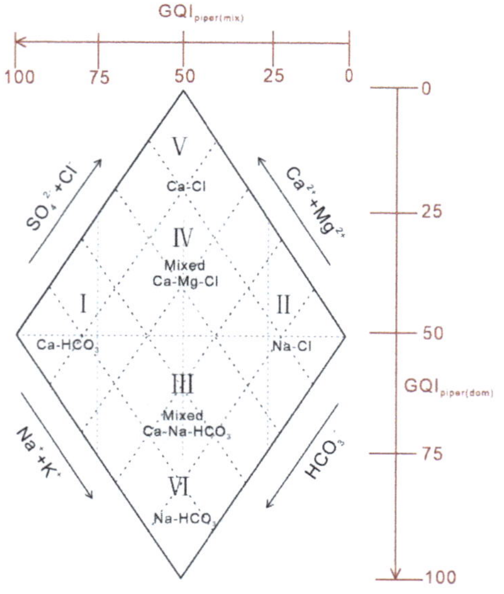

Figure 3. Development of the $GQI_{Piper(mix)}$ and $GQI_{Piper(dom)}$ resultant domains.

$GQI_{Piper(mix)}$ varies from 0 (high-concentration saline water in domain II) to 100 (freshwater in domain I). $GQI_{Piper(mix)}$ and $GQI_{Piper(dom)}$ can be utilized in conjunction to enhance the definition and description of the other domains. $GQI_{Piper(dom)}$ varies from 0, representing CaCl-type water (domain V), to 100, denoting $NaHCO_3$-type water (domain VI). The $GQI_{Piper(dom)}$ spans from 0, representing CaCl-type water (domain V), to 100, indicating $NaHCO_3$-type water (domain VI):

$$GQI_{Piper(dom)} = \left[\frac{(Na^+ + K^+)}{\text{Total cations}} + \frac{(HCO_3^-)}{\text{Total anions}}\right] \times 50\ (\text{meq/L}) \quad (2)$$

The range of $GQI_{Piper\ (mix)}$ and $GQI_{Piper\ (dom)}$ and the hydrochemical domains used in pairs are shown in Table 2.

Table 2. $GQI_{Piper(mix)}$ and $GQI_{Piper(dom)}$ to determine hydrogeochemical domains.

Domain	$GQI_{Piper(mix)}$	$GQI_{Piper(dom)}$
I	50–100	25–75
II	0–50	25–75
III	25–75	50–75
IV	25–75	25–50
V	25–75	0–25
VI	25–75	75–100

A straightforward method for detecting seawater intrusion is the seawater contribution proportion (*fsea*), which ranges from 0 to 100, yielding an index, GQI_{fsea}, where freshwater has a lower *fsea* value:

$$GQI_{fsea} = (1 - fsea) \times 100 \quad (3)$$

Cl^- measurements in specific places can be utilized to compute *fsea*, and in the absence of such observations, the freshwater to seawater ratio may be designated to range from 0 to 566 meq/L. The integration of the $GQI_{Piper\ (mix)}$ and *fsea* produces a novel index, the Groundwater Quality Index for Seawater Intrusion (GQI_{SWI}). The formulas are presented below:

$$GQI_{SWI} = \frac{GQI_{Piper(mix)} - GQI_{fsea}}{2} \quad (4)$$

The GQI_{SWI} spans from 0 to 100, with 0 representing seawater and 100 representing freshwater. Indicator values for freshwater typically exceed 75, whereas those for seawater generally fall below 50. GQI_{SWI} scores for mixed groundwater fluctuate between 50 and 75 [48].

3.2.5. Hydrogen and Oxygen Stable Isotopes

The thousandths deviation method is utilized to denote the isotopes of hydrogen and oxygen. The isotope ratios of the collected samples are compared to those of the standard samples based on the kilo deviation, specifically the kilo deviation of the $R_{samples}$ from the $R_{standards}$. The δ-value of a sample based on this ratio is defined as:

$$\delta\ (‰) = \frac{\text{Rsample} - \text{R standard}}{\text{R standard}} \times 1000 \quad (5)$$

where R_{sample} and $R_{standard}$ are the abundance ratios of the water sample and standard. The more enriched the heavy isotopes, the higher the δ-value. In 1961, Craig found that the composition of δ^2H and $\delta^{18}O$ in atmospheric precipitation in North America showed a linear relationship of $\delta^2H = 8\delta^{18}O + 10$, known as the Global Atmospheric Precipitation

Line (GMWL), or Craig line. In addition, each region also has a regional atmospheric precipitation line (LMWL), and the expression of the local atmospheric precipitation line (LMWL) in the Gulf of Lehigh is: $\delta^2H = 7.8\delta^{18}O + 6.3$ [24].

4. Results and Discussion

Groundwater chemistry in fresh groundwater and seawater intrusion areas has different characteristics, and analyzing the characteristics of groundwater chemistry is of great significance for seawater intrusion studies. Groundwater was categorized into freshwater TDS (g/L) < 1, brackish water TDS (g/L) 1–3, and saline water TDS (g/L) > 3 based on TDS values [49]. The groundwater samples were categorized. Among the 115 samples in the 2020 water-abundance period, 55 samples were fresh water, accounting for 47.83%, 52 samples were brackish water, accounting for 45.22%, and eight samples were saline water, accounting for 6.95%.

4.1. Hydrochemical Composition

The pH value (Table 3) of the groundwater samples varied from 6.18 to 8.88, with an average value of 7.07. The pH value in the study area tended to be stable in general, and showed a pattern of change from weakly acidic to weakly alkaline; the TDS value varied from 276 mg/L to 6674.9 mg/L, with an average value of 1346.16 mg/L, and the maximum value was about 24.2 times the minimum value.

Table 3. Groundwater chemical parameters (N = 115 in mg/L).

	Ca^{2+}	Mg^{2+}	Na^+	K^+	HCO_3^-	SO_4^{2-}	Cl^-	TDS	pH
Min	25.7	6.9	21.2	0.4	76.27	10	26.6	276	6.18
Max	786.57	218.81	2100	90	705	2912	3190.77	6674.9	8.88
Avg	213.13	52.47	165.75	8.84	291.81	370.43	358.37	1346.16	7.07
Med	197.19	42.55	69.4	2.3	283.73	240	252	1023	6.98
Sd	132.34	39.33	279.99	13.72	107.78	465.55	427.46	1162.02	0.48
CV(%)	62.00	75.00	169.00	155.00	37.00	126.00	119.00	86.00	7.00

The changes of cation concentration in the groundwater samples in the study area were as follows: TDS < 1 g/L (Table 4), $Ca^{2+} > Na^+ > Mg^{2+} > K^+$ in descending order, with Ca^{2+} occupying the main advantage; 1 g/L < TDS < 3 g/L (Table 5), $Ca^{2+} > Na^+ > Mg^{2+} > K^+$ in descending order, with Ca^{2+} occupying the main advantage; some samples with $Na^+ > Ca^{2+} > K^+$, and some samples with $Na^+ > Ca^{2+} > K^+$. $Ca^{2+} > Na^+ > Mg^{2+} > K^+$, with Ca^{2+} dominating, and $Na^+ > Ca^{2+}$ in some of the samples; the concentrations of cations in groundwater with TDS > 3 g/L (Table 6) were, in descending order, $Na^+ > Ca^{2+} > Mg^{2+} > K^+$, and $Mg^{2+} > Ca^{2+}$ in some of the samples.

Table 4. Freshwater chemical parameters (N = 55 in mg/L).

	Ca^{2+}	Mg^{2+}	Na^+	K^+	HCO_3^-	SO_4^{2-}	Cl^-	TDS	pH
Min	25.70	6.90	21.20	0.70	76.27	10.00	26.60	276.00	6.43
Max	278.40	62.48	235.00	31.00	493.02	395.00	304.90	999.71	7.94
Avg	145.75	31.48	69.40	4.96	247.90	135.33	167.94	693.06	7.04
Med	160.32	31.12	49.40	2.20	247.12	139.00	177.27	729.00	6.82
Sd	69.27	14.80	52.10	6.37	81.51	96.75	79.24	204.56	0.44
CV(%)	47.53	47.00	75.07	128.46	32.88	71.50	47.18	29.52	6.24

Table 5. Brackish chemical parameters (N = 52 in mg/L).

	Ca^{2+}	Mg^{2+}	Na^{+}	K^{+}	HCO_{3-}	$SO_4{}^{2-}$	Cl^{-}	TDS	pH
Min	42.10	22.40	34.70	0.40	111.00	16.00	129.00	1007.66	6.18
Max	560.10	167.80	684.00	47.20	622.37	1278.00	957.23	2827.93	8.88
Avg	253.04	59.24	156.40	9.18	317.36	431.85	373.65	1486.79	7.07
Med	241.84	55.25	129.00	2.20	305.40	359.00	324.31	1354.50	7.02
Sd	112.16	26.99	140.06	12.75	102.06	259.36	190.39	492.62	0.55
CV(%)	44.33	45.56	89.55	138.84	32.16	60.06	50.96	33.13	7.80

Table 6. Saline chemical parameters (N = 8 in mg/L).

	Ca^{2+}	Mg^{2+}	Na^{+}	K^{+}	HCO_{3-}	$SO_4{}^{2-}$	Cl^{-}	TDS	pH
Min	74.10	69.29	249.00	2.20	272.00	280.00	719.70	3112.33	6.85
Max	786.57	218.81	2100.00	90.00	705.00	2912.00	3190.77	6674.90	7.65
Avg	417.04	152.75	888.88	33.33	427.57	1587.50	1568.37	4922.10	7.24
Med	445.25	167.40	628.00	27.95	433.45	1450.00	1322.72	5217.54	7.27
Sd	248.02	53.82	664.58	27.62	144.79	901.27	838.65	1336.78	0.29
CV(%)	59.47	35.23	74.77	82.89	33.86	56.77	53.47	27.16	4.62

The anion concentration of groundwater samples varied as follows: TDS < 1 g/L (Table 4), HCO_{3-} > Cl^{-} > $SO_4{}^{2-}$, $SO_4{}^{2-}$, in which the concentration of some samples was higher than that of Cl^{-} or HCO_{3-}; TDS > 1 g/L, Cl^{-} and $SO_4{}^{2-}$ > HCO_{3-}, $SO_4{}^{2-}$ and Cl^{-} concentration of different samples had their own advantages compared to each other; $SO_4{}^{2-}$ and Cl^{-} concentrations of different samples had their own advantages.

4.2. Water Chemistry Type

From Figure 4, it can be seen that the cations of freshwater samples of groundwater samples are dominated by Ca^{2+}, and there is no obvious dominant anion; the cations of brackish water samples are dominated by Ca^{2+}, and the anions are dominated by Cl^{-} and $SO4^{2-}$; and the cations of saline water samples are dominated by Na^{+} and Ca^{2+}, and the anions are dominated by Cl^{-} and $SO_4{}^{2-}$.

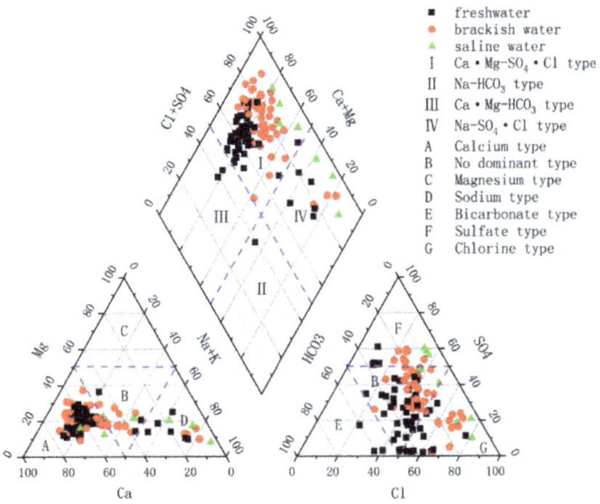

Figure 4. Piper triple line map of groundwater hydrochemistry [22].

Combined with the results of previous research and the results of the present analysis, it is believed that under the influence of aerosol diffusion by the sea breeze [40,50], the chemical type of the unpolluted natural groundwater in the northern part of Laizhou City is the $Ca\text{-}HCO_3{\cdot}Cl$ type. The Piper diagram reveals that cations in freshwater and brackish water samples are dominated by Ca^{2+}, with groundwater samples concentrated in Zone

I. Based on the characteristics of major ions, it is inferred that the higher Ca^{2+} concentration in brackish water may result from cation exchange [40]. Combined with Shukarev classification, from freshwater to saline water, the groundwater chemistry transitions from $Ca\text{-}HCO_3{\cdot}Cl$-type water to $Ca{\cdot}Na\text{-}SO_4{\cdot}Cl$-type water, and finally to Na-Cl-type water.

Groundwater samples occupy 40 percent of the freshening phase and 60 percent of the intrusion phase, with intrusion dominating. Groundwater samples occupy 13 of the 16 phases in the hydrochemical evolution diagram (Figure 5). The main hydrochemical facies of 115 groundwater samples are Ca-MixCl, $Ca\text{-}MixHCO_3/MixSO_4$, $Ca\text{-}HCO_3/SO_4$, Ca-Cl, and Na-Cl, which accounted for 83.48% of the total samples. There were 27 samples of Ca-MixCl phase, 26 samples of $Ca\text{-}MixHCO_3/MixSO_4$ phase, 11 samples of $Ca\text{-}MixHCO_3$ phase, and 15 samples of $Ca\text{-}MixSO_4$ phase. Also, 46.09% of the total samples were $Ca\text{-}MixHCO_3/MixSO_4$ and Ca-MixCl facies samples, 46.09% of the total samples were $Ca\text{-}HCO_3/MixSO_4$ facies samples of 17 samples, 15 samples were $Ca\text{-}SO_4$ facies, and only two samples were $Ca\text{-}HCO_3$ facies. There were 15 Ca-Cl facies samples and 11 Na-Cl facies samples. Groundwater hydrochemical facies are characterized by complexity and variability.

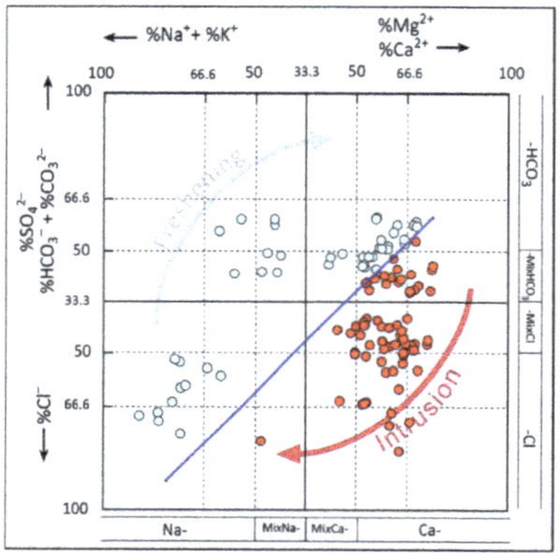

Figure 5. The hydrochemical evolution facies diagram.

As can be seen from Figure 6, most of the groundwater samples in the study area are evolving from $Ca\text{-}HCO_3$-type water to Ca-Cl-type water, with mixing dominating the invasion phase. During the desalination stage, positive cation exchange has a greater influence on the groundwater chemistry type. Only very few areas of the study area exhibit $Na\text{-}HCO_3$-type water, indicating that there is no obvious trend of desalination of brackish water in the study area, which may be attributed to insufficient groundwater recharge.

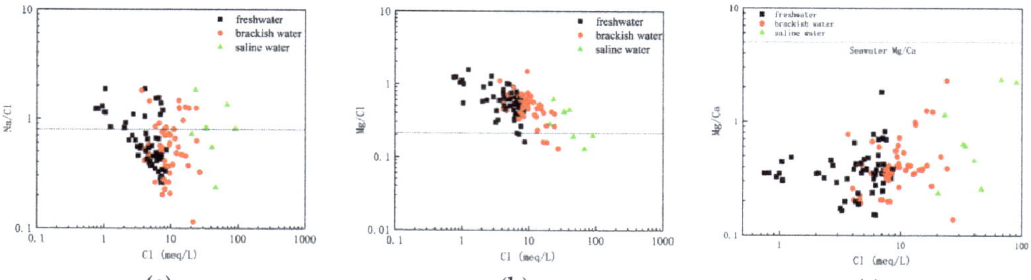

(a) (b) (c)

Figure 6. The relationship between main ion ratios to Cl^- and Cl^-: (**a**) Na^+/Cl^- and Cl^-; (**b**) Mg^{2+}/Cl^- and Cl^-; (**c**) Mg^{2+}/Ca^{2+} and Cl^-. The horizontal dotted line indicates the seawater ratio line.

4.3. Hydrogeochemical Variations

The Na/Cl ratio has no specific pattern in the scatter plot (Figure 6a). The range of Na/Cl ratios in the study area is 0.11–1.88, and 75.65% of the groundwater samples have Na/Cl ratios less than 0.86, indicating that the study area is mainly affected by seawater intrusion [51]. Furthermore, 21.74% of the groundwater samples have Na/Cl ratios equal to or greater than 1, and areas with high Na/Cl ratios (>1) may be contaminated by anthropogenic activities [51].

The Mg/Cl ratios in the scatter plots showed a decrease with the increase of Cl^- concentration and converged around the seawater ratio line (Figure 6b). Typical Mg/Ca ratios for seawater range from 4.5 to 5, while freshwater Mg/Ca ratios are less than 1 [52]. The Mg/Ca ratios in the study area ranged from 0.14 to 2.35 and did not tend to be close to the seawater Mg/Ca ratio line with increasing Cl^- concentration (Figure 6c), indicating that a considerable number of samples had excess Ca^{2+} and deficient Mg^{2+}, which was presumably affected by the mixing of seawater and freshwater, as well as by the influence of Mg^{2+} and the influence of seawater and freshwater mixing. It is assumed that Mg^{2+} is affected by the mixing of seawater and freshwater, as well as by the cation exchange.

The K/Cl ratio in the scatter plot shows an increase with Cl^- concentration and a certain degree of proximity to the seawater dilution line (Figure 7a), suggesting that the K^+ may be primarily sourced from invaded seawater and influenced by cation exchange.

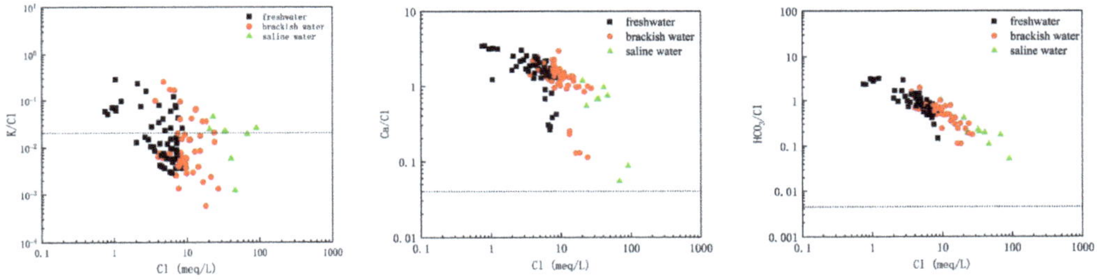

(a) (b) (c)

Figure 7. The relationship between main ion ratios to Cl^- and Cl^-: (**a**) K^+/Cl^- and Cl^-; (**b**) Ca^{2+}/Cl^- and Cl^-; (**c**) HCO_{3-}/Cl^- and Cl^-. The horizontal dotted line indicates the seawater ratio line.

The Ca/Cl ratios all show a pattern of decreasing with increasing Cl^- concentration in the scatter plots, and the Ca/Cl ratios of some samples were higher than the seawater ratio line (Figure 7b). Ca^{2+} is inferred to be affected by seawater intrusion and cation exchange in conjunction with water chemistry components and the location of groundwater Piper diagrams.

The HCO_3/Cl ratios in the scatter plots all show a pattern of decreasing with increasing Cl^- concentration, and HCO_{3-} is higher than the seawater ratio line in the scatter plots of the groundwater samples and shows a certain linear pattern (Figure 7c), which indicates that HCO_{3-} is mainly derived from freshwater and mineral dissolution, and that the content is affected by seawater intrusion mainly because of the mixing of seawater and freshwater from the groundwater in the aquifer.

The SO_4/Cl ratios in the scatter plot show an increase in Cl^- concentration and a certain degree of closeness to the seawater ratio line of the pattern (Figure 8a), indicating that $SO_4{}^{2-}$ originated from invaded seawater. Some samples fell above the 1:1 line, suggesting that dissolution of gypsum is one of the sources of $SO_4{}^2$; in addition, some samples were far above the 1:1 line (Figure 8b), suggesting that they may be affected by human activities (mariculture).

From Figure 9, it can be seen that most of the groundwater samples fall within the rock-dominated and evaporation-dominated zones. Freshwater and brackish water are dominated by rock, and saline water is mainly affected by evaporation dominance. As TDS increases, groundwater changes from rock-dominated to evaporation-dominated. It shows the same pattern as the southern coast of Leizhou Bay [33].

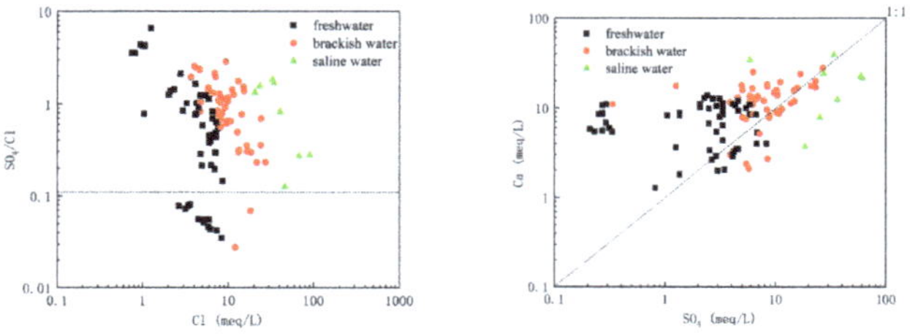

(**a**) (**b**)

Figure 8. The relationship between main ion ratios: (**a**) SO_4^{2-}/Cl^- and Cl^-; (**b**) Ca^{2+} / SO_4^{2-}. The horizontal dotted line indicates the seawater ratio line.

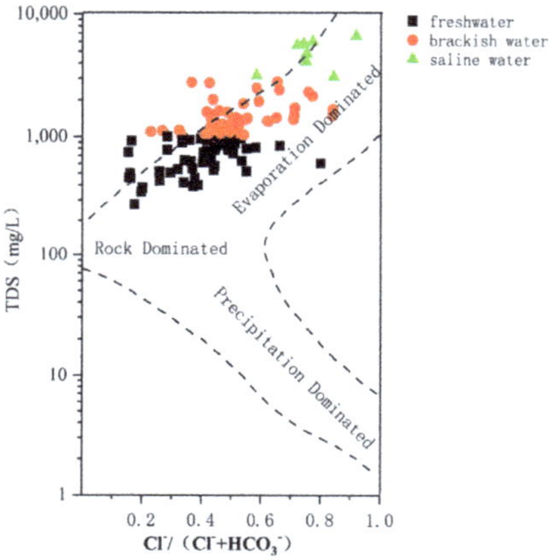

Figure 9. Gibbs diagram.

4.4. Scope of Seawater Intrusion

The distribution of GQI_{SWI} values for the 115 samples ranged from 49.34 to 85.34, with a mean value of 74.18 and a median value of 76.33. The main types of groundwater in the study area are freshwater and mixed groundwater (Figure 10), and only one area with GQI_{SWI} values lower than 50 exists in the southwestern part of Cangshang. Mixed groundwater is mainly distributed in the western part of the city of Laizhou, along the area from Zhuyueyi village, over the west to the northern part of Zhuqiao town.

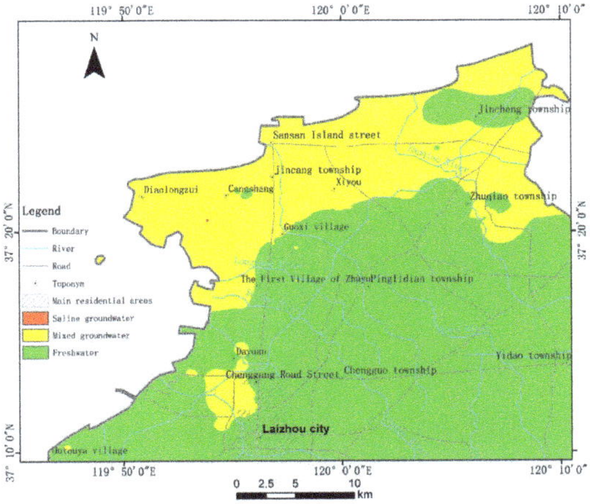

Figure 10. Spatial distribution of the extent of saltwater intrusion.

4.5. Sources of Groundwater Recharge

The theoretical values of δ^2H and $\delta^{18}O$ of seawater are both 0, but the near-shore surface seawater is affected by land surface water and groundwater, which will deviate from the theoretical values. Two surface seawater hydrogen and oxygen isotope samples were collected: T-XY-01 had a δ^2H value of −19.07 ‰ and a $\delta^{18}O$ value of −2.18 ‰; T-YX-01 had a δ^2H value of −19.10 ‰ and a $\delta^{18}O$ value of −2.07 ‰.

The δ^2H value of groundwater samples: −9.12 ‰~−5.94 ‰, mean value −7.85 ‰; $\delta^{18}O$ value: −62.40 ‰~−46.08 ‰, mean value −56.76 ‰. It can be seen that the hydrogen and oxygen isotope values of the groundwater samples all fall near or below the global atmospheric precipitation line and the regional atmospheric precipitation line (Figure 11), which indicates that atmospheric precipitation is the main source of groundwater.

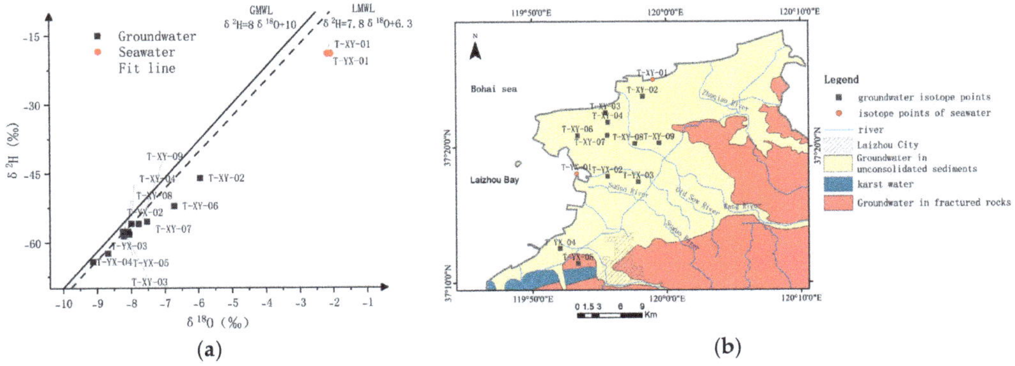

Figure 11. δ^2H and $\delta^{18}O$ relation: (**a**) The relationship between δ^2H and $\delta^{18}O$; (**b**) Map of isotope sampling point locations.

In addition, points 7, 6, and 2 deviate from the regional atmospheric precipitation line, and, according to the location of the sampling points, this is probably due to the fact that the groundwater in the region is subjected to the effect of evaporation and fractionation.

5. Conclusions

The distribution of groundwater samples in Piper's trilinear map indicates the complexity of groundwater chemistry types in the study area. From freshwater to saline water, the groundwater chemistry transitions from Ca-HCO_3·Cl-type water to Ca·Na-SO_4·Cl-type water, and finally to Na-Cl-type water. The hydrochemical evolution diagram shows that most of the groundwater samples in the study area are in the invasive stage, and the hydrochemical phases of the groundwater samples are complex. Ca-MixCl and Ca-Cl phases are the major hydrochemical phases of the groundwater samples.

The main ions in the groundwater in the study area are controlled by seawater intrusion: Na^+, Cl^-, K^+, Mg^{2+}, $SO_4{}^{2-}$ are mainly derived from seawater intrusion, and are affected by human activities, cation exchange, and water–rock interactions; HCO_{3^-} is mainly derived from freshwater and mineral dissolution, and its content is affected by seawater intrusion; changes in Ca^{2+} content are affected by seawater intrusion, cation exchange, and mineral dissolution. As TDS increases, groundwater changes from rock-dominated to evaporation-dominated.

The current status of seawater intrusion in the study area was evaluated using the seawater intrusion groundwater quality index (GQI_{SWI}), and seawater intrusion was mainly concentrated in the northern part of the study area. In view of the extent of seawater intrusion in the past, it is recommended that the monitoring of seawater intrusion in the northern part of the study area be strengthened and that groundwater recharge to the study area be increased.

Hydrogen and oxygen isotope data of groundwater and seawater showed that the main source of recharge in the study area is atmospheric precipitation.

Author Contributions: Software, X.H. and D.Z.; Investigation, Z.Z., Y.L. and H.Z.; Data curation, Y.Z. and K.S.; Writing—original draft, X.H.; Writing—review and editing, X.H. and L.K.; Project administration, L.Z. All authors have read and agreed to the published version of the manuscript.

Funding: This study was supported by Open Research Fund of Observation and Research Station of Seawater Intrusion and Soil Salinization, Laizhou Bay, Ministry of Natural Resources (2024LZORS003), and the China Geological Survey program (DD20242961, DD20243124, and DD20208080).

Data Availability Statement: The original contributions presented in this study are included in the article. Further inquiries can be directed to the corresponding author.

Acknowledgments: The authors give their most sincere thanks to the editors and reviewers for their contributions to the improvement of this article.

Conflicts of Interest: The authors declare no conflicts of interest.

Abbreviations

The following abbreviations are used in this manuscript:

GQI_{SWI}	The groundwater quality index for seawater intrusion

References

1. Cao, T.; Han, D.; Song, X. Past, present, and future of global seawater intrusion research: A bibliometric analysis. *J. Hydrol.* **2021**, *603 Pt A*, 126844. [CrossRef]
2. Jeen, S.-W.; Kang, J.; Jung, H.; Lee, J. Review of Seawater Intrusion in Western Coastal Regions of South Korea. *Water* **2021**, *13*, 761. [CrossRef]
3. Shi, L.; Jiao, J.J. Seawater intrusion and coastal aquifer management in China: A review. *Environ. Earth Sci.* **2014**, *72*, 2811–2819. [CrossRef]
4. Zhang, Z.; Liu, C.; Yi, L.; Ren, H.; Li, S.; Liu, N.; Ren, L.; Liu, J.; Li, R.; Dong, Y. Hydrochemical fingerprints and health risk assessment of groundwater contamination in the Bohai Sea region, China. *Mar. Pollut. Bull.* **2025**, *212*, 117539. [CrossRef] [PubMed]
5. Yamanaka, M.; Bottrell, S.H.; Wu, J.H.; Kumagai, Y.; Mori, K.; Satake, H. Chlorine stable isotope evidence for salinization processes of confined groundwater in southwestern Nobi Plain aquifer system, central Japan. *J. Hydrol.* **2014**, *519*, 295–306. [CrossRef]

6. V, S.P.; S.K, A.; S, R.A.; R.J, J.J. An assessment of vulnerability in aquifer using GALDIT model in aquifer zones between Mookkaiyur to Tharuvaikulam, Southeast coast of India. *J. Coast. Conserv.* **2025**, *29*, 5. [CrossRef]
7. Hossain, M.I.; Reza, A.H.M.S.; Shafiuzzaman, S.M.; Islam, M.S.-U.; Rahman, M.A. The effects of seawater intrusion on sustainable coastal areas: A comprehensive study on Bagerhat district, Bangladesh. *Reg. Stud. Mar. Sci.* **2025**, *82*, 104038. [CrossRef]
8. Souid, F.; Telahigue, F.; Agoubi, B.; Kharroubi, A. Isotopic behavior and self-organizing maps for identifying groundwater salinization processes in Jerba Island, Tunisia. *Environ. Earth Sci.* **2020**, *79*, 175. [CrossRef]
9. Mountadar, S.; Younsi, A.; Hayani, A.; Siniti, M.; Tahiri, S. Groundwater salinization process in the coastal aquifer Sidi Abed-Ouled Ghanem (Province of El Jadida, Morocco). *J. Afr. Earth Sci.* **2018**, *147*, 169–177. [CrossRef]
10. Zha, X.; Hernández Anguiano, J.H.; Padilla Benítez, F.; Cruz-Falcón, A.; Miranda-Avilés, R.; Mendoza Cantú, M.E.; Perera-Burgos, J.A.; Liao, X.; Navarro-Céspedes, J.M.; Acosta-Reyes, M.A.; et al. Status of seawater intrusion in Mexico: A review. *J. Hydrol.-Reg. Stud.* **2025**, *57*, 102189. [CrossRef]
11. Cruz, J.V.; Andrade, C. Groundwater salinization in Graciosa and Pico islands (Azores archipelago, Portugal): Processes and impacts. *J. Hydrol.-Reg. Stud.* **2017**, *12*, 69–87. [CrossRef]
12. Gil-Márquez, J.M.; Barberá, J.A.; Andreo, B.; Mudarra, M. Hydrological and geochemical processes constraining groundwater salinity in wetland areas related to evaporitic (karst) systems. A case study from Southern Spain. *J. Hydrol.* **2017**, *544*, 538–554. [CrossRef]
13. Wang, Y.; Tong, J.; Hu, B.X.; Dai, H. Combining Isotope and Hydrogeochemistry Methods to Study the Seawater Intrusion: A Case Study in Longkou City, Shandong Province, China. *Water* **2022**, *14*, 789. [CrossRef]
14. Xiong, G.; Chen, G.; Wu, J.; Wang, Z.; Yu, H.; Fu, T.; Liu, W.; Xu, X.; Hou, G.; Yang, Y.; et al. Identifying the characteristics and potential risk of seawater intrusion for southern China by the SBM-DEA model. *Sci. Total Environ.* **2022**, *844*, 157205. [CrossRef]
15. Al-Taliby, W.; Dekhn, H. Modeling of Groundwater Pollution by Former Seawater in Submarine Groundwater Discharge Driven by Saltwater Intrusion. *J. Ecol. Eng.* **2022**, *23*, 166–175. [CrossRef]
16. Zang, Y.; Li, M. Numerical assessment of compressed air injection for mitigating seawater intrusion in a coastal unconfined aquifer. *J. Hydrol.* **2021**, *595*, 125964. [CrossRef]
17. Yamanaka, M.; Kumagai, Y. Sulfur isotope constraint on the provenance of salinity in a confined aquifer system of the southwestern Nobi Plain, central Japan. *J. Hydrol.* **2006**, *325*, 35–55. [CrossRef]
18. Wang, Y.; Jiao, J.J. Origin of groundwater salinity and hydrogeochemical processes in the confined Quaternary aquifer of the Pearl River Delta, China. *J. Hydrol.* **2012**, *438–439*, 112–124. [CrossRef]
19. Zhao, Q.; Su, X.; Kang, B.; Zhang, Y.; Wu, X.; Liu, M. A hydrogeochemistry and multi-isotope (Sr, O, H, and C) study of groundwater salinity origin and hydrogeochemcial processes in the shallow confined aquifer of northern Yangtze River downstream coastal plain, China. *Appl. Geochem.* **2017**, *86*, 49–58. [CrossRef]
20. Zeynolabedin, A.; Ghiassi, R.; Norooz, R.; Najib, S.; Fadili, A. Evaluation of geoelectrical models efficiency for coastal seawater intrusion by applying uncertainty analysis. *J. Hydrol.* **2021**, *603*, 127086. [CrossRef]
21. Balasubramanian, M.; Sridhar, S.G.D.; Ayyamperumal, R.; Karuppannan, S.; Gopalakrishnan, G.; Chakraborty, M.; Huang, X. Isotopic signatures, hydrochemical and multivariate statistical analysis of seawater intrusion in the coastal aquifers of Chennai and Tiruvallur District, Tamil Nadu, India. *Mar. Pollut. Bull.* **2022**, *174*, 113232. [CrossRef] [PubMed]
22. Wang, Z.; Su, Q.; Wang, S.; Gao, Z.; Liu, J. Coupling hydrochemistry and stable isotopes (δ^2H and δ^{18}O) to identify the major factors affecting the hydrochemical evolution of groundwater in the Western Yellow Sea Coast, China. *Appl. Geochem.* **2022**, *138*, 105221. [CrossRef]
23. Zhang, Z.; Yi, L. Research methods for seawater intrusion in China and recommendations for novel radium-radon technologies. *Mar. Environ. Res.* **2024**, *198*, 106530. [CrossRef]
24. Han, D.; Kohfahl, C.; Song, X.; Xiao, G.; Yang, J. Geochemical and isotopic evidence for palaeo-seawater intrusion into the south coast aquifer of Laizhou Bay, China. *Appl. Geochem.* **2011**, *26*, 863–883. [CrossRef]
25. Yin, Z.; Geng, S.; Zou, B.; Lin, W.; Cai, Z.; Yang, M.; Cheng, R.; Li, B.; Ning, P. *Study on Seawater Intrusion in the Coastal Area of Laizhou City*; Ocean Press: Beijing, China, 1992.
26. Zhang, Z.; Wang, L. Driving Mechanism of Land Use/Coverage Change in Southern Laizhou Bay. *Sci. Geogr. Sin.* **2007**, *5*, 40.
27. Du, Y.; Ma, T.; Chen, L.; Shan, H.; Xiao, C.; Lu, Y.; Liu, C.; Cai, H. Genesis of salinized groundwater in Quaternary aquifer system of coastal plain, Laizhou Bay, China: Geochemical evidences, especially from bromine stable isotope. *Appl. Geochem.* **2015**, *59*, 155–165. [CrossRef]
28. Chen, Q.; Wei, J.; Wang, H.; Shi, L.; Gao, Z.; Liu, S.; Ning, F.; Jia, C.; Ji, Y.; Dong, F.; et al. Discussion on the Fluorosis in Seawater-Intrusion Areas Along Coastal Zones in Laizhou Bay and Other Parts of China. *Int. J. Environ. Res.* **2019**, *13*, 435–442. [CrossRef]
29. Xue, Y.; Xie, C.; Wu, J.; Liu, P.; Shi, H. *Seawater Intrusion and Saltwater-Freshwater Interface Migration Mechanism Research*; Nangjing University Press: Nanjing, China, 1991.

30. Fu, T.; Fu, Y.; Li, C.; Dong, M.; Qi, C.; Wang, Z.; Chen, G.; Su, Q.; Xu, X.; Yu, H. Seawater intrusion-triggered high fluoride groundwater development on the eastern coast of China. *Environ. Sci. Pollut. Res.* **2024**, *31*, 11307–11320. [CrossRef]
31. Sun, Q.; Gao, M.; Wen, Z.; Hou, G.; Dang, X.; Liu, S.; Zhao, G. Hydrochemical evolution processes of multiple-water quality interfaces (fresh/saline water, saline water/brine) on muddy coast under pumping conditions. *Sci. Total Environ.* **2023**, *857*, 159297. [CrossRef]
32. Yang, F.; Liu, S.; Jia, C.; Gao, M.; Chang, W.; Wang, Y. Hydrochemical characteristics and functions of groundwater in southern Laizhou Bay based on the multivariate statistical analysis approach. *Estuar. Coast. Shelf Sci.* **2021**, *250*, 107153. [CrossRef]
33. Wang, Z.; Wang, S.; Liu, W.; Su, Q.; Tong, H.; Xu, X.; Gao, Z.; Liu, J. Hydrochemical Characteristics and Irrigation Suitability Evaluation of Groundwater with Different Degrees of Seawater Intrusion. *Water* **2020**, *12*, 3460. [CrossRef]
34. Liu, S.; Tang, Z.; Gao, M.; Hou, G. Evolutionary process of saline-water intrusion in Holocene and Late Pleistocene groundwater in southern Laizhou Bay. *Sci. Total Environ.* **2017**, *607–608*, 586–599. [CrossRef]
35. Han, D.M.; Song, X.F.; Currell, M.J.; Yang, J.L.; Xiao, G.Q. Chemical and isotopic constraints on evolution of groundwater salinization in the coastal plain aquifer of Laizhou Bay, China. *J. Hydrol.* **2014**, *508*, 12–27. [CrossRef]
36. Xue, Y.W.; Xie, J.; Xie, C.; Zhang, Y. Research on Seawater Intrusion and Saline Intrusion along the Coast of Laizhou Bay. *Chin. Sci. Bull.* **1997**, *42*, 2360–2368.
37. Wang, T.; Dai, W.; Bai, Y. Analysis of Seawater Intrusion Status and Development Trends in the Eastern Coast of Laizhou Bay. *West. Resour.* **2020**, *5*.
38. Shu, L.; Huang, L.; Chen, H.; Lu, C.; Liu, B. Risk Assessment and Zoning of Seawater Intrusion Hazard in Coastal Region of Laizhou City Based on AHP-EWM Method. *J. Jilin Univ. (Earth Sci. Ed.)* **2023**, *53*, 16. [CrossRef]
39. Wang, W.; Feng, L.; Lin, Y. This study presents an analysis of seawater intrusion status in Yantai City and discusses engineering and ecological prevention measures. *Ground Water* **2018**, *40*, 3.
40. Wu, J.; Xue, Y.; Liu, P.; Wang, J.; Jiang, Q.; Shi, H.; Ning, P. Development and Hydrochemical Characteristic of Seawater Intrusion in Longkou-Laizhou District. *J. NanJing Univ. (Nat. Sci.)* **1994**, *30*, 98–110.
41. Li, F. *Monitoring and Numerical Simulation of Saltwater Intrusion in the Eastern Coast of Laizhou Bay, China*; Ocean University of China: Qingdao, China, 2005.
42. Giménez-Forcada, E. Space/time development of seawater intrusion: A study case in Vinaroz coastal plain (Eastern Spain) using HFE-Diagram, and spatial distribution of hydrochemical facies. *J. Hydrol.* **2014**, *517*, 617–627. [CrossRef]
43. Giménez-Forcada, E. Dynamic of Sea Water Interface using Hydrochemical Facies Evolution Diagram. *Groundwater* **2010**, *48*, 212–216. [CrossRef]
44. Giménez-Forcada, E.; Sánchez San Román, F.J. An Excel Macro to Plot the HFE-Diagram to Identify Sea Water Intrusion Phases. *Groundwater* **2015**, *53*, 819–824. [CrossRef] [PubMed]
45. Aladejana, J.A.; Kalin, R.M.; Sentenac, P.; Hassan, I. Groundwater quality index as a hydrochemical tool for monitoring saltwater intrusion into coastal freshwater aquifer of Eastern Dahomey Basin, Southwestern Nigeria. *Groundw. Sustain. Dev.* **2021**, *13*, 100568. [CrossRef]
46. Le, T.T.V.; Lertsirivorakul, R.; Bui, T.V.; Schulmeister, M.K. An Application of HFE-D for Evaluating Sea Water Intrusion in Coastal Aquifers of Southern Vietnam. *Groundwater* **2020**, *58*, 1012–1022. [CrossRef]
47. Gibbs, R.J. Mechanisms Controlling World Water Chemistry. *Science* **1970**, *170*, 1088–1090. [CrossRef]
48. Tomaszkiewicz, M.; Abou Najm, M.; El-Fadel, M. Development of a groundwater quality index for seawater intrusion in coastal aquifers. *Environ. Model. Softw.* **2014**, *57*, 13–26. [CrossRef]
49. Gao, M.; Zhang, Y.; Liu, S.; Wang, S.; Kong, X.; Zhao, J.; Guo, F. Palaeogeographic Condition for Origin of Underground Brine in Southern Coast of Laizhou Bay, Bohai Sea. *Geol. Rev.* **2015**, *61*, 393–400. [CrossRef]
50. Bouzourra, H.; Bouhlila, R.; Elango, L.; Slama, F.; Ouslati, N. Characterization of mechanisms and processes of groundwater salinization in irrigated coastal area using statistics, GIS, and hydrogeochemical investigations. *Environ. Sci. Pollut. Res.* **2015**, *22*, 2643–2660. [CrossRef]
51. Asare, A.; Appiah-Adjei, E.K.; Ali, B.; Owusu-Nimo, F. Assessment of seawater intrusion using ionic ratios: The case of coastal communities along the Central Region of Ghana. *Environ. Earth Sci.* **2021**, *80*, 307. [CrossRef]
52. Jimmy, J.; Vincentive, P. *Coastal Hydrogeology*; Cambridge University Press: Cambridge, UK, 2019. [CrossRef]

water

MDPI

Article

Analysis of Soil Salinization Characteristics in Coastal Area of Panjin City, Liaoning Province

Jiaquan Sun [1,2,3], Wanbing Song [1], Xiubo Sun [1], Jian Cui [1,2,3,*] and Hongwei Ma [1,2,3,*]

1 Shenyang Center, China Geological Survey, Shenyang 110034, China; sunjiaquan2025@163.com (J.S.); yswb1228@163.com (W.S.); hljsunxb@126.com (X.S.)
2 Key Laboratory of Black Soil Evolution and Ecological Effect, Ministry of Natural Resources, Shenyang 110034, China
3 Key Laboratory of Black Soil Evolution and Ecological Effect, Liaoning Province, Shenyang 110034, China
* Correspondence: cj119@sina.com (J.C.); mahw83@163.com (H.M.); Tel.: +86-186-4004-6120 (J.C.); +86-186-9887-5616 (H.M.)

Abstract

Soil salinization is one of the major geological environmental issues in the coastal area of Panjin City, Liaoning Province. By collecting soil samples from the upper 0–40 cm and conducting total salt content tests, this study summarizes the statistical characteristics of soil salinity content, analyzes the correlations between soil salinity ions and total salt content, explores the spatial distribution characteristics of soil salinization in the study area, evaluates soil salinization, and identifies the driving factors of soil salinization in the region. The results show that the total salt content in the study area ranges from 1.3 to 9.0 g kg^{-1}, with Na^+, Cl^-, and $SO_4{}^{2-}$ being the dominant ions. Total salt content is positively correlated with Mg^{2+}, Na^+, Cl^-, and $SO_4{}^{2-}$, indicating that the main forms of salinity are sulfates and chlorides. The degree of soil salinization is classified as mild to moderate. The types of soil salinization change from the center to the sides in the following order: sulfate type → chloride–sulfate type → sulfate–chloride type. The degree of soil salinization shows distinct zonality, gradually decreasing from the coastal area to the inland. The main driving factor of soil salinization is the groundwater level, while the evapotranspiration ratio and groundwater salinity are secondary factors that jointly control the process.

Keywords: soil salinization; total salt content; correlation; salinization analysis; driving factors

1. Introduction

Soil salinization is a common environmental geological issue in coastal areas, characterized by the continuous accumulation of soluble salts in the soil surface, forming salt crusts and salt patches [1]. Soil salinization is a global ecological problem that directly leads to a reduction in arable land area, soil erosion, and soil desertification, thereby triggering a series of vicious cycles. It severely affects human life and hinders economic development, potentially leading to food security issues. Globally, 932 million hm^2 of land is affected by salinization, accounting for about one-third of the world's usable land [2,3]. In China, saline soils account for about 5% of the country's usable land area [4]. Chinese saline soils are characterized by their diversity, large area, and wide distribution, mainly found in the arid and semi-arid regions of the northwest, such as Xinjiang, Qinghai, Inner Mongolia, and Ningxia [5–8], the inland areas of the northeast, including Jilin, Inner Mongolia, and Heilongjiang [9,10], and the eastern coastal regions, such as Shandong, Jiangsu, and Hebei [11]. The trend of salinization is still expanding nationwide [12].

With the development of society and the increasing population, the importance of arable land resources is becoming more significant, and the issue of soil salinization is gradually attracting global attention. Since President Xi Jinping emphasized the comprehensive transformation and utilization of saline-alkali land, several hot topics related to soil salinization have been intensively studied. In recent years, research on soil salinization has mainly focused on the following aspects: (1) monitoring methods for salinization; (2) factors influencing salinization; and (3) spatial distribution characteristics of salinization. Yuan et al. [13–15] conducted in-depth research on the methods for accurately extracting soil salinization information. The results showed that although determining soil salinity by measuring soluble salt content is accurate, it is costly and difficult to implement on a large scale. In contrast, using hyperspectral remote sensing to extract salinization information is relatively accurate and has advantages such as wide monitoring range, strong continuous monitoring capability, and fast information updates. In recent years, some domestic scholars have also used isotopic techniques to reflect the degree of soil salinization, but this method is still immature and not yet widely applied. Fu et al. [16–19] studied the factors causing soil salinization and their distribution characteristics. The results indicated that climatic conditions, groundwater depth, and topography are closely related to the spatial distribution of soil salinization in the study area, and human activities also influence the formation of soil salinization to a certain extent. Li et al. [20–22] investigated the spatial distribution characteristics of soil salinization. The results showed that the influence of various factors on soil salinity at different soil depths is complex, and the interaction between any two factors has a more significant impact on the spatial distribution of soil salinity.

The southeastern coastal area of Liaoning Province is a concentrated distribution area of coastal saline soils. Due to the long-term influence of seawater infiltration [23], the groundwater level is shallow, and the groundwater salinity is high. Coupled with unreasonable land use by humans, soil salinization is widespread in this region. The spatial distribution of salinization is uneven, and the variation of water and salt in space is complex, severely restricting local agricultural development. The southeastern coastal area of Liaoning Province, as an important reserve land resource in the northeast, has great potential for development and utilization [24]. Correctly and effectively analyzing the spatial differentiation characteristics of soil salinity and its driving factors is of practical significance for the scientific use of saline land resources.

This study, supported by the China Geological Survey project, focuses on the southern coastal area of Panjin City, a core region of the southeastern coastal area of Liaoning Province. It aims to reveal the distribution characteristics and patterns of soil salinity ions, analyze the correlations between soil salinity ions and total salt content, explore the spatial distribution characteristics of soil salinization in the study area, evaluate soil salinization, and identify the driving factors of soil salinization. This study is intended to actively respond to national strategic needs, clarify the development direction of comprehensive utilization of saline-alkali land, provide theoretical basis and technical support for formulating precise comprehensive improvement measures and management systems for coastal saline land agriculture, offer geological basis for national land space planning in coastal areas and soil improvement in coastal areas, promoting zero growth of land degradation in coastal zones [25], and constructing a regional integrated monitoring network for resources and environment.

2. Materials and Methods

2.1. Study the General Situation of the Region

The study area is located in the coastal area of Panjin City, Liaoning Province, situated in the southwest of Liaoning Province, on the east coast of the Bohai Sea, downstream of the Liao River. It includes Dawang District, Panshan County, and Xinglongtai District, covering a total area of 1757.21 km^2. It is an important convergence area between land and sea in the Bohai Rim region of Northeast China. The terrain of the study area slopes from north to south, gradually tilting towards the south with relatively low-lying areas. The main geomorphological type is the Holocene alluvial–marine plain. The study area has a warm temperate continental semi-humid monsoon climate, with distinct seasons and concurrent heat and rainfall. The average annual high temperature is 17 °C, the average annual low temperature is 6 °C, and the average annual precipitation is over 700 mm, mainly concentrated in July and August, accounting for more than 50% of the annual precipitation. The average annual evapotranspiration ranges from 1000 to 1400 mm. The main soil types are Fluvisols, Paddy soils, and Coastal Saline soils. The dominant lithology in the area is silt loam and silt. The groundwater level is high, and the salinity is strong, which easily leads to soil salinization.

2.2. Sample Collection

The sampling period was chosen during the non-fertilization period of agricultural land to avoid the influence of fertilizers on soil test indicators. The sampling locations were in the southern coastal area of Panjin City, south of the Jingha and Danshen Expressways. Nine sampling profiles were set up in the study area, perpendicular to the southern coastline of Panjin City, with five sampling points along each profile, resulting in a total of 45 soil samples collected. Soil samples were collected from the surface layer of plots with obvious salinization, with a sampling depth of 0–40 cm. The double diagonal method was used to collect cubic mixed samples with a sampling shovel. During the sampling process, the sampling amount was kept consistent vertically, and efforts were made to keep the soil samples fresh.

2.3. Sample Testing

The sample tests included the analysis of 10 water-soluble salts in the soil: fluoride (F^-), bicarbonate (HCO_3^-), chloride (Cl^-), calcium (Ca^{2+}), magnesium (Mg^{2+}), sulfate (SO_4^{2-}), potassium (K^+), sodium (Na^+), nitrate (NO_3^-), and phosphate (PO_4^{3-}). Potassium and sodium ions were determined by flame photometry (The equipment used in the flame photometry is FP6410 produced by Shanghai INESA, Shanghai, China), calcium and magnesium ions by EDTA complexometric titration, bicarbonate ions by double indicator neutralization method, sulfate ions by EDTA indirect complexometric titration, chloride ions by standard silver nitrate direct titration, fluoride, nitrate and phosphate by ion chromatography, and total salt content by the electrical conductivity method. The sample tests were completed by the Shenyang Geological Survey Center of the China Geological Survey.

2.4. Data Sources and Processing

The data on soil water-soluble salt analysis in this study were obtained from laboratory tests of soil samples. The data on groundwater depth and salinity in the study area were sourced from the Liaoning Provincial Geological Environment Monitoring Station, and the evapotranspiration ratio data were obtained from the China Meteorological Data Center.

SPSS 27 was used to perform statistical analysis, correlation analysis, and principal component analysis on the soil salinity test data. ArcGIS 10.6 was employed for interpolation to draw isopleths and create maps.

3. Results

3.1. Statistical Characteristics of Soil Salinity Content

The study area has diverse ion types (Table 1), and the content of each ion shows significant spatial variation, reflecting the diversity of salinity sources and influencing factors in the study area. The pH value in the study area ranges from 5.93 to 8.97, with a relatively small variation and an average value of 7.89, indicating slight alkalization of the soil. The total salt content ranges from 1.3 to 9.0 g/kg, with significant variation. Among the 10 measured ions, Ca^{2+}, Na^{+}, SO_4^{2-} Cl^{-}, and HCO_3^{-} account for 90.84% of the total ion sum. The cation concentration shows the order $Na^{+} > Ca^{2+} > Mg^{2+} > K^{+}$, with Na^{+} and Ca^{2+} being the dominant cations in the saline soils of the study area. Na^{+} has a strong toxic effect on plants, and the hydrolysis of exchangeable sodium ions can make the soil alkaline, further deteriorating soil quality and becoming a key environmental factor limiting vegetation survival and growth. The anion concentration shows the order $SO_4^{2-} > Cl^{-} > HCO_3^{-} > NO_3^{-} > F^{-} > PO_4^{3-}$, with the sum of sulfate and chloride ions accounting for 46.44%. Cl^{-} and SO_4^{2-} are the dominant anions in the study area. Excessive Cl^{-} and SO_4^{2-} can affect plants by inhibiting dehydrogenase [26], influencing plant metabolism, impacting microbial community development, disrupting soil nutrient cycling, and causing soil pollution.

Table 1. Characteristic values of soil salt composition in the monitoring area.

Feature Value	pH	Total Salt/(g kg^{-1})	Ion Concentration/(mg kg^{-1})									
			Mg^{2+}	Ca^{2+}	Na^{+}	K^{+}	F^{-}	Cl^{-}	SO_4^{2-}	NO_3^{-}	HCO_3^{-}	PO_4^{3-}
Maximum Value	8.97	9.0	225.00	921.00	1760.0	134.00	67.40	3043	2155	473.00	755	317
Minimum Value	5.93	1.3	14.30	31.20	113.00	10.30	0.00	13.10	27.00	0.00	0.00	0.06
Average Value	7.89	3.9	61.78	173.79	486.22	43.96	14.70	533.30	619.99	84.60	304.62	8.60
Standard Deviation	0.71	1.8	49.09	173.57	347.09	28.97	10.89	581.35	540.25	116.23	163.00	46.57
CV	-	46.15	79.46	99.87	71.39	65.90	74.08	109.01	87.14	137.39	53.51	541.51
Ion Proportion (%)	-	-	2.65	7.45	20.85	1.89	0.63	22.87	26.59	3.63	13.07	0.37

3.2. Correlation Analysis of Soil Salinity Ions and Total Salt Content

By analyzing the correlations between the contents of various ions in the soil, we can infer the sources of soil salinity and reflect the movement trends of salts. As shown in Table 2, total salt content is positively correlated with Mg^{2+}, Na^{+}, Cl^{-}, and SO_4^{2-}, indicating that the total salt content increases with the increase in these ions. Cl^{-} and SO_4^{2-} are positively correlated with Mg^{2+}, Ca^{2+}, Na^{+}, and K^{+}, suggesting that the salts in the soil may exist in the forms of $MgCl_2$, $CaCl_2$, NaCl, KCl, $MgSO_4$, $CaSO_4$, Na_2SO_4, and K_2SO_4. The correlation between Cl^{-} and Na^{+} and between SO_4^{2-} and Mg^{2+} is greater than 0.85, indicating that the salts in the soil are most likely in the forms of NaCl and $MgSO_4$. The presence of NaCl also suggests that soil salinity ions may be influenced by shallow brackish water and seawater.

Principal component analysis was applied to the soil salinity ions using the dimensionality reduction approach. The data were condensed to obtain representative soil salinization factors and to quantitatively describe the main characteristics of soil salinity ions based on the comprehensive evaluation of principal component analysis. The soil data were standardized to unify the units of multiple indicators and subjected to the Kaiser–Meyer–Olkin (KMO) test and Bartlett's test of sphericity. The total variance was interpreted, and the orthogonal rotation method was used to derive the final component matrix.

Based on the total salt content and major salinity ions in the study area, three principal components were extracted. The cumulative contribution rate of the first, second, and third principal components is 72.09%, which exceeds 70% and can basically represent the main information of the original variable factors (Table 3).

Table 2. Correlation matrix of soil salinity parameters.

	Total Salt Content	Mg^{2+}	Ca^{2+}	Na^+	K^+	F^-	Cl^-	SO_4^{2-}	NO_3^-	HCO_3^-	PO_4^{3-}
Total Salt Content	1										
Mg^{2+}	0.627 **	1									
Ca^{2+}	0.291	0.555 **	1								
Na^+	0.687 **	0.711 **	0.206	1							
K^+	0.282	0.568 **	0.205	0.544 **	1						
F^-	0.230	−0.229	−0.180	−0.118	−0.199	1					
Cl^-	0.536 **	0.681 **	0.474 **	0.868 **	0.458 **	−0.186	1				
SO_4^{2-}	0.510 **	0.863 **	0.562 **	0.590 **	0.637 **	−0.220	0.454 **	1			
NO_3^-	0.296 *	0.672 **	0.590 **	0.236	0.243	−0.257	0.239	0.657 **	1		
HCO_3^-	0.326 *	−0.260	−0.237	0.041	−0.115	0.485 **	−0.107	−0.266	−0.187	1	
PO_4^{3-}	−0.016	0.058	−0.047	0.142	0.143	0.044	0.071	0.077	−0.124	−0.263	1

Note: **. Significant correlation at the 0.01 level (double tail); *. Significant correlation at the 0.05 level (single tail).

Table 3. Explanation of total variance by principal component analysis.

	Initial Eigenvalue			Extracted Sum of Squared Loadings		
Component	**Initial Eigenvalue**	**Initial Eigenvalue Variance Percentage**	**Cumulative%**	**Total**	**Extracted Variance Percentage**	**Cumulative%**
1	4.387	43.868	43.868	4.387	43.868	43.868
2	1.157	15.165	59.034	1.157	15.165	59.034
3	1.305	13.055	72.089	1.305	13.055	72.089
4	0.906	9.058	81.147			
5	0.733	7.328	88.475			
6	0.454	4.539	93.014			
7	0.404	4.036	97.050			
8	0.197	1.968	99.018			
9	0.089	0.893	99.911			
10	0.009	0.089	100.000			

The factor loadings of the salinity ion indicators are shown in Table 4. In the first principal component, Mg^{2+}, SO_4^{2-}, Cl^-, and Na^+ have the highest loadings, indicating that these four indicators contain the most comprehensive original variable information and have the greatest impact on soil salinization. The high loadings of Mg^{2+}, SO_4^{2-}, Cl^-, and Na^+ (all greater than 0.7) suggest that these ions are closely related to soil salinization. They also reflect that the main components of soil salinity are sulfates and chlorides, which corroborates the results of the ion correlation analysis. The salts in the soil are most likely in the forms of NaCl and $MgSO_4$. The second and third principal components contain fewer factors with lower loadings, indicating less reference value and a smaller impact on soil salinization.

3.3. Spatial Distribution Characteristics of Soil Salinization

According to the Coastal Soil Salinization Monitoring and Evaluation Technical Regulations, the total salt content in the southern coastal area of Panjin City ranges from 1.3 to 9.0. Based on the frequency distribution histogram of soil total salt content, the degree of soil salinization in the study area is classified as mild to moderate salinization (Figure 1).

Table 4. Factor load coefficient after rotation.

Indicator	Factor Load Coefficient		
	1	2	3
Mg^{2+}	0.940	0.043	0.080
Na^{+}	0.754	0.541	−0.146
Ca^{2+}	0.651	−0.292	0.333
K^{+}	0.672	0.245	−0.221
F^{-}	−0.359	0.580	0.251
Cl^{-}	0.762	0.367	−0.094
${SO_4}^{2-}$	0.888	−0.049	0.083
${NO_3}^{-}$	0.668	−0.361	0.436
${HCO_3}^{-}$	−0.321	0.683	0.497
${PO_4}^{3-}$	0.099	0.081	−0.776

Notes: Extraction method: principal component analysis. Three components were extracted.

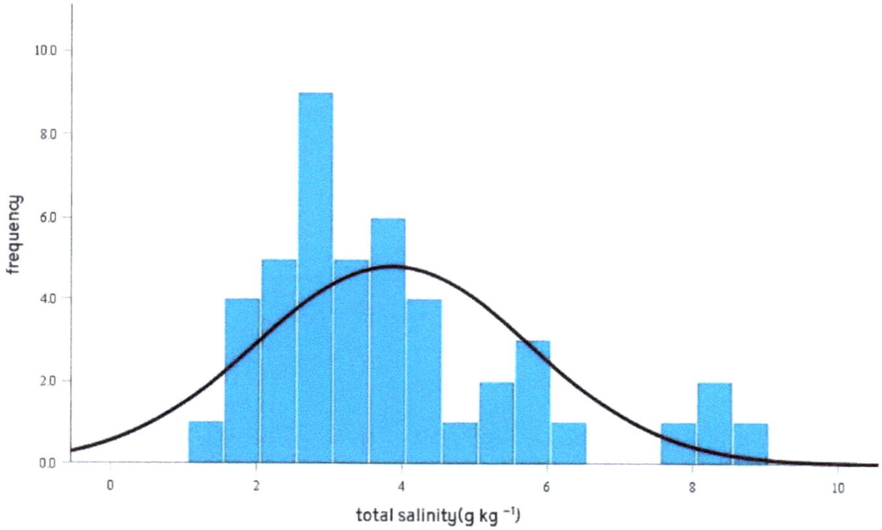

Figure 1. Histogram of soil total salinity frequency distribution.

According to the standard of the Chinese Soil Science Society for soil classification [27], the type of soil salinization in the area is determined based on the ratio of chloride ion content to sulfate ion content. The area of chloride-sulfate type soil salinization in the study area is 765.74 km^2, accounting for 43.06% of the total area, mainly distributed in the central part of the study area, the Liaohe Estuary Wetland. The area of sulfate–chloride type soil salinization is 895.40 km^2, accounting for 50.35% of the total area, mainly distributed on both sides of the study area. The area of sulfate type soil salinization is 6.42 km^2, accounting for 3.61% of the total area, distributed sporadically in the southern part of Dawang District. The area of chloride type soil salinization is 3.28 km^2, accounting for 1.84% of the total area, distributed sporadically in the southeastern part of the study area (Figure 2). The ratio of chloride ions to sulfate ions in the study area ranges from 0.17 to 4.60. The types of soil salinization in the southern coastal area of Panjin City have a distinct zonal distribution. The overall trend of salinization type change from the center to the sides is: sulfate type → chloride-sulfate type → sulfate-chloride type. The degree and intensity of salinization gradually decrease, and the area gradually increases. The most widely distributed types of soil salinization are chloride–sulfate type and sulfate–chloride type.

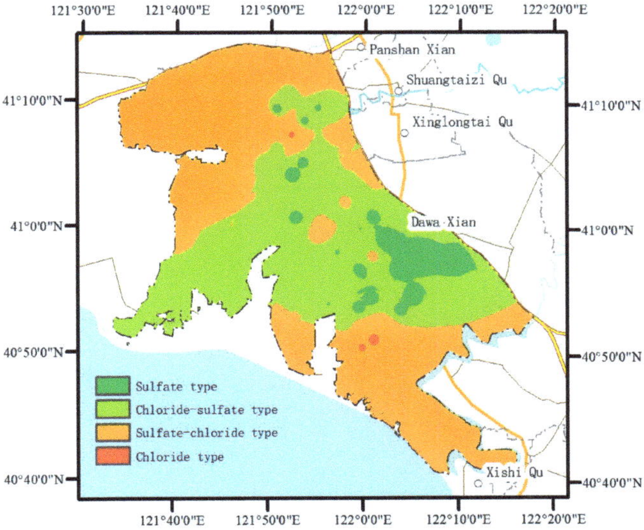

Figure 2. Spatial distribution characteristics of soil salinization in the study area.

3.4. Evaluation of Soil Salinization and Analysis of Driving Factors

Referring to the "Technical Guidelines for Environmental Impact Assessment—Soil Environment," the soil salinization in the study area was evaluated using the soil salinity comprehensive scoring method. According to the evaluation results, the soil salinization in the southern coastal area of Panjin City has a distinct zonal distribution, with concentrated distribution ranges for each grade of salinization. The degree of soil salinization generally decreases from the coastal area to the inland (Figure 3). The area of mild salinization is 116.69 km^2, accounting for 6.64% of the study area, mainly distributed in the western part of the study area (Dongguo Town–Fule Village–Changtun Village–Miaogou–Shixin Town) and the southwestern part (Yuanzhiyi–Xiaohe Branch Farm). The area of moderate salinization is 1377.23 km^2, accounting for 78.38% of the study area, mainly concentrated in the central part (Yueyaz–Lujiatou–Xinxing Town–Zhaoquanhe Town–Yanchang Village–Nandianzi–Gunlonggang–Sunjialiuzi–Luojiapu–Xinsheng Farm Third Brigade Third Squadron). The area of severe salinization is 263.23 km^2, accounting for 14.98% of the study area, mainly distributed in the eastern part (Dawang District–Xudong Village–Qingnian Village–Shuangjingzi Village) and the area around Dougouzi Village–Liaobin Township.

The main factors affecting soil salinization, such as groundwater salinity, groundwater level, and evapotranspiration ratio were selected for the evaluation. The soil texture was not included in the analysis because the soil types in the study area are mainly silt loam and silt, which have consistent soil texture attributes. The relationship between these factors and soil salinization was further revealed using SPSS 27 for stepwise regression analysis.

As shown in Table 5, all regression coefficients are less than 0.05 and pass the significance test. The factor most strongly correlated with soil total salt content is groundwater level (0.399). The factor most strongly correlated with groundwater level is the evapotranspiration ratio (0.757), followed by groundwater salinity (0.536). The most significant correlation with groundwater salinity is the evapotranspiration ratio (−0.521). This indicates that the groundwater level has the most direct impact on soil salinization in the area. The significant vertical migration and exchange between soil and groundwater suggest that the groundwater level affects the movement and accumulation of salts, which is the main way of salt migration in the area. The significant correlation between groundwater level and groundwater salinity and evapotranspiration ratio indicates that these two factors also indirectly affect the soil total salt content. In summary, soil salinization in the area is mainly controlled by the groundwater level, with evapotranspiration ratio and groundwater salinity as secondary factors.

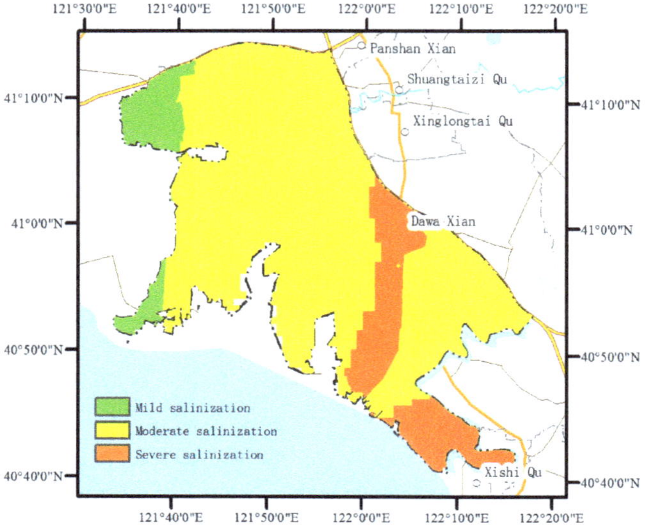

Figure 3. Distribution of soil salinization in the study area.

Table 5. Correlation coefficient of influencing factors of soil total salinity.

	Soil Total Salinity	Evapotranspiration Ratio	Groundwater Salinity	Groundwater Level
Soil Total Salinity	1	0.195	−0.127	0.399 **
Evapotranspiration Ratio	0.195	1	−0.521 **	0.757 **
Groundwater Salinity	−0.127	−0.521 **	1	−0.536 **
Groundwater Level	0.399 **	0.757 **	−0.536 **	1

Note: **. Significant at the 0.01 level.

Based on the evaluation results and regression analysis, the driving factors of soil salinization in the study area are as follows: (1) the shallow groundwater level in the study area ranges from 1.38 to 10.31 m. The groundwater level shows a distinct zonal decrease from northwest to southeast. The spatial distribution of soil salinization is highly similar to the spatial distribution of the groundwater level. As the groundwater level decreases, the soil salt content increases. The shallower the groundwater level, the more convenient the conditions for capillary water to rise to the surface, leading to soil salinization. (2) The average annual evaporation in the study area is generally high, with an evaporation–precipitation ratio of 1.68 to 2.32. The highest annual evaporation is 1669.6 mm, which is 2.3 times the annual precipitation. Strong evaporation leads to significant water loss from the soil, with the salinization process exceeding the leaching process, resulting in soil salinization. (3) The total dissolved solids in groundwater directly affect the formation and distribution of saline soils. The concentration of total dissolved solids in the study area ranges from 344 to 7504 mg/L. The higher the concentration of total dissolved solids in groundwater, the more salts are carried by capillary water, leading to higher salt accumulation in the soil and increased likelihood of soil salinization. (4) The lithology in the study area is mainly silt loam and silt. These soil characteristics include fine particles, high homogeneity, tight structure, small pores, weak permeability, and strong adsorption capacity. These characteristics cause soluble ions to adhere to the soil surface, making it difficult to remove salts and leading to soil salinization.

4. Discussion

This study comprehensively analyzes the characteristics, spatial distribution, and driving factors of soil salinity in the southern coastal area of Panjin City, Liaoning Province, revealing the formation mechanism and dominant factors of soil salinization in the region. The discussion is as follows.

4.1. Causes of Soil Salinization and Dominant Driving Factors

The results indicate that soil salinization in the study area results from the combined effects of natural factors and intrinsic mechanisms. Groundwater depth is the most direct and critical factor controlling the spatial variation of soil salinity. Shallower groundwater depth enhances capillary action, facilitating the migration and accumulation of salts from highly mineralized groundwater into the topsoil. The evaporation–concentration effect, characterized by a high evapotranspiration ratio, serves as the core driving force for salt accumulation in the surface layer. Its intensity far exceeds that of precipitation leaching, leading to a strong salt accumulation process. Groundwater mineralization provides the direct source of salts, and its concentration determines the flux of upward salt migration. Furthermore, the widespread distribution of silty loam and silt textures in the study area, due to their poor permeability, strong capillary action, and good adsorption properties, significantly promotes salt ascent and retention while inhibiting the desalination process. These four factors—groundwater level, evaporation, water quality, and soil texture—constitute a complete "source–pathway–sink–dynamics" causation system for salinization.

4.2. Soil Ion Composition and Source Indicators

Both soil ion correlation and principal component analysis show that Na^{+}, Mg^{2+}, Cl^{-}, and $SO_4{}^{2-}$ are the dominant ions in the study area, and these ions exhibit strong homology, suggesting that salts primarily exist in the form of chlorides and sulfates such as NaCl and $MgSO_4$. This ion combination strongly indicates a marine origin, confirming the continuous influence of seawater immersion, marine sedimentation, and saline groundwater on modern soil salt accumulation. This is highly consistent with the coastal geographical location and geological background of the area. It is worth noting that the relative increase in Ca^{2+} and $HCO_3{}^{-}$ content in some areas may reflect the influence of terrestrial salt sources, possibly related to activities such as groundwater extraction and agricultural irrigation. In summary, the characteristics of soil salt ion composition fully demonstrate that the salts in the study area mainly originate from seawater immersion, marine sedimentary parent material, and saline groundwater recharge. This understanding provides important guidance for formulating targeted salinization prevention and control measures. During the treatment process, special attention should be paid to blocking seawater intrusion pathways and controlling groundwater levels to prevent salts from rising to the topsoil through capillary action.

4.3. Spatial Distribution Patterns and Their Significance

The degree and type of soil salinization exhibit a clear zonal distribution pattern, decreasing from the coast to the inland areas. The salinization types transition from chloride or sulfate–chloride types in the coastal areas to chloride–sulfate and sulfate types in the inland areas. This pattern is consistent with the spatial trends of groundwater depth and mineralization, further confirming the controlling role of groundwater on salt distribution. It provides a scientific basis for zonal management with significant economic benefits. In severely salinized coastal areas, the construction of drainage engineering and the cultivation of salt-tolerant pioneer crops (e.g., Suaeda salsa) can effectively reduce soil salinity, creating conditions for subsequent agricultural use. In moderately salinized areas, the promotion of integrated rice-reed-fishery ecological models can improve soil quality while increasing comprehensive income per mu. In lightly salinized inland areas, the development of salt-tolerant economic crops (e.g., salt-tolerant rice and wolfberry) and facility agriculture can significantly enhance land output efficiency. The zonal management model not only reduces treatment costs but also enables efficient utilization of saline-alkali land resources, which is of great significance for expanding cultivated land area, ensuring food security, and promoting agricultural income growth.

4.4. Practical Value and Limitations of the Study

This study combines traditional geoscientific surveys with modern statistical and spatial analysis techniques to identify the dominant factors influencing salinization in the Panjin coastal area and their weights, providing a specific scientific basis and geoscientific support for local land spatial planning, precise improvement of saline-alkali land, and sustainable agricultural development. The management measures proposed based on the research results offer important scientific guidance for the local government to formulate effective measures to alleviate groundwater salinization, improve saline-alkali land, and expand cultivated land area. Through saline-alkali land improvement and the cultivation of suitable crops, the scale of rice and salt-tolerant special economic crops can be effectively expanded, driving the development of agricultural product processing and eco-tourism, thereby forming new economic growth points.

However, this study is primarily based on sampling data from a specific period, and the understanding of dynamic salinization processes (e.g., seasonal fluctuations and long-term evolution trends) remains insufficient. Future research could incorporate time-series remote sensing monitoring and long-term positioning observations, and further quantify the impact of human activities (e.g., irrigation methods and land use changes) on water-salt migration to construct more comprehensive prediction and evaluation models.

5. Conclusions

(1) The soil salt composition characteristics in the area are significant and indicative of typical marine origins. The total salt content in the soil varies widely (1.3–9.0 g/kg), and the ion composition is dominated by Na^+, Cl^-, and SO_4^{2-}. Their combination fully reflects the compound effects of seawater immersion, marine sedimentation, and saline groundwater recharge. Principal component analysis further shows that Mg^{2+}, Na^+, Cl^-, and SO_4^{2-} are key indicators characterizing the degree of salinization, confirming that chlorides and sulfates are the main salt forms.

(2) Soil salinization exhibits significant spatial heterogeneity and zonal patterns. Spatial analysis shows that from the coast to the inland, salinization types transition in a zonal sequence from sulfate → chloride–sulfate → sulfate–chloride types, with salinization intensity gradually decreasing. Moderately salinized soil dominates (78.38%), overall consistent with the water-salt migration of coastal alluvial–marine plains.

(3) Groundwater depth is the dominant driving factor of salinization, with significant multivariate synergistic effects. Through stepwise regression and mechanistic analysis, groundwater depth is identified as the most important factor affecting the spatial differentiation of soil salinity. Together with the evapotranspiration ratio and groundwater mineralization, it constitutes a "source–pathway–sink–dynamics" coupling mechanism.

Shallow groundwater levels, strong evaporation, high mineralization, and silt texture synergistically promote salt surface accumulation and inhibit natural desalination processes.

(4) The research results provide a theoretical and technical basis for the precise management of coastal saline-alkali land. This study integrates geoscientific surveys, statistical analysis, and GIS spatial modeling to systematically identify key driving factors and their weights. The constructed zonal management strategy can provide theoretical and technical support for land ecological restoration, improvement of saline-alkali farmland, and agricultural adaptation planning in the southeastern coastal areas of Liaoning Province. It is recommended to strengthen dynamic monitoring and quantitative assessment of human activity interference in the future and develop a multi-scenario salinization risk warning system to achieve sustainable use of regional land resources.

Author Contributions: Conceptualization, J.S. and H.M.; methodology, J.S. and H.M.; software, H.M.; resources, J.C. and W.S.; data curation, J.S.; writing—original draft, X.S. and W.S.; writing—review and editing, X.S., J.S., J.C. and H.M. All authors have read and agreed to the published version of the manuscript.

Funding: This study was supported by the China Geological Survey projects (DD20240061, DD20230461, DD20221730).

Data Availability Statement: The original contributions presented in the study are included in the article, and further inquiries can be directed to the first author.

Conflicts of Interest: The authors declare no conflicts of interest.

References

1. Yu, Y. The Characteristics of Saline-Alkaline Patches in Xinjiang Farmalnd Oasis Soil. Master's Thesis, Xinjiang Agricultural University, Ürümqi, China, 2015.
2. Ivits, E.; Cherlet, M.; Tóth, T.; Lewińska, K.E.; Tóth, G. Characterisation of productivity limitation of salt-affected lands in different climatic regions of Europe using remote sensing derived productivity indicators. *Land Degrad. Dev.* **2013**, *24*, 438–452. [CrossRef]
3. Wang, S.P.; Chen, Y.; Pan, D.W.; Xue, W.D.; Zhou, L.; Zhao, N.; Gong, Z.Q.; Zhang, X.R. Review on salt marshes management: Status, problems and countermeasures. *Ind. Miner. Process.* **2023**, *52*, 59–68. [CrossRef]
4. Yang, J.S. Development and prospect of the research on salt-affected soils in China. *Acta Soil Sci.* **2008**, *45*, 837–845. Available online: https://kns.cnki.net/KCMS/detail/detail.aspx?dbcode=CJFQ&dbname=CJFD2008&filename=TRXB200805010 (accessed on 31 August 2025).
5. Zhu, S.B.; Wueruli, T.E.X.; Tang, G.M.; Zhang, Y.S.; Xu, W.L.; Ma, H.G. Research Progress on the Changes and Control Measures of Saline-Alkali Land in Xinjiang. *Shandong Agric. Sci.* **2023**, *55*, 158–165. [CrossRef]
6. Ren, X.L.; Zhang, W.; Liu, X.; Luo, Y.Z.; Wei, K. The analysis of dynamic spatial distribution characteristic of saline soil salinity in Xining basin, Qinghai., Qinghai. *J. Qinghai Univ. (Nat. Sci. Ed.)* **2016**, *34*, 9–17+30. [CrossRef]
7. Pan, P.; Zhao, T.; Mao, L.M.; Feng, X.W. Dynamic Changes and Control Approaches of Soil Salinization in Ningxia. *J. Ningxia Agric. For. Sci. Technol.* **2001**, *5*, 38–39+42. Available online: http://nxnlkj.nxaas.com.cn/thesisDetails?columnId=63697257&Fpath=home&index=0&lang=zh (accessed on 31 August 2025).
8. Liu, S.B.; Arong, Q.Q.G.; Wang, L. Study on dynamically monitoring the land salinization in the typical area of Inner Mongolia. *Environ. Monit. China* **2001**, *4*, 24–27. Available online: https://www.cnemce.cn/ch/reader/view_abstract.aspx?file_no=200104100&flag=1 (accessed on 31 August 2025).
9. Lan, T. Geochemical Evaluation and Cause Analysis of Soil Salinization in Western Jilin Province. Ph.D. Thesis, Jilin University, Changchun, China, 2023. [CrossRef]
10. Ma, B.; Yi, J.G. Analysis on Sensitivity of Soil Salinization in the Plain Area of Heilongjiang Province. *Environ. Sci. Manag.* **2011**, *36*, 133–135+146. Available online: https://vpn.cgl.org.cn/https/77726476706e69737468656265737421fbf952d2243e635930068cb8/nzkhtml/xmlRead/trialRead.html?dbCode=CJFD&tableName=CJFDTOTAL&fileName=BFHJ201105035&fileSourceType=1&appId=KNS_BASIC_PSMC&invoice=F9jS3Rrcnjtf2+wB4FxheNMDp7ggtGgiQmfaDIiX8lbqj799BKHuHYnzQEahLoLhoIpen/o1s4xDBajP4pSo8xbd2ViQlhPAyimmYKaEdyxaSsOpeXrn2EgdG0xEcmqfXqurRk3EgXtaieMp5JXZ47zKVI4cAjhlh8hR2ZY/Hm0= (accessed on 31 August 2025).

11. Fu, E.G.; Liu, H.G.; Ma, Z.; Guo, X. Characteristics and status assessment of soil salinization in the northern Shouguang area, Shandong province. *Geol. Surv. Res.* **2016**, *39*, 300–304. Available online: https://vpn.cgl.org.cn/https/77726476706e69737468656265737421fbf952d2243e635930068cb8/nzkhtml/xmlRead/trialRead.html?dbCode=CJFD&tableName=CJFDTOTAL&fileName=QHWJ201604012&fileSourceType=1&appId=KNS_BASIC_PSMC&invoice=HHLC5lcArKW6OWEUJGg9vzyjH4NmZj2t+yMsiqFIELUU6G4DQCm85seWgony7kG1fF1VdkUtkIEfsj5odBgtUO3QHmxoLuE0522NFnpnd9xLcRE245u5Xhqo9Bi74h6c1cpDUL8GHbuG8ZSfOINGKZ35tt+2bSUcLJWMpmHQUNc= (accessed on 31 August 2025).
12. Ma, Y.; Tashpolat, N. Current Status and Development Trend of Soil Salinity Monitoring Research in China. *Sustainability* **2023**, *15*, 5874. [CrossRef]
13. Yuan, G.X.; Chen, D.W.; Xu, Y.Y.; Meng, D.Z.; Zhang, Y.Y.; Wang, X.D. Summary of Methods for Extracting Soil Salinization Information. *J. North China Univ. Water Resour. Electr. Power (Nat. Sci. Ed.)* **2022**, *43*, 95–101. [CrossRef]
14. Wang, Y.J.; Gao, X.M.; Bao, S.M.; Li, W.; Fu, Y.; Wu, R.; Wang, J.Y.; Guo, L.Y. Prediction model of salt content in salinized soil based on hyperspectral data. *Agric. Eng.* **2024**, *14*, 113–120. [CrossRef]
15. Chen, P.; Wang, W.; Wang, G.M.; Ma, J.C. Distribution characteristics of soil salinization in the east coastal plain of Cangzhou. *N. China Geol.* **2022**, *45*, 36–43. [CrossRef]
16. Fu, X.; Zhang, H.R.; Wang, Y.B.; Huang, C.; Liu, X.Y.; Zhang, H.C.; Xu, Z.H. Three-dimensional Spatial Differentiation of Soil Salinity and Its Influencing Factors in the Saline Soil Areas of the Yellow River Delta. *J. Geo-Inf. Sci.* **2024**, 1–16. Available online: https://link.cnki.net/urlid/11.5809.p.20241114.1704.008 (accessed on 31 August 2025).
17. Chen, Y.Q.; Xu, L.Y.; Wang, Z.W.; Wang, X.P.; Yao, R.J.; Zhao, X. Stoichiometric Characteristics of Nutrient-Microbial Biomass-Extracellular Enzyme Activity in Different Salt-affected Soils. *Acta Pedol. Sin.* **2024**, *62*, 811–824. [CrossRef]
18. Niu, X.X.; Gao, S.S.; Li, W.C.; Jia, Z.J.; Wu, Y.Y.; Chen, J.H.; Ma, D.X. Research Progress and Prospects on Impacts of Groundwater Dynamics on Soil Water and Salt Based on Bibliometrics. *Bull. Soil Water Conserv.* **2024**, *44*, 122–133+143. [CrossRef]
19. Zhao, L. Spatiotemporal Variation Characteristics and Influencing Factors of Soil Salinization in Typical Yellow River Irrigation Areas of Shandong Province. Master's Thesis, Jinan University, Changchun, China, 2024.
20. Li, Z.; Du, Z.; Bi, S.T.; Ye, T.; Zhang, Q.; Chen, Y. Prediction of Soil Salinity and Analysis of Influencing Factors in Coastal Plains Based on the Geographically Weighted Random Forests. *Environ. Sci.* **2024**, *46*, 4982–4992.
21. Wang, J.Z.; Ding, J.L.; Ge, X.Y.; Peng, J.; Hu, Z.W. Monitoring soil salinization on the basis of remote sensing and proximal soil sensing: Progress and perspective. *Natl. Remote Sens. Bull.* **2024**, *28*, 2187–2208. [CrossRef]
22. Luo, Z.H.; Zhang, C.; Feng, S.Y.; Tang, M.; Liu, R.; Kong, J.Y. Advances in research on methods for optical remote sensing monitoring of soil salinization. *Remote Sens. Nat. Resour.* **2024**, *36*, 9–22.
23. Fu, T.F.; Zhang, Y.; Gao, J.W.; Chen, G.Q.; Liu, W.Q.; Su, Q. Study on the spatio-temporal variability of saline soil salinity in the Yellow River Delta. *Period. Ocean Univ. China* **2017**, *47*, 50–60. [CrossRef]
24. Qiao, W.H. Breakthrough, awakening the dormant cultivated land resources. *Econ. Dly.* **2023**, 1–4. Available online: https://vpn.cgl.org.cn/https/77726476706e69737468656265737421fbf952d2243e635930068cb8/nzkhtml/xmlRead/trialRead.html?dbCode=CCND&tableName=CCNDTOTAL&fileName=JJRB202305230090&fileSourceType=1&appId=KNS_BASIC_PSMC&invoice=ZtWcW4l/CJRStXiS08OcUptt/JXagkt7Rqq7iuGnLb6TX0d88CwGjJ/zpzlmhP26jX056XgFArXzTgQ5T/X4+J+/G620oX1S9ZeZ/bbQELSxpMeRhvlWUU3WBdcq9xaXnW54RgK6yUVRCvS5FG0Xr6L9A0G5NpH5Bq/b3IG0DvI= (accessed on 31 August 2025).
25. Liu, Q.P.; Li, H.; Tian, H.Z.; Wei, N.; Wu, Z.; Li, L.; Chang, Q. Land use and cover change and sustainable development in the coastal zone of China's mainland. *Sci. Geogr. Sin.* **2024**, *44*, 1695–1705. [CrossRef]
26. Li, J.L. Research on the Effect of Saline Wetland from Pulp Wastewater Irrigation in Yellow River Dleta. Ph.D. Thesis, Ocean University of China, Qingdao, China, 2008.
27. Liu, L. Constitute and distributing character of salinity in soil in Xinjiang. *Arid Environ. Monit.* **2009**, *23*, 227–229. Available online: https://vpn.cgl.org.cn/https/77726476706e69737468656265737421fbf952d2243e635930068cb8/reader/flowpdf?invoice=ml1B51KVNkpP66i5VynCeXcbtm7XJL9TGreLbDDkmzVfWPxfLEFsuyeFuflFDQP49rXNlOaNNLeK9ZmnHX%2BieBDvVnYnN9gWKrDshZ52lLOv8w%2B5Dc3pGzsQO7CnLykv7xjOwnZPhYGqudJGZ%2FTsaFvXpGtNgMYXRgq4QQgpT8o%3D&platform=NZKPT&product=CJFQ&filename=GHJC200904010&tablename=cjfd2009&type=JOURNAL&scope=trial&dflag=pdf&pages=&language=CHS&trial=&nonce=473B6A9F34564D1E9227AE8E3140DBAF&cflag=pdf (accessed on 31 August 2025).

water

MDPI

Article

Spatial–Temporal Change and Dominant Factors of Coastline in Zhuhai City from 1987 to 2022

Tao Ma [1,†], Haolin Li [2,†], Yandi She [1], Yuanyuan Zhao [1], Xueke Feng [1] and Feng Zhang [3,*]

1 State Key Laboratory of Plateau Ecology and Agriculture, Qinghai University, Xining 810016, China; matao2438@163.com (T.M.); sheyandi0530@163.com (Y.S.); 2015990030@qhu.edu.cn (Y.Z.); fengxk9537@163.com (X.F.)
2 School for Environment and Sustainability, University of Michigan, Ann Arbor, MI 48109, USA; haolinli@umich.edu
3 Technology Center of Natural Resources and Planning in Zhuhai, Zhuhai 519015, China
* Correspondence: zhzhangfeng@126.com
† These authors contributed equally to this work.

Abstract

Understanding the spatiotemporal variations and driving mechanisms of coastlines is crucial for their adequate protection, utilization, and sustainable development. In this study, the changes in various coastline types in Zhuhai from 1987 to 2022 were analyzed by using long-term Landsat and GaoFen satellite imagery. The Index of Coastline Type Diversity (ICTD), Index of Coastline Utilization Degree (ICUD) and the Digital Shoreline Analysis System (DSAS) analysis indicators were employed to investigate coastline change. Both quantitative and qualitative analyses were integrated to comprehensively elucidate the impacts of various driving factors. The results indicate that the total length of Zhuhai coastline increased from 761.50 km in 1987 to 798.91 km in 2022, with natural coastlines decreasing by 89.82 km and artificial coastlines increasing by 153.40 km. The rapid expansion of artificial coastlines since 2007 led to a marked decline in the ICTD indicator, while the ICUD indicator increased from 146.42 in 1987 to 216.37 in 2022, reflecting the intensified and continuous influence of anthropogenic activities. Additionally, the end point rate (EPR) and Weighted Linear Regression Rate (WLR) changed by 8.09 m/yr and 6.62 m/yr, respectively. The Shoreline Change Envelope (SCE) and Net Shoreline Movement (NSM) exhibited average changes of 331.42 m and 224.32 m, respectively. Gray correlation and regression analyses further revealed that climate factors exhibited the strongest association with natural coastline changes, while economic development indicators showed the strongest correlation with artificial coastline dynamics. The relationship of Number of Berths in Main Ports (Nb) with coastline changes strongly suggests that human activities are the primary driver of these changes. These findings provide a robust scientific basis for coastal zone management in Zhuhai.

Keywords: Zhuhai; coastline; DSAS; gray correlation; anthropogenic activities

1. Introduction

Coastlines, as critical transitional zones between terrestrial and marine environments, function not only as ecologically sensitive corridors but also as pivotal hubs for anthropogenic activities and resource exploitation [1–3]. Globally, coastal zones accommodate approximately 50% of the world's population [4,5]. In China, more than 50% of medium-sized and larger cities are located in coastal areas, contributing over 60% of the nation's

GDP [6], thereby intensifying the conflict between anthropogenic exploitation and ecological conservation. In recent decades, accelerated sea-level rise (SLR) and coastal erosion have further heightened the risks of coastline retreat. For example, projections suggest that by 2050, SLR-driven erosion may submerge up to 70% of Italy's beaches [7]. Concurrently, anthropogenic interventions (e.g., land reclamation and port construction) have led to significant morphological fragmentation. In China, artificial coastlines expanded at an annual rate of 14.5% before 2010 [8]. These synergistic effects of natural and anthropogenic stressors necessitate a systematic investigation into the spatiotemporal evolution and multidimensional driving mechanisms of coastlines to reconcile development-conservation dilemmas, enhance coastal resilience, and achieve sustainable governance.

Accurate coastline mapping is fundamental to studying the spatial–temporal change and driving factors of the coastline. Traditional coastline mapping methods based on field surveys have gradually been supplanted by remote sensing technologies, owing to the high cost and labor intensity of ground-based approaches [9,10]. The change in coastline mapping has progressed from the use of single aerial or satellite images to the fusion of hyperspectral and LiDAR data, and from manual interpretation and band statistics to machine learning algorithms [11–13]. The research has also expanded from localized key areas to global applications [14,15]. It can be seen that the current research on the spatial–temporal change in the coastline has made significant progress in methodology and application scale. However, there remains a significant gap in the long-term analysis of morphological differentiation, spatial heterogeneity, and multidimensional driving mechanisms in Zhuhai—a core coastal city within China's Special Economic Zone and the Guangdong-Hong Kong-Macao Greater Bay Area. Existing studies on the coastline of Zhuhai mostly focus on short periods [16–19], and although some studies have analyzed the long time-series changes in the Guangdong-Hong Kong-Macao Greater Bay Area, most of them are based on administrative divisions, which makes it difficult to reveal the changes in a refined way [20–22].

Understanding the spatial–temporal change in coastlines is essential for achieving sustainable development. Previous studies have identified both natural processes and anthropogenic activities as the principal factors influencing coastline changes, with natural factors traditionally regarded as the primary drivers [23]. However, in recent years, intensified anthropogenic activities have emerged as the predominant force behind these changes. Zhang et al. demonstrated through layer superposition and impact analysis that anthropogenic interventions, along with variations in material composition and landforms, play a critical role in shaping the coastlines of Southeast Asian islands [24]. Amrouni et al. found that the retreat of North African coastlines, coupled with rapid urban growth, has accelerated inland seawater intrusion by approximately 5 km, thereby negatively affecting crop production and food security [25]. Above results indicate that anthropogenic processes, rather than natural forces, primarily drive coastline retreat. In addition, Lansu et al. reported that coastal infrastructure development restricts the available space for coastal ecosystems, producing irreversible effects on coastline morphology [26]. Most existing research, however, tends to analyze driving factors using either qualitative or quantitative methods in isolation. Yang et al. combined tidal station data and reclamation project statistics to show that anthropogenic activities have dominated the evolution of mainland China's coastline over the past 30 years [27]. Wang et al. qualitatively evaluated 50 years of coastline change in Huizhou city and concluded that anthropogenic factors play the dominant role, with natural factors having minimal influence [28]. Relying solely on a single analytical approach may underestimate the complexity and relative influence of various driving factors. Therefore, integrating both qualitative and quantitative analyses is necessary to comprehensively assess the factors influencing coastline changes.

To address the shortcomings in existing research on the spatiotemporal changes and driving factors of Zhuhai coastline, this study establishes a research framework comprising spatiotemporal change analysis, dynamic assessments of utilization and locational shifts, and driving factor evaluation. Utilizing methodologies such as utilization evaluation, the Digital Shoreline Analysis System (DSAS), gray correlation analysis, and regression analysis, we extract detailed spatiotemporal data on the coastline, assess changes in utilization and position, and quantify the relative influences of natural and socioeconomic factors to determine the dominant drivers. The objective of this study is to provide a robust scientific basis for coastal zone planning in Zhuhai and offer insights applicable to other estuarine cities.

2. Materials and Methods

2.1. Study Area

Zhuhai is located on the western margin of the Pearl River Estuary (21°48′–22°33′ N, 113°03′–114°19′ E) (Figure 1), which contains multiple islands and embayments, including Hengqin, Gaolan, Hebao and Qi'ao, parts of the Wanshan Archipelago, Xianglu Bay and Jiuzhou Bay. Coastal dynamics are governed by semi-diurnal tides and associated residual currents. Seasonal discharge from the Pearl River drives estuarine circulation and sediment plumes. Wave action and typhoon-driven storm surge reshape the shoreline during extreme events. Longshore sediment transport controls the morphology of beaches, tidal flats and shoals. Changes in river sediment supply and human activities have modified local hydrodynamics and sediment budgets. Human activities include large-scale land reclamation, port construction around Gaolan and Jiuzhou and the Hong Kong–Zhuhai–Macao Bridge with its artificial structures. The region has a subtropical climate with warm temperatures year-round and distinct wet and dry seasons [29]. This climate together with the geographic setting increases vulnerability to sea-level rise, coastal erosion and other climate-change stresses. Recent coastal urbanization has substantially altered the natural landscape and expanded artificial shorelines, particularly around Hengqin and Gaolan islands [30].

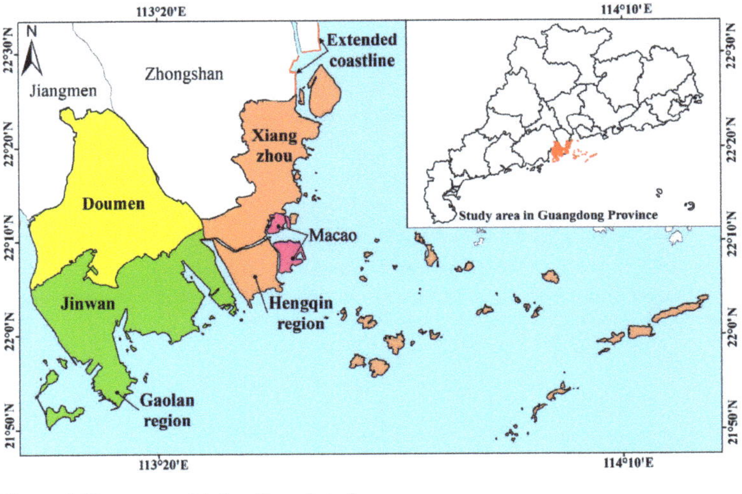

Figure 1. The geographic location of study area.

2.2. Data

In this study, a combination of Landsat series satellite imagery and GaoFen satellite data was employed as the primary data source (Table 1). Landsat imagery was acquired from the United States Geological Survey Earth Explorer platform (https://earthexplorer.usgs.gov (accessed on 12 December 2024)), and GaoFen data were obtained from ChinaGEOSS Data Sharing Network (https://noda.ac.cn (accessed on 13 December 2024)). Following standardized procedures, the imagery underwent radiometric calibration, mosaic merging and geometric correction to ensure spatial and spectral consistency. These preprocessed images formed the basis for coastline extraction. Coastline features were identified through the visual interpretation approach, in which remote sensing data were analyzed based on the spectral signatures of typical coastline elements. The final delineation of the coastline was conducted in accordance with established coastline classification criteria and definitional principles, ensuring both accuracy and consistency across the temporal dataset.

Table 1. Remote sensing information used in the study area.

No.	Remote Sensing Data Type	Time
1	Landsat5 TM	8 December 1987
2	Landsat5 TM	1 November 1997
3	Landsat5 TM	29 January 2007
4	Landsat7 ETM	18 December 2017
5	GF-2, GF-7	3 February 2022

2.3. *Coastline Classification System*

Based on the current development status of Zhuhai coast and considering features such as image scale, shape, shadow, color, and the surrounding environment derived from Landsat and high-fraction imagery, the continental coastline in the study area is classified into two primary categories: natural and artificial coastlines [8,31,32]. The natural coastline is further subdivided into bedrock, sandy and biological coastline. In contrast, the artificial coastline is categorized according to the marine functional area classification established by the State Oceanic Administration in China, which includes industrial, port terminal, road and breeding. Table 2 presents the complete classification system for the coastline in the study area.

2.4. *Visual Interpretation*

Given the characteristics of Zhuhai's coastline, characterized by an intricate configuration, numerous embayments, extensive muddy shores, and ambiguous land-water boundaries in specific sectors, this study utilized visual interpretation for coastline delineation. This methodology integrated manual interpretation of remote sensing imagery with field verification. Initial interpretation leveraged key features within false-color composite imagery, including tone, texture, geomorphology, and surrounding features. Coastline positions were subsequently determined based on specific type, as detailed in Table 2. Uncertain shoreline segments were confirmed via field surveys to generate the final coastline dataset.

Table 2. Coastline classification system description in the study area.

Category	Class	Description	Image Example
Natural	Bedrock	Bedrock coastlines typically follow the water's edge line as the shoreline reference, which is often irregular and jagged due to the resistance of exposed rock formations. On remote sensing imagery, the bedrock appears light-toned with striping patterns, and the shores are frequently dotted with reefs, boulders, sea cliffs, and other erosional landforms.	
	Sandy	Sandy coastlines are typically delineated using the upper beach ridge line or the vegetation line as the shoreline reference. They exhibit a smoother water's edge line; bright white, uniform hue in areas not reached by the tide, and a darker hue in areas wetted by the tide. Beaches often show striping patterns on remote sensing imagery.	
	Biological	Biological coastline is a special kind of coastal space formed by the special development of a certain kind of organism. Among the types of biological coastlines visible in Zhuhai City, mangrove coastlines are generally mangrove forests, which are distributed in numerous independent patches in the intertidal zone and are dark green in true-color images and red in false-color images.	
Artificial	Industrial	Industrial: Coastline used for waterfront industries for port construction development.	
	Port Terminal	Coastline means coastline used for the construction of port terminals, including coastline used for piers, jetties, and other construction function uses.	
	Road	Coastline used to connect land-linked islands, generally straight in shape.	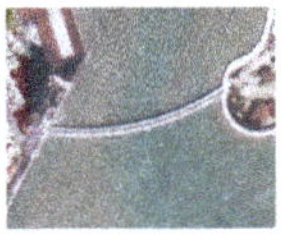
	Breeding	Breeding generally refers to the use of aquaculture, fisheries production pond dike coastline, generally located in estuaries, shallow, fish ponds around the perimeter, mostly regular block target.	
Island		The water–land boundary of atoll far from land or an island.	

2.5. Methods

2.5.1. Coastline Utilization Index

In this study, the index of coastline type diversity (ICTD) and the index of coastline utilization degree (ICUD) were used to quantify the diversity and utilization intensity of coastline types, respectively. ICTD measure the diversity of coastline types based on their relative lengths [33]. The calculation formula is as follows:

$$\text{ICTD} = 1 - \frac{\sum_{i=1}^{n} L_i^2}{\left(\sum_{i=1}^{n} L_i\right)^2}, \quad (1)$$

where ICTD represents the diversity of coastline types in the study area in a certain year, n denotes the total number of coastline types; i indexes each type; L_i represents the length of the ith coastline type. The ICTD value ranges from 0 to 1, with values closer to 0 indicating lower diversity and those nearer to 1 signifying a greater complexity and diversity in coastline types. This index is closely related to the proportionate lengths of the various coastline types. The ICUD quantifies the intensity of anthropogenic utilization along the coastline by incorporating both the proportion and the weight of different coastline types [34]. The calculation formula is as follows:

$$\text{ICUD} = \sum_{i=1}^{n} (A_i \times w_i) \times 100, \quad (2)$$

where ICUD indicates the utilization intensity of the coastline in the study area in a certain year; i indexes each type; A_i stands for the proportion of ith length relative to the total coastline; W_i indicates the weight reflecting the degree of anthropogenic activity impact on that type. The higher ICUD value implies that anthropogenic activities exert a stronger influence on the coastline. The weighting factors (W_i) for each coastline type are determined by referring to previous studies (see Table 3 for detailed information) [35,36].

Table 3. Influence weight of anthropogenic activities on coastlines.

Class	Island	Bedrock	Sandy	Biological	Artificial	Breeding
Weight	1	1	2	3	4	4

2.5.2. DSAS Analysis

Before performing the DSAS analysis (developed by United States Geological Survey, Reston, VA, USA), transect generation was required. Considering the spatial characteristics of the study area and the resolution of the imagery, transects were generated at 100 m intervals and extended to a length of 2000 m. As a result, a total of 1736 transects intersecting the coastline were obtained; their spatial distribution is further detailed in the DSAS index analysis. The Net Shoreline Movement (NSM), End Point Rate (EPR), Weighted Linear Regression Rate (WLR) and Shoreline Change Envelope (SCE) were used to evaluate the shoreline dynamics in the study area. NSM represents the distance between the oldest and the most recent coastlines, thereby quantifying overall coastline change. EPR indicates the rate and direction of coastline movement [37]. WLR—an extension of the Linear Regression Rate (LRR)—determines the best-fit line that characterizes the coastline trend [38]. SCE is defined as the maximum distance between all coastlines intersecting a transect. For a detailed description of the DSAS metrics, see the DSAS Version 5.0 User Guide (https://pubs.usgs.gov/publication/ofr20181179 (accessed on 5 February 2025)).

2.5.3. Gray Correlation Analysis

Gray correlation analysis quantifies the degree of association between variables by examining the similarity of their trend curves (i.e., the more similar the trends, the higher the correlation). This method is particularly suitable for small sample sizes and data that do not conform to a normal distribution. The analysis proceeds through the following steps [39]:

Step 1: Mean normalization

$$x_i'(k) = \frac{x_i(k)}{\frac{1}{n}\sum_{k=1}^{n} x_i(k)}, \tag{3}$$

where $x_i'(k)$ represent the normalized value of the ith variable in year k. Each raw data series is normalized to remove dimensional effects.

Step 2: Calculation of absolute differences

$$a = \underset{i}{min}\,\underset{k}{min}\,\Delta_i k, \tag{4}$$

$$b = \underset{i}{max}\,\underset{k}{max}\,\Delta_i k, \tag{5}$$

where a and b are the minimum and maximum difference between the ith variable and the length of the study area coastline in year k, respectively. Δik is the absolute difference between the ith variable and the length of the study area coastline in year k.

Step 3: Calculation of correlation coefficient

$$y_i k = \frac{a+\rho b}{\Delta_i k + \rho b}, \tag{6}$$

where $y_i k$ denotes the correlation coefficient of the ith variable with the length of the coastline of the study area in year k, and ρ is the discrimination coefficient, which usually takes the value of 0.5.

Step 4: Gray correlation calculation

$$Y = \frac{1}{n}\sum_{k=1}^{n} y_i k, \tag{7}$$

where Y denotes the average correlation coefficient between the ith variable and the length of the coastline in the study area over k years.

3. Results

3.1. Spatial–Temporal Change in Coastline

Figure 2 illustrates the dynamic evolution of the coastline in the study area from 1987 to 2022. Significant spatiotemporal changes were observed over the study period. In 1987, the total coastline length was 761.50 km, comprising 184.63 km (24.24%) of natural coastlines and 103.91 km (13.64%) of artificial coastlines. Among the natural segments, bedrock coastlines dominated, accounting for 83.49% (154.32 km) of the natural coastline, while aquaculture dikes constituted the major component (67.13%, 69.71 km) of the artificial coastline. By 1997, the overall coastline length had increased to 768.62 km; however, the natural coastline decreased to 118.62 km (15.43%), whereas the artificial coastline expanded to 202.27 km (26.31%). Within the natural category, bedrock coastlines continued to predominate (88.67%, 105.23 km). In contrast, the composition of the artificial coastline shifted, with artificial construction becoming the dominant component (62.20%, 125.82 km). In 2007, the total coastline slightly decreased to 750.50 km. Natural coastlines further declined to 107.80 km (14.36%), maintaining bedrock as the primary type (84.00%, 90.55 km), while the artificial coastline exhibited substantial growth, reaching 270.52 km (36.05%), of which artificial construction accounted for 54.54% (147.57 km). By 2017, the total coastline had expanded to 785.15 km—an overall increase of approximately 5.00% since 1987. During this period, natural coastlines experienced a marked reduction to 91.85 km (11.70%), with bedrock coastlines notably declining by 69.39%. Conversely, the artificial coastline surged to 277.89 km (35.39%), with artificial construction comprising 66.53% (184.72 km) of this total. In 2022, the total coastline reached 798.91 km. Natural coastlines were reduced further to 94.81 km (11.87%), reflecting an overall decline of 48.64% since 1987; bedrock coastlines suffered a particularly severe reduction (−69.39%), while biological coastlines paradoxically increased by 158.82%. In contrast, artificial coastlines grew markedly (147.62%), with artificial construction exhibiting an extraordinary rise of 466.73%. Notably, the aquaculture dikes showed an initial increase followed by a subsequent decline over the study period. Spatial analysis revealed a pattern of natural coastline retreat concurrent with artificial coastline expansion, predominantly concentrated in the central and southern regions of the study area (Figure 3). The spatial overlap strongly shows that intensive human activities are the primary driver of coastal transformation, with significant ecological implications arising from the accelerated loss of natural habitats and the proliferation of anthropogenic infrastructure.

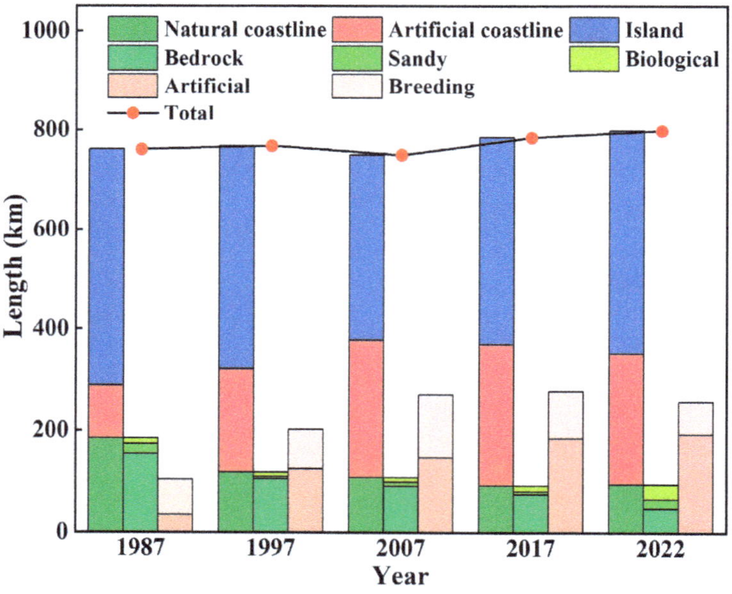

Figure 2. Change trend of coastline length in the study area from 1987 to 2022.

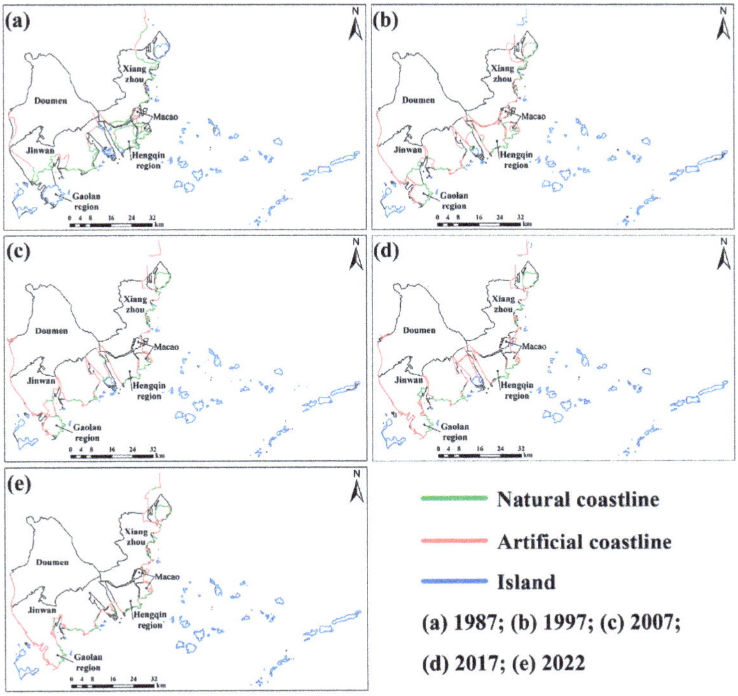

Figure 3. Spatial distribution of coastline in the study area from 1987 to 2022.

3.2. Dynamic Changes in Coastline Utilization and Location

The dynamic evolution of coastline utilization within the study area reveals distinct temporal patterns (Figure 4). ICTD exhibited a notable increase between 1987 and 2007, followed by a significant decline from 2007 to 2022. This temporal trend reflects the trans-formation in the proportional distribution of coastline types over time. During the 1987–2007 period, the proportions of major coastline categories—such as bedrock coastlines (ranging from 83.49% to 88.67% of natural coastlines), artificial construction coastlines (rising from 54.54% to 67.13% of artificial coastlines), and aquaculture dikes (from 62.20% to 75.22%)—exhibited only modest fluctuations. These relatively stable proportions contributed to a higher diversity of coastline types. However, post-2007, substantial structural changes occurred. Bedrock coastlines experienced a sharp decline, decreasing by 69.39% by 2022, while aquaculture dike coastlines also significantly contracted. In contrast, artificial construction coastlines expanded dramatically, with an increase of 466.73%. This rapid expansion of a single artificial type led to a reduction in the overall diversity of coastline types, resulting in the observed decline in ICTD. Simultaneously, ICUD exhibited a steady upward trend, increasing at an average rate of 1.97 units per year from 1987 to 2022. This continuous rise indicates a growing intensity of anthropogenic intervention and utilization along the coastline. The opposite trend of ICTD and ICUD reflects that the coastal development in the study area has shifted from relative balance to increasingly intensive utilization dominated by anthropogenic in the past 35 years.

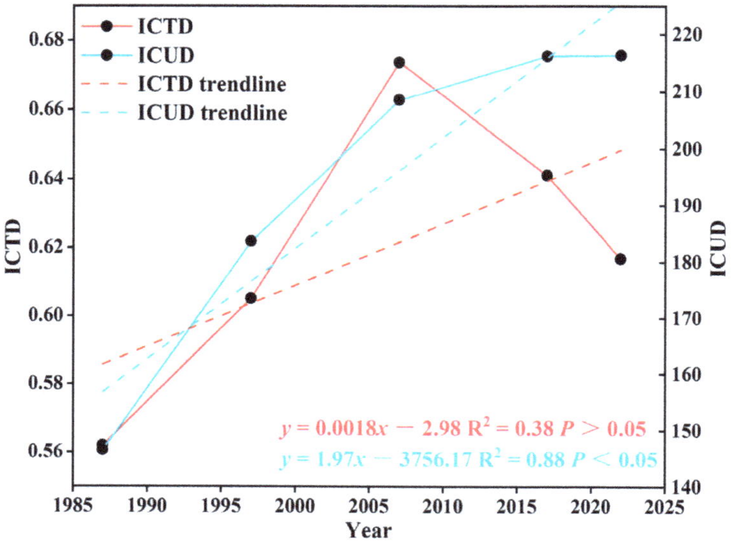

Figure 4. Change trend of ICTD and ICUD in the study area from 1987 to 2022.

The DSAS analysis results for the study area from 1987 to 2022, as illustrated in Figure 5, revealed pronounced spatiotemporal changes in the coastline. EPR analysis indicated an average coastline change rate of 8.09 ± 0.01 m/yr, with 26.27% of transects exhibiting statistically significant erosion (mean rate: −7.32 m/yr) and 72.69% showing statistically significant accretion (mean rate: 13.64 m/yr). These results reflect a prevailing trend of coastline progradation over the study period. WLR analysis indicated a slightly lower average rate of 6.62 ± 3.34 m/yr. Among all transects, 26.04% were classified as erosional (only 5.23% statistically significant, with an average rate of −3.29 m/yr) and 73.96% as accretional (of which 13.83% were statistically significant, averaging 10.10 m/yr). These findings further support the general trend of coastline advancement, though localized erosional processes persist. SCE represents the maximum displacement among all observed coastlines at each transect, averaged 331.42 m, ranging from 0.65 m to 1001.95 m, reflecting the spatial heterogeneity of coastline responses across the study area. NSM represents the linear distance between the earliest and most recent coastlines, recording an average change of 224.32 m. Notably, 26.48% of transects showed negative displacement, with a maximum retreat of −859.40 m and an average of −111.68 m, whereas 73.52% of transects experienced positive displacement, with a maximum advancement of 1001.54 m and an average of 345.33 m. Spatially, the most pronounced EPR reductions were concentrated in the eastern region, while localized areas in the south exhibited the highest accretion rates. The spatial patterns of WLR, SCE, and NSM closely aligned with those observed in the EPR analysis, reinforcing the consistency of regional coastline dynamics across multiple indicators.

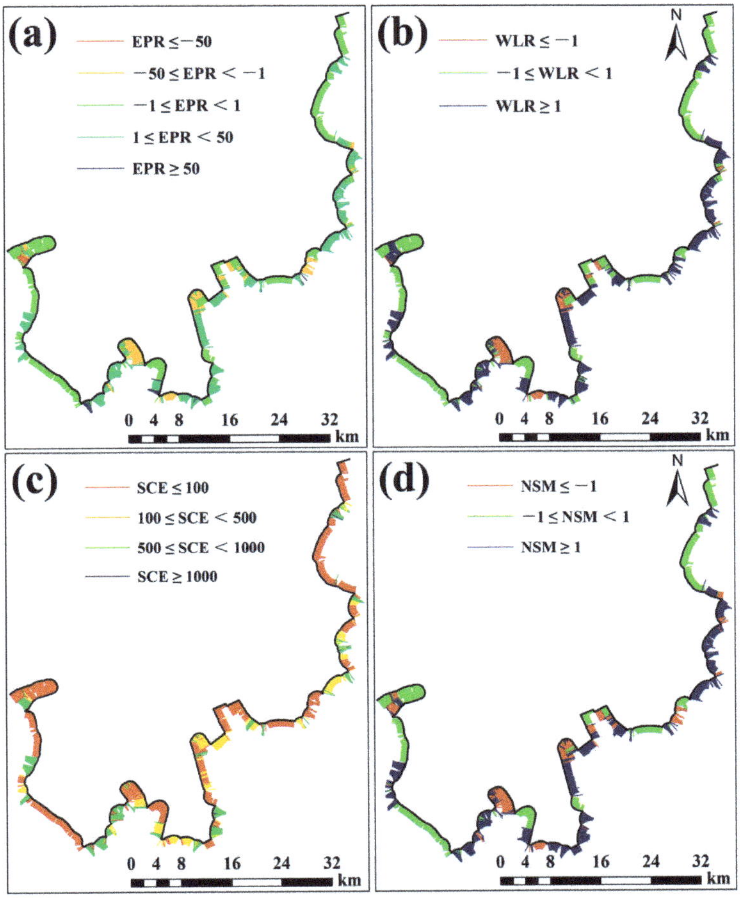

Figure 5. Spatial change trend of ICTD and ICUD in the study area from 1987 to 2022. (**a**) End Point Rate; (**b**) Weighted Linear Regression; (**c**) Shoreline Change Envelope; (**d**) Net Shoreline Movement.

3.3. Influencing Factors of Spatial–Temporal Change in Coastline

The results of the gray relational analysis revealed that the highest gray relational coefficients with natural coastline were associated with MF (0.85), MAT (0.83), and AP (0.81), followed by PRP (0.65), AA (0.65), Nb (0.65), QL (0.61), and APO (0.61). The lowest coefficients were observed for GDP (0.49) and Ex (0.47), both below the critical threshold of 0.5, indicating weak correlations (Figure 6). In contrast, the highest gray relational coefficients with artificial coastline were observed for Ex (0.64), PT (0.63), TFT (0.61), and GDP (0.60). Coefficients below 0.5 were recorded for PRP (0.49), MF (0.49), MAT (0.49), AA (0.47), and Nb (0.46). Regarding the total coastline, the strongest associations were observed with MAT (0.98) and AP (0.85), while Ex (0.46) was the only factor with a coefficient below 0.5. Overall, climatic variables (specifically MAT and AP) demonstrated the strongest associations with natural coastline changes, whereas economic development indicators such as GDP, TFT, and Ex showed the strongest correlations with artificial coastline dynamics. Further regression analysis revealed that MF and Nb were significantly associated with natural coastlines (Figure 7). Specifically, MF exhibited a significant positive correlation (slope = 1.05, p = 0.01), whereas Nb showed a significant negative correlation (slope = −1.03, p = 0.002). Regarding the artificial coastline, MF, TFT, AA, and Nb all displayed statistically significant correlations. Among these, MF had a significant negative correlation (slope = −1.03, $p = 0.03$), while TFT, AA, and Nb showed significant positive correlations. No variables were found to be significantly correlated with the total coastline length. However, Nb exhibited an especially strong relationship with coastline change, demonstrating an inverse association with natural coastline and a direct association with artificial coastline. This suggests that anthropogenic activities was the most influential factor driving changes in coastline length.

Similar conclusions can be drawn from typical regional coastline changes. In the southwestern part of the study area, Gaolan Island exhibited a clear trend of seaward expansion from 1987 to 1997, during which road infrastructure connecting the island to the mainland was completed (Figure 8). This expansion trend weakened between 1997 and 2007, although land reclamation began in the northern part of Gaolan Island in 2003, accompanied by the initiation of engineering construction on its western side. By 2022, land enclosure in the northern region had been completed, and construction activities had extended to the western and southern sections connected to the mainland, eventually forming a port. As of 2022, the northern part of the island was fully reclaimed, with ongoing infrastructure development in the west and south, contributing to continued westward expansion of the harbor. Hengqin Island, located in the southern portion of the study area, experienced particularly significant seaward expansion during the 1987–1997 period (Figure 9). The eastern islands underwent widespread outward growth due to polder farming and the development of coastline engineering projects. These efforts resulted in the enclosure of the original natural coastline and the creation of new artificial coastline, leading to a substantial increase in coastal zone area. Consequently, the proportion of artificial coastline steadily approached that of natural coastline. Between 1997 and 2007, this expansion trend diminished, with reclamation and aquaculture mainly concentrated in the Nanwan and Jinwan areas. After 2007, the rate of coastline expansion further declined compared to the previous periods. Expansion became primarily driven by engineering construction projects, while the contribution from reclamation and aquaculture decreased.

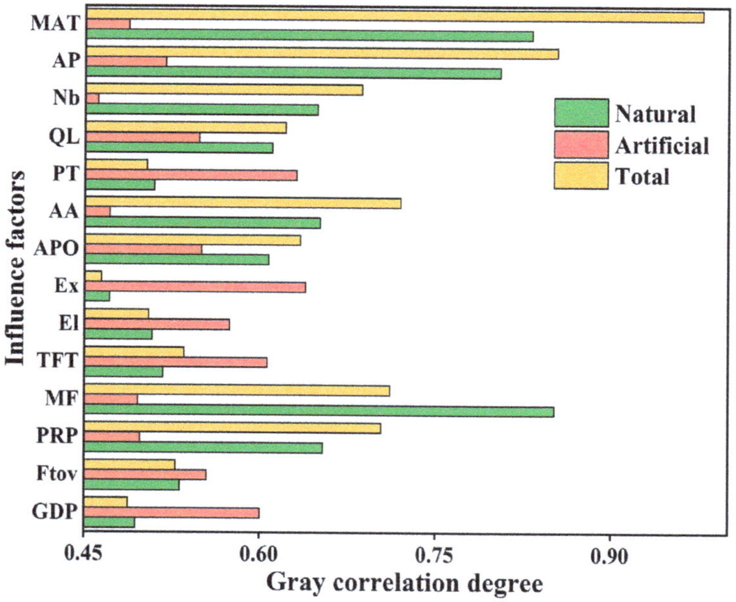

Figure 6. Gray correlation coefficients of different influencing factors in the study area from 1987 to 2022.

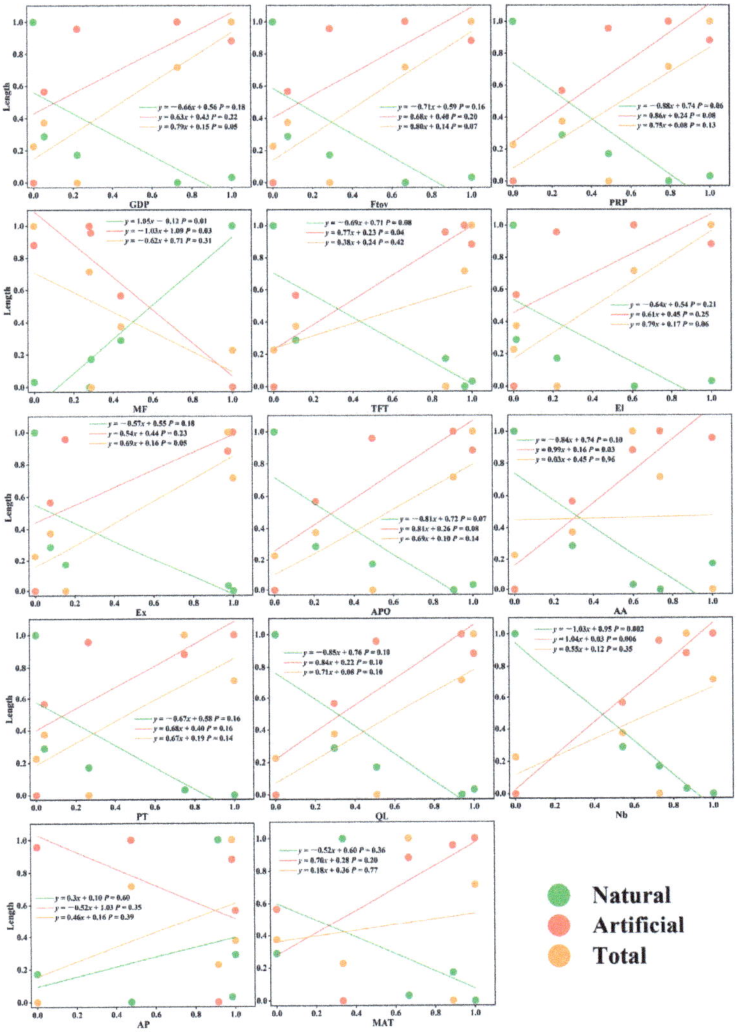

Figure 7. Linear regression of coastline and different influencing factors in the study area.

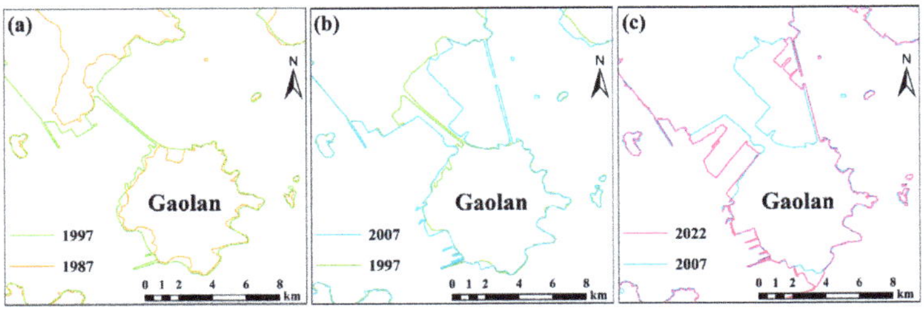

Figure 8. Spatial change in coastline in Gaolan region from 1987 to 2022. (**a**) Change from 1987 to 1997; (**b**) Change from 1997 to 2007; (**c**) Change from 2007 to 2022.

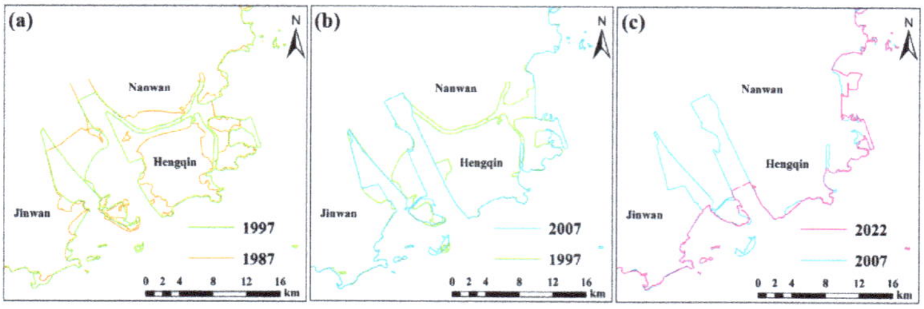

Figure 9. Spatial change in coastline in Hengqin region from 1987 to 2022. (**a**) Change from 1987 to 1997; (**b**) Change from 1997 to 2007; (**c**) Change from 2007 to 2022.

4. Discussion

The coastline of Zhuhai has undergone marked spatiotemporal changes from 1987 to 2022, characterized by a dominant trend of artificial coastline expansion and natural coastline retreat. DSAS analysis revealed a prevailing accretional trend, particularly in the southern parts of Zhuhai, such as Gaolan and Hengqin Islands, where extensive land reclamation and coastal infrastructure construction have been concentrated. The average EPR reached 8.09 ± 0.01 m/yr, suggesting strong anthropogenic influence over natural geomorphic processes. In the early period (1987–1997), infrastructure development such as road construction and initial reclamation projects marked the beginning of rapid coastline advancement. Between 1997 and 2007, land enclosure expanded northward and westward in Gaolan Island, while large-scale polder farming and coastline engineering in Hengqin significantly altered the coastline configuration. After 2007, the pace of expansion slowed, replaced by more localized engineering projects. This change aligns with observations in other rapidly urbanizing Chinese coastal cities. Han et al. (2023) found similar patterns of morphological change driven by anthropogenic activity in estuarine environments [40]. Li and Dong (2023) documented phased and spatially uneven ecological transformations in the Pearl River Delta, closely tied to urban development intensity [41]. While the morphological gain from artificial coastline construction supports urban growth and port development, it has also led to a marked reduction in ICTD and ICUD, signaling a shift from balanced land–sea interaction to anthropogenic coastal systems.

The coastal transformation in Zhuhai is shaped by both climatic variables and socio-economic development. Gray relational analysis revealed that natural coastline change was most strongly associated with climate factors, particularly mean annual temperature (MAT, 0.83) and annual precipitation (AP, 0.81). These results suggest that hydroclimatic dynamics affect natural coastline morphology through influencing runoff, erosion, and sedimentation, although their influence is increasingly overshadowed by anthropogenic activity. In contrast, artificial coastlines showed strong correlations with economic indicators such as Ex, PT and TFT, which are closely tied to infrastructure expansion. Regression analysis also highlighted the number of berths (Nb) as a key indicator: negatively correlated with natural coastlines ($p = 0.002$) and positively with artificial coastlines ($p = 0.01$), consistent with the narrative of land use transformation. These findings reinforce prior studies on the role of economic growth in coastal change. Han et al. (2023) emphasized how anthropogenic activities have restructured estuarine geomorphology [40]. Li and Dong (2023) identified GDP-driven expansion as a primary contributor to ecological decline in urban coastal zones [41]. In Zhuhai, the shift from natural to engineered coastlines reflects broader national trends, but also suggests an urgent need to balance development with ecological resilience.

However, this study operates within medium-resolution remote sensing data. Firstly, Landsat imagery was established as the robust tool for regional coastline monitoring, imposes inherent limitations on the detection of sub-pixel features such as narrow beach berms or minor scarps [14,42,43]. In addition, the observed reduction in rocky coastline and the variability of biological coasts (including aquaculture areas) likely reflect both human and natural drivers. Reduced riverine sediment supply caused by upstream water diversion and intensive agriculture, together with port construction and other engineering works, can prevent estuarine replenishment of adjacent beaches and alter local hydrodynamics. Enhanced wave attack on exposed promontories and typhoon-driven storm surge also contribute to shoreline change. Disentangling these influences requires higher-resolution imagery, in situ sediment and hydrodynamic measurements, and explicit mapping of aquaculture and reclamation activities; future work should integrate these data with tidal and wave records to better separate anthropogenic from natural controls.

To ensure sustainable coastal development in Zhuhai, it is essential to address the growing tension between economic expansion and environmental conservation. Based on the identified trends and driving mechanisms, several recommendations are proposed. Firstly, coastal ecological zoning should be strengthened to protect remaining natural coastline types, particularly bedrock and biological coastlines that serve vital ecological functions [44]. Designating conservation zones can help prevent irreversible habitat loss caused by unchecked urban encroachment. Secondly, the city should adopt an Integrated Coastal Zone Management (ICZM) framework to coordinate land use planning, engineering development, and ecological protection. The integration of nature-based solutions such as mangrove restoration and green buffer zones alongside necessary gray infrastructure has proven effective in enhancing coastal resilience [45]. Thirdly, a long-term dynamic monitoring system should be implemented using remote sensing and coastline analysis tools (e.g., DSAS) to detect early warning signs of erosion, coastline degradation, or ecological decline. Recent studies recommend coupling remote sensing data with hydrodynamic models for improved decision-making. Finally, stakeholder participation and policy coordination between municipal departments, developers, and local communities are crucial. Public awareness campaigns and transparent coastline usage regulations can increase compliance and foster a shared responsibility for coastal stewardship.

5. Conclusions

This study reveals that anthropogenic drivers dominate coastline evolution in rapidly urbanizing coastal zones, with implications for global deltaic and bay cities. Analysis of Zhuhai (1987–2022) demonstrates that land reclamation and infrastructure development drove extensive coastline accretion, particularly in southern areas like Hengqin and Gaolan Islands. Crucially, economic forces (GDP, Ex, TFT) outweighed climatic factors (MAT, AP) in steering artificial coastline changes, while Nb correlations confirm port development as a primary morphological driver. These findings highlight a global-scale paradigm: where coastal urbanization intensifies, economic priorities systematically override natural dynamics. The irreversible transition from natural to anthropogenic coastlines underscores an urgent need for strategic balancing of development and ecological resilience—a critical challenge for coastal cities worldwide.

Author Contributions: Conceptualization, T.M., H.L. and F.Z.; methodology, T.M.; software, H.L.; validation, T.M., H.L. and F.Z.; formal analysis, Y.S., Y.Z. and X.F.; investigation, H.L. and F.Z.; resources, F.Z.; data curation, F.Z.; writing—original draft preparation, T.M., H.L. and F.Z.; writing—review and editing, T.M., H.L. and F.Z.; visualization, Y.S., Y.Z. and X.F.; supervision, T.M. and F.Z.; project administration, F.Z.; funding acquisition, T.M. All authors have read and agreed to the published version of the manuscript.

Funding: This research was funded by the Open Project of State Key Laboratory of Plateau Ecology and Agriculture, Qinghai University, grant number 2024-ZZ-07.

Data Availability Statement: The data presented in this study are available on request from the corresponding author. The data are not publicly available due to privacy or ethical restrictions.

Conflicts of Interest: The authors declare no conflicts of interest.

Abbreviations

The following abbreviations are used in this manuscript:

GDP	Gross Domestic Product
Ftov	The total output value of fishery
PRP	Permanent Residents Population
MF	Mobile Fishermen
TFT	Turnover of Foreign Trade
El	Electricity of agriculture, forestry, animal husbandry and fishery
Ex	Expenditures in local government general public budgets/Agriculture, forestry and water affairs
APO	Aquatic Product Output
AA	Aquaculture Area
PT	Port Throughput
QL	Quay Length
Nb	Number of Berths in Main Ports
AP	Annual Precipitation
MAT	Mean Annual Temperature

References

1. Luo, Y.; Han, G.; Qin, W.; Chen, L.; Gao, Z. *Eco-Environment and Sustainable Management in the Coastal Zone*; Science Press: Beijing, China, 2021.
2. Liu, L.; Xu, W.; Yue, Q.; Teng, X.; Hu, H. Problems and countermeasures of coastline protection and utilization in China. *Ocean. Coast. Manag.* **2018**, *153*, 124–130. [CrossRef]
3. Hou, X.; Wu, T.; Hou, W.; Chen, J.; Wang, Y.; Yu, L. Characteristics of coastline changes in mainland China since the early 1940s. *Sci. China Earth Sci.* **2016**, *59*, 1791–1802. [CrossRef]
4. Boye, C.; Appeaning Addo, K.; Wiafe, G.; Dzigbodi-Adjimah, K. Spatio-temporal analyses of shoreline change in the Western Region of Ghana. *J. Coast. Conserv.* **2018**, *22*, 769–776. [CrossRef]
5. Primavera, J.H. Overcoming the impacts of aquaculture on the coastal zone. *Ocean. Coast. Manag.* **2006**, *49*, 531–545. [CrossRef]
6. Ren, C.; Wang, Z.; Zhang, Y.; Zhang, B.; Chen, L.; Xi, Y.; Xiao, X.; Doughty, R.B.; Liu, M.; Jia, M. Rapid expansion of coastal aquaculture ponds in China from Landsat observations during 1984–2016. *Int. J. Appl. Earth Obs. Geoinf.* **2019**, *82*, 101902. [CrossRef]
7. Celata, F.; Gioia, E. Resist or retreat? Beach erosion and the climate crisis in Italy: Scenarios, impacts and challenges. *Appl. Geogr.* **2024**, *169*, 103335. [CrossRef]
8. Li, K.; Zhang, L.; Chen, B.; Zuo, J.; Yang, F.; Li, L. Analysis of China's coastline changes during 1990–2020. *Remote Sens.* **2023**, *15*, 981. [CrossRef]
9. Liang, L.; Liu, Q.; Liu, G.; Li, X.; Huang, C. Review of coastline extraction methods based on remote sensing images. *J. Geo-Inf. Sci.* **2018**, *20*, 1745–1755.
10. Toure, S.; Diop, O.; Kpalma, K.; Maiga, A.S. Shoreline detection using optical remote sensing: A review. *ISPRS Int. J. Geo-Inf.* **2019**, *8*, 75. [CrossRef]
11. Ohenhen, L.O.; Shirzaei, M.; Ojha, C.; Sherpa, S.F.; Nicholls, R. Disappearing cities on US coasts. *Nat. Commun.* **2024**, *627*, 108–115. [CrossRef]
12. Sunny, D.S.; Islam, K.A.; Mullick, M.R.A.; Ellis, J.T. Performance study of imageries from MODIS, Landsat 8 and Sentinel-2 on measuring shoreline change at a regional scale. *Remote Sens. Appl. Soc. Environ.* **2022**, *28*, 100816. [CrossRef]
13. Rostami, E.; Sharifi, M.; Hasanlou, M. Shoreline extraction using time series of sentinel-2 Satellite images by google earth engine platform. *ISPRS Ann. Photogramm. Remote Sens. Spat. Inf. Sci.* **2023**, *10*, 653–659. [CrossRef]
14. Zuo, J.; Zhang, L.; Xiao, J.; Chen, B.; Zhang, B.; Hu, Y.; Mamun, M.A.A.; Wang, Y.; Li, K. GCL_FCS30: A global coastline dataset with 30-m resolution and a fine classification system from 2010 to 2020. *Sci. Data* **2025**, *12*, 129. [CrossRef]

15. Zhang, Q.; Zhang, Z.; Xu, N.; Li, Y. Fully automatic training sample collection for detecting multi-decadal inland/seaward urban sprawl. *Remote Sens. Environ.* **2023**, *298*, 113801. [CrossRef]
16. Wang, H. Research on Land Use Spatial-time Evolution Based on Geographical Conditions Monitoring in Development Zone. *Geomat. Spat. Inf. Technol.* **2018**, *41*, 151–159.
17. Lin, J.; Tang, D.; Gao, Y.; Li, X. Remote Sensing of the Classification, Development and Utilization of Coastline in Zhuhai City. *Ocean. Dev. Manag.* **2019**, *36*, 69–73, 93.
18. Yang, L.; Sun, W.; Ma, Y.; Ren, G. Remote Sensing Analyses of the Spatial and Temporal Changes in Zhuhai Shoreline. *Mar. Sci.* **2017**, *41*, 20–28.
19. Zhang, H. Monitoring of Coastline Change in Zhuhai Based on High Resolution Remote Sensing. *Bull. Surv. Mapp.* **2016**, *11*, 55–59, 71.
20. Yin, N.; Tang, J.; Yang, Y.; Gao, X.; Song, S.; Hu, Q. Variations of shoreline and land use in Guangdong-Hong Kong-Macao Greater Bay Area from 1989 to 2021. *Mar. Geol. Front.* **2023**, *39*, 1–11.
21. Huang, Z.; Fang, L.; Wen, H.; Zhang, K.; Wang, X.; Chen, T. Responses of Indo-Pacific humpback dolphins (Sousa chinensis) to construction of the Hong Kong–Zhuhai–Macao Bridge. *Front. Mar. Sci.* **2024**, *11*, 1407937. [CrossRef]
22. Hu, R.; Yao, L.; Yu, J.; Chen, P.; Wang, D. Remote sensing of the coastline variation of the guangdong–hongkong–macao greater bay area in the past four decades. *J. Mar. Sci. Eng.* **2021**, *9*, 1318. [CrossRef]
23. Almar, R.; Boucharel, J.; Graffin, M.; Abessolo, G.O.; Thoumyre, G.; Papa, F.; Ranasinghe, R.; Montano, J.; Bergsma, E.W.; Baba, M.W. Influence of El Niño on the variability of global shoreline position. *Nat. Commun.* **2023**, *14*, 3133. [CrossRef]
24. Zhang, Y.; Li, D.; Fan, C.; Xu, H.; Hou, X. Southeast Asia island coastline changes and driving forces from 1990 to 2015. *Ocean Coast. Manag.* **2021**, *215*, 105967. [CrossRef]
25. Amrouni, O.; Hzami, A.; Heggy, E. Photogrammetric assessment of shoreline retreat in North Africa: Anthropogenic and natural drivers. *ISPRS J. Photogramm. Remote Sens.* **2019**, *157*, 73–92. [CrossRef]
26. Lansu, E.M.; Reijers, V.C.; Höfer, S.; Luijendijk, A.; Rietkerk, M.; Wassen, M.J.; Lammerts, E.J.; van der Heide, T. A global analysis of how human infrastructure squeezes sandy coasts. *Nat. Commun.* **2024**, *15*, 432. [CrossRef]
27. Yang, G.; Huang, K.; Zhu, L.; Sun, W.; Chen, C.; Meng, X.; Wang, L.; Ge, Y. Spatio-temporal changes in China's mainland shorelines over 30 years using Landsat time series data (1990–2019). *Earth Syst. Sci. Data Discuss.* **2024**, *2024*, 1–26. [CrossRef]
28. Wang, T.; Zhang, G.; Zhang, K.; Fu, Q. Dynamic changes of the Huizhou Coastline in nearly 50 years based on Landsat images and DSAS. *Natl. Remote Sens. Bull.* **2024**, *28*, 689–703. [CrossRef]
29. Ai, B.; Lai, Z.; Zeng, J.; Jian, Z.; Zhao, J.; Sun, S. Detection of wetland degradation and restoration in urbanizing Zhuhai City based on google earth engine. *Ocean Coast. Manag.* **2025**, *261*, 107518. [CrossRef]
30. Zeng, Z.; Lai, C.; Wang, Z.; Chen, Y.; Chen, X. Future sea level rise exacerbates compound floods induced by rainstorm and storm tide during super typhoon events: A case study from Zhuhai, China. *Sci. Total Environ.* **2024**, *911*, 168799. [CrossRef]
31. Zhu, L.; Huang, Y.; Yang, G.; Sun, W.; Chen, C.; Huang, K. Information extraction and spatio-temporal evolution analysis of the coastline in Hangzhou Bay based on Google Earth Engine and remote sensing technology. *Remote Sens. Nat. Resour.* **2023**, *35*, 50–60.
32. Wang, J.; Wu, Z.; Li, S.; Wang, S.; Zhang, X.; Gao, Q. Coastline and Land Use Change Detection and Analysis with Remote Sensing in the Pearl River Estuary Gulf. *Sci. Geogr. Sin.* **2016**, *36*, 1903–1911.
33. Wu, T. Analysis of Spatio-Temporal Characteristics of Mainland Coastline Changes in China in Nearly 70 Years. Ph.D. Thesis, Yantai Institute of Coastal Zone Research, Chinese Academy of Sciences, Yantai, China, 2015.
34. Wu, T.; Hou, X.; Xu, X. Spatio-temporal characteristics of the mainland coastline utilization degree over the last 70 years in China. *Ocean Coast. Manag.* **2014**, *98*, 150–157. [CrossRef]
35. Yang, F.; Zhang, L.; Chen, B.; Li, K.; Liao, J.; Mahmood, R.; Hasan, M.E.; Mamun, M.A.A.; Raza, S.A.; Sutrisno, D. Long-term change of coastline length along selected coastal countries of Eurasia and African continents. *Remote Sens.* **2023**, *15*, 2344. [CrossRef]
36. Ning, Z.; Jiang, C.; Chen, J.; Wu, Z.; Yao, Z.; Ma, Y.; Deng, T.; Chen, Y. Long-term spatiotemporal analysis of coastline morphological evolutions and their underlying mechanisms in the Pearl River Delta region of China. *Ocean Coast. Manag.* **2024**, *258*, 107426. [CrossRef]
37. Crowell, M.; Honeycutt, M.; Hatheway, D. Coastal erosion hazards study: Phase one mapping. *J. Coast. Res.* **1999**, 10–20.
38. Keyes, T.K. Applied regression analysis and multivariable methods. *Technometrics* **2001**, *43*, 101. [CrossRef]
39. Deng, J. Introduction to grey systems theory. *J. Grey Syst.* **1989**, *1*, 1–24.
40. Han, Z.; Wang, H.; Xie, H.; Li, H.; Li, W. How Does Human Activity Shape the Largest Estuarine Bay of the Pearl River Estuary, South China (1964–2019). *Water* **2023**, *15*, 4143. [CrossRef]
41. Li, T.; Dong, Y. Phased and polarized development of ecological quality in the rapidly-urbanized Pearl River Delta, China. *Environ. Sci. Pollut. Res.* **2023**, *30*, 36176–36189. [CrossRef]

42. Cai, H.; Li, C.; Luan, X.; Ai, B.; Yan, L.; Wen, Z. Analysis of the spatiotemporal evolution of the coastline of Jiaozhou Bay and its driving factors. *Ocean Coast. Manag.* **2022**, *226*, 106246. [CrossRef]
43. Liu, Y.; Feng, J.; Cheng, Q.; Tsou, J.Y.; Huang, B.; Ji, C.; Yang, Y.; Zhang, Y. Investigating spatiotemporal coastline changes and impacts on coastal zone management: A case study in Pearl River Estuary and Hong Kong's coast. *Ocean Coast. Manag.* **2024**, *257*, 107354. [CrossRef]
44. Barbier, E.B.; Hacker, S.D.; Kennedy, C.; Koch, E.W.; Stier, A.C.; Silliman, B.R. The value of estuarine and coastal ecosystem services. *Ecol. Monogr.* **2011**, *81*, 169–193. [CrossRef]
45. Temmerman, S.; Meire, P.; Bouma, T.J.; Herman, P.M.; Ysebaert, T.; De Vriend, H.J. Ecosystem-based coastal defence in the face of global change. *Nature* **2013**, *504*, 79–83. [CrossRef]

water

MDPI

Article

Numerical Simulations of Extratropical Storm Surge in the Bohai Bay Based on a Coupled Atmosphere–Ocean–Wave Model

Yong Li [1,2,*], Xuezheng Liu [1,2], Junjie Liu [1,2] and Guangsen Xiong [1,2]

1 Ecological Environment Research Center of Estuarine and Coastal Zone, Hebei University of Environmental Engineering, Qinhuangdao 066102, China
2 Hebei Engineering Research Center for Ecological Restoration of Seaward Rivers and Coastal Waters, Qinhuangdao 066102, China
* Correspondence: liyong@hebuee.edu.cn

Abstract

The Bohai Bay is particularly vulnerable to storm surges triggered by extratropical storms or cold-air outbreaks. A coupled atmosphere–ocean–wave model with high resolution is presented and applied to simulate a cold-air outbreak that happened in late November 2004. The surge dynamics are examined in detail. Each model component is separately validated, demonstrating that the triply coupled system can reproduce intense winds, storm surge amplitudes, and significant surface waves with high fidelity. The potential coupling effects on the simulation results are investigated. Six experiments are performed covering various coupling models, and a two-way nesting technique is utilized during simulation. After comparison it shows that there is little difference in wind speed between the three numerical models and that the reanalysis data may significantly underestimate extreme winds. The evident improvements are obtained for peak values of water level when using the atmosphere–ocean coupled configuration versus uncoupled model simulation. It also can be found that the negative surge can be captured by each of the coupled and uncoupled models. The ocean–wave coupled configuration yields significant wave heights that closely match in situ measurements, underscoring the critical role of ocean–wave interaction in storm wave prediction. Our findings confirm that the fully coupled model is well-suited for forecasting extratropical storm surge in Bohai Bay. Northeast winds emerge as the primary driver, with the western coast of Bohai Bay bearing the greatest impact.

Keywords: Bohai bay; cold-air outbreak; extratropical storm surge; coupled atmosphere–ocean–wave model; WRF; ROMS; SWAN

1. Introduction

Bohai Bay, one of three embayments comprising the Bohai Gulf, is highly susceptible to storm surges driven by cold-air outbreaks, which occur primarily in spring, autumn, and winter. These surges profoundly affect a severe impact on the topography and landforms near the coast of Bohai Bay [1,2]. Furthermore, storm surges may cause significant human life losses and severe damage to coastal structures, and thus it is very important to get acquainted with the extratropical storm surges occurring in the Bohai Bay [3,4].

As a semi-enclosed, funnel-shaped inner sea at mid-latitudes, the Bohai Sea features shallow depths (average $\sim$18 m) and gentle slopes (Figure 1). It connects to the Yellow Sea via the Bohai Strait and consists of Bohai Bay, Liaodong Bay, and Laizhou Bay. During cold-air surges, strong easterly winds drive open seawater westward into the bay, causing rapid sea-level rise due to the bay's restricted geometry.

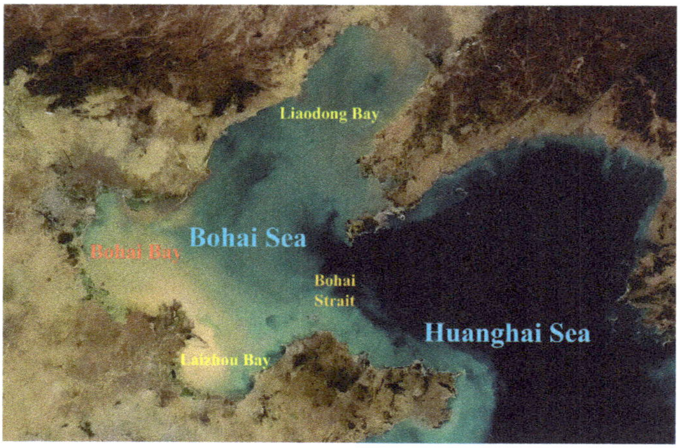

Figure 1. A geographical map of Bohai Bay.

Extratropical storm surge in Bohai Bay manifests as long-period waves primarily induced by cold-air outbreaks, with surge intensity governed by wind strength and direction. Numerous studies have aimed to observe, interpret, and predict severe storm surges [5–11], and numerical simulations have become increasingly vital for coastal planning and disaster mitigation.

Numerical models for storm surge analysis fall into three main categories. The first category, widely adopted, couples ocean models with prescribed wind fields [12–14]. Here, ocean models compute water levels, while wind field data, typically from reanalysis products or analytical wind models, serves as surface forcing. For example, Zhao and Jiang [12] developed 20 cold-air scenarios to evaluate the influence of storm tracks, pressure fields, and wind durations on maximum surge in the Bohai Sea using the Finite-Volume Coastal Ocean Model (FVCOM). Mo et al. [13] employed a three-dimensional model driven by Climate Forecast System Reanalysis (CFSR) data to investigate the roles of wind direction, speed, and cyclone–tide interactions.

The second category uses uncoupled atmospheric and ocean models [6,15–17], where winds from atmospheric simulations force separate ocean runs. Li et al. [6] used Fifth-Generation Mesoscale Model (MM5) winds to drive surge simulations, identifying wind stress as the dominant surge mechanism. Ding and Ding [15] forced FVCOM with outputs from the Weather Research and Forecasting (WRF) system to reproduce the October 2003 surge and later assessed land reclamation impacts [16]. Li et al. [17] integrated the Regional Ocean Modeling System (ROMS) with Quick Scatter Meter/National Center for Environmental Prediction (QSCAT/NCEP) wind stress to model surge and inundation along the western Bohai Gulf.

The third category encompasses fully multi-field coupled models [5,18–23]. Warner et al. [5] developed a Coupled Ocean–Atmosphere–Wave–Sediment Transport (COAWST) modeling system, combining the ocean model (ROMS), the atmosphere model (WRF), the wave model (Simulating WAves Nearshore, SWAN), and the sediment model (Community Sediment Transport Model, CSTM), demonstrating improved storm intensity prediction. The COAWST system was originally developed to evaluate model sensitivity during the simulation of Hurricane Isabel in 2003. Subsequently, it was applied to hindcast Hurricane Ivan in 2004 [18]. A series of sensitivity experiments then assessed the effects of coupling on the atmospheric, oceanic, and wave predictions during the tropical cyclone (TC). Results revealed a marked improvement in storm intensity forecasts when using the fully atmosphere–ocean–wave coupled configuration compared to uncoupled simulations.

Further, to examine atmosphere–ocean–wave interactions during tropical cyclones, Hurricane Ida (November 2009) and Typhoon Kai-tak (1213) (August 2012) were simulated with COAWST [19,20]. Li et al. [21] employed a coupled atmosphere–ocean model to analyze the mid-October 2003 extratropical surge in Bohai Bay but omitted wave dynamics. However, waves play a crucial role in reshaping the bay's seabed and coastline by lifting and redistributing sediment. Incorporating a wave component into the coupled framework is therefore essential for accurately capturing seabed and nearshore morphological changes during storm surges [3,24].

As noted above, the Bohai Bay remains highly prone to extratropical storm surges and, while numerous studies have approached the topic from various angles [3,4,6,12,15–17], few have used a fully coupled atmosphere–ocean–wave model. To enhance surge forecasting and mechanistic insight for Bohai Bay, we introduce such a three-way coupled system, in which each component exchanges data fields via two coupling packages. This coupled model is applied to simulate an extratropical storm surge that happened in November 2004. The potential coupling effects on the simulation results are investigated. Six experiments are performed covering different coupling ways. The calculated results are compared with observation data, which identifies whether the fully coupled model can simulate the winds and ocean hydrodynamics accurately. Section 2 describes component models, including atmospheric models, ocean models, and wave models. In addition, the coupling packages and methods are introduced. Section 3 outlines the experiment design and model settings. Section 4 presents the calculated results and discusses the impact of different coupling ways on storm surge simulation. Finally, Section 5 summarizes our findings and offers conclusions.

2. Numerical Models

2.1. WRF

The atmospheric component of the coupled system is the WRF model, a next-generation mesoscale numerical weather prediction model developed by the National Centre for Atmospheric Research (NCAR, Boulder, CO, USA) and collaborating institutes. This model has been widely employed for both real-world and idealized simulations. Numerous studies demonstrate that WRF performs reliably across scales ranging from meters to thousands of kilometers [25–27].

The WRF model solves the fully compressible Eulerian equations using terrain-following hydrostatic-pressure coordinates to simulate various atmospheric variables. The modeling framework in the WRF model offers flexible configuration options, allowing users to choose different parameterizations for key atmospheric processes, including microphysics, cumulus schemes, boundary-layer treatments, and radiation physics. In the present study, the WRF v3.6 is coupled with the ocean model ROMS and wave model SWAN.

The wave-induced surface roughness at the coupling interface is computed in WRF using the following relation [28]:

$$z_0 = 3.35 H_s \left(\frac{u^*}{c_w} \right)^{3.4} + 0.11 \frac{\nu}{u^*} \quad (1)$$

where z_0 is the roughness length, H_s is the significant wave height, u^* is the wind friction speed, c_w is the phase speed at the peak period, and ν is the kinematic viscosity.

2.2. ROMS

The ROMS model is a free-surface, sigma-coordinate, primitive equations ocean model and widely used for numerical simulations in physical oceanography [29,30]. The

ROMS ocean model solves the Reynolds-averaged Navier–Stokes equations based on the hydrostatic and Boussinesq approximations [31]. Here, ROMS v3.6 is integrated into the coupled framework.

The governing equations in Cartesian coordinates are shown as follows. The Reynolds-averaged formulation of the momentum equations in the x- and y-directions are:

$$\frac{\partial u}{\partial t} + \vec{v} \cdot \nabla u - fv = -\frac{\partial \phi}{\partial x} - \frac{\partial}{\partial z}\left(\overline{u'w'} - \nu\frac{\partial u}{\partial z}\right) + F_u + D_u \tag{2}$$

$$\frac{\partial v}{\partial t} + \vec{v} \cdot \nabla v + fu = -\frac{\partial \phi}{\partial y} - \frac{\partial}{\partial z}\left(\overline{v'w'} - \nu\frac{\partial v}{\partial z}\right) + F_v + D_v \tag{3}$$

The equation of state is:

$$\rho = \rho(T, S, P) \tag{4}$$

Hydrostatic balance yields:

$$\frac{\partial \phi}{\partial z} = -\frac{\rho g}{\rho_0} \tag{5}$$

Continuity for an incompressible fluid is:

$$\frac{\partial u}{\partial x} + \frac{\partial v}{\partial y} + \frac{\partial w}{\partial z} = 0 \tag{6}$$

Closure is achieved by parameterizing Reynolds stresses and turbulent tracer fluxes:

$$\overline{u'w'} = -K_m\frac{\partial u}{\partial z} \tag{7}$$

$$\overline{v'w'} = -K_m\frac{\partial v}{\partial z} \tag{8}$$

where u, v, and w are the components of speed vector $\vec{v}$ in the x, y, and z directions, respectively, t is time, T is the temperature, S is the salinity and P is the pressure, g is gravity acceleration, ρ is the water density, ϕ is the dynamic pressure, f is the Coriolis parameter, D is the diffusive term, F is the forcing term, including surface wind stress and wave radiation stress, and K_m is the eddy viscosity for momentum.

2.3. SWAN

Accurate wave prediction is essential for storm surge modeling. In this study, we employ the third-generation spectral wave model SWAN (v40.91A). SWAN simulates wind-wave generation and propagation in coastal waters, incorporating key processes such as refraction, diffraction, shoaling, nonlinear wave–wave interactions, and energy dissipation [32,33]. Specifically designed for shallow-water environments, SWAN solves the action balance equation:

$$\frac{\partial N}{\partial t} + \frac{\partial c_x N}{\partial x} + \frac{\partial c_y N}{\partial y} + \frac{\partial c_\sigma N}{\partial \sigma} + \frac{\partial c_\theta N}{\partial \theta} = \frac{S_w}{\sigma} \tag{9}$$

where N is the action density spectrum, σ is the relative radian frequency, θ is direction normal to the wave crest, x and y are coordinate space, and c_x and c_y are the group velocities in x- and y-directions, respectively. c_σ is the propagation speed in frequency space and c_θ is a speed in directional space, and S_w is the combined source and sink terms of wave energy density.

2.4. Coupled WRF–ROMS–SWAN Model

The WRF atmosphere model, ROMS ocean model, and SWAN wave model are coupled via the open-source Model Coupling Toolkit (MCT, https://web.cels.anl.gov/projects/climate/mct/, accessed on 1 July 2025) and the Spherical Coordinate Remapping and Interpolation Package (SCRIP, https://github.com/SCRIP-Project/SCRIP, accessed on 1 July 2025) [34]. MCT provides a fully parallelized framework using MPI for exchanging model variables, mesh regridding, and time averaging [35]. In addition, the grid points have to be interpolated for data exchange between different model grids and the SCRIP can be competent for this job. The SCRIP computes addresses and weights for remapping and interpolating fields between different grids [36].

During the ocean–atmosphere–wave coupling, model variables are exchanged as follows: ROMS receives wind stress and heat flux parameters from WRF and conversely supplies WRF with updated sea surface temperature values; the wave height, period, length, direction, and so on are passed from SWAN to ROMS; the sea surface height and depth averaged currents are passed from ROMS to SWAN; the 10 m winds are passed from WRF to SWAN; and the sea surface roughness in WRF is computed by significant wave height, length, period, and so on. Figure 2 illustrates the data-exchange pathways. All three models support parallel execution via MPI on distributed-memory architectures.

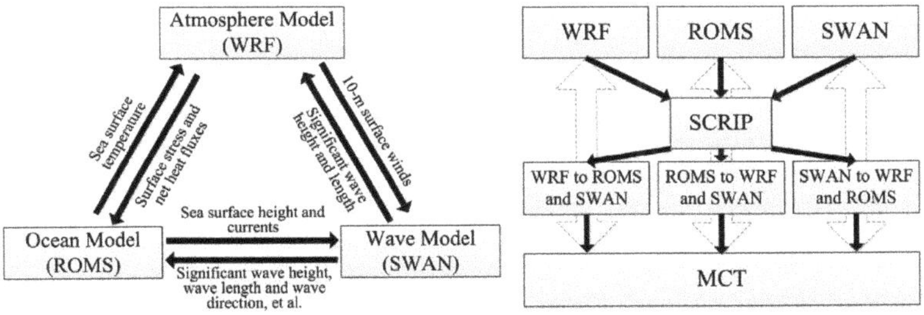

Figure 2. Schematic of the 3-way coupled model.

3. Model Application

We simulate the storm surge induced by a cold-air outbreak in November 2004. Arctic air masses from Siberia advected across the Bohai Sea. The coupled system runs from 20 to 27 November 2004, with all times in Coordinated Universal Time (UTC). To assess the contribution of each component, we conduct six experiments (Table 1). The first (EXP1) implements fully coupled WRF–ROMS–SWAN integration (three-way), exchanging fields every 10 min. Experiments 2 and 3 perform two-way coupling between WRF–ROMS and WRF–SWAN (two-way). The final three (EXP4–EXP6) run each model independently (one-way), driven by ECMWF's ERA-Interim reanalysis.

Table 1. Modeling experiments.

Experiment	Component
EXP1 (3-Way)	WRF–ROMS–SWAN coupling
EXP2 (2-Way)	WRF–ROMS coupling
EXP3 (2-Way)	WRF–SWAN coupling
EXP4 (1-Way)	SWAN
EXP5 (1-Way)	WRF
EXP6 (1-Way)	ROMS

3.1. Atmospheric Model Setup

The WRF computational domain and associated topography are depicted in Figure 3. A coarse domain of 150 × 150 grid points at 10 km spacing was configured to fully encompass the storm and minimize open-boundary effects. The domain extends roughly from 29.0° N to 42.5° N and 112.5° E to 130.5° E. A nested grid is employed at three times finer horizontal resolution. Vertical discretization consists of 44 non-uniform σ levels. Time steps of 60 s (coarse) and 20 s (nest) are applied.

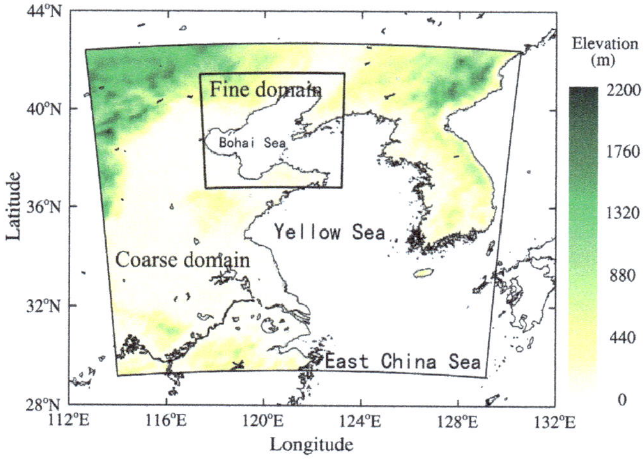

Figure 3. Domain configuration and topography of the region used in WRF model.

The following WRF parameterizations were selected: time discretization was achieved with a third-order Runge–Kutta scheme and spatial derivatives were handled by a fifth-order scheme, Lin microphysics [37], Noah land-surface model [38], Monine–Obukhov surface-layer scheme, Rapid Radiative Transfer Model (RRTM) longwave radiation [39], Dudhia shortwave radiation [40], Yonsei University (YSU) planetary boundary-layer scheme [41], and Kain–Fritsch cumulus parameterization [42]. Initial and lateral boundary conditions were obtained from ERA-Interim reanalysis data (0.5°/6 h). Further details are provided by Skamarock et al. [43].

3.2. Ocean Model Setup

The ROMS computational domain and seafloor topography are shown in Figure 4, with open southern boundary and closed boundaries on the remaining sides. The ROMS domain is configured to encompass the entire Bohai Sea and most of the Yellow Sea (117° E–127° E, 35° N–41° N), with grid spacings of 0.04° in longitude and 0.03° in latitude. There are 250 × 200 grid points for coarse domain and the horizontal resolution ratio between the nest grid and the coarse grid is 3. Six sigma levels are used in the vertical direction. Time steps of 60 s are used in the coarse domain and one-third of that value in the nested domain.

A third-order upstream scheme is used for horizontal momentum advection, together with the Mellor-Yamada 2.5 turbulence closure for vertical mixing. Bottom friction is represented by a quadratic parameterization with a drag coefficient of 0.012. Bathymetric data were compiled from high-resolution (1 min) National Geophysical Data Center of National Oceanic and Atmospheric Administration (NOAA) and Sung Kyun Kwan University's (SKKU) digital elevation models [44], then smoothed via a second-order Shapiro filter. Coastline information was obtained from NOAA's World Vector Shoreline database.

Both open and closed lateral boundary conditions are implemented; open boundaries incorporate free surface dynamics, 2D momentum conservation, and 3D momentum radiation physics. Tidal forcing includes eight principal constituents (M2, S2, N2, K2, K1, O1, P1, and Q1), with water surface elevations derived from Oregon State University's OTPS (OSU Tidal Prediction Software, https://www.tpxo.net/otps, accessed on 1 July 2025) tidal prediction system. Closed boundaries employ no-slip wall conditions for surface elevation and normal speed components. Initial conditions assume quiescent flow (zero speed) with water levels set to mean sea level.

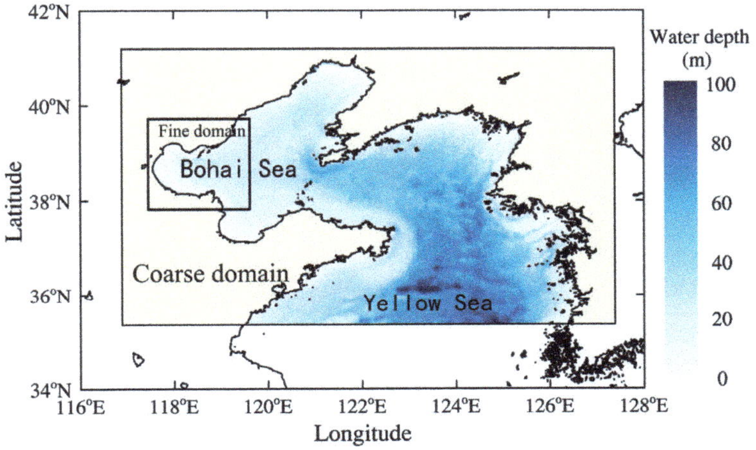

Figure 4. Domain configuration and topography of the region in ROMS model and SWAN model.

3.3. SWAN Model Setup

SWAN executed the same grid and bathymetry as ROMS (Figure 4). Initial conditions were generated via a stationary spin-up using reanalysis wind data. The JONSWAP spectrum served as the boundary condition for the initial stationary steady-state simulation and the non-stationary simulation. The relevant parameters are derived from the solutions of the global WaveWatch 3 (WW3) model.

In this configuration, nonlinear quadruplet wave interactions were activated, and wind-wave growth was parameterized following the Komen formulation [45]. Additional wave parameters used in this study are summarized in Table 2. The model ran with a 120 s time step, and outputs of significant wave height, wave period, and wavelength were recorded every 10 min.

Table 2. SWAN parameters.

Parameters	Value
Frequency range (Hz)	0.05–1.0
Frequency bins	24
Breaking constant	0.75
Direction Full	circle
Direction bins	72
Bottom friction	Madsen formulation
Friction parameter	0.05
Minimum water depth (m)	0.05

4. Results and Discussion

4.1. Atmosphere Results

Accuracy representation of the wind field is essential for realistic storm surge and wave state simulations. Wind vectors at 4 h intervals from 2100 UTC 23 November 2004, as produced by EXP1, are presented in Figure 5. At 2100 UTC on 23 November, easterly to northeasterly winds dominated the northern and central Bohai Sea. The wind speed increased rapidly, reaching a peak of 16.0 m/s at approximately 0100 UTC on 24 November (Figure 5b). Thereafter, wind speeds declined gradually to about 12 m/s by 0900 UTC on 24 November. Winds exceeding Beaufort force 6 persisted for over 10 h, with a prevailing northeast direction across most of the Bohai Sea (Figure 5d).

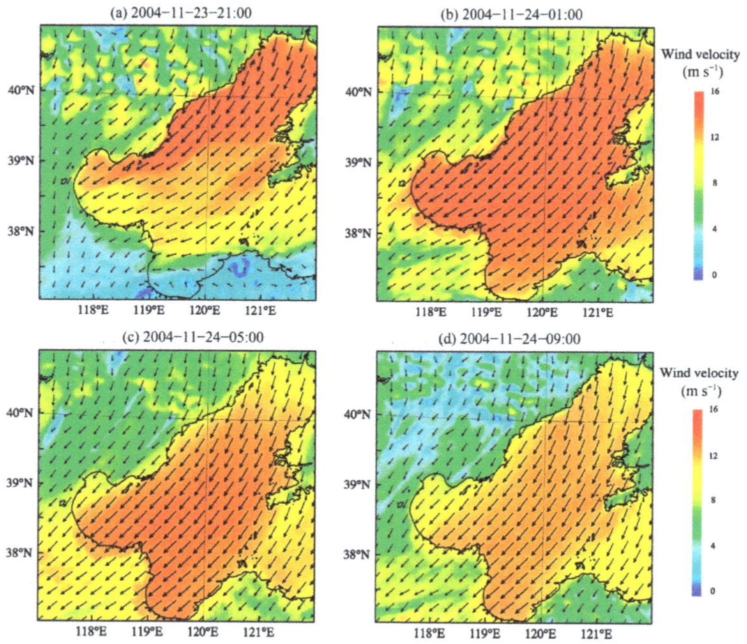

Figure 5. Wind snapshots every 4 h from 2100 UTC 23 November 2004 as simulated by EXP1. The arrows and colors represent the wind speed direction and intensity.

Sea level temperature and pressure at 4 h intervals are shown in Figure 6. A pronounced temperature drop was simulated, with surface temperatures in Bohai Bay falling by nearly 10 °C over 12 h. Sea level pressure decreased from north to south as a cold high-pressure system advanced southward. Prior to 2100 UTC on 23 November (Figure 6b), the intensifying pressure gradient was associated with increasing wind speeds. In summary, the southeastward movement of the northwest cold high pressure caused abrupt cooling and sustained strong winds, triggering the extratropical storm surge.

Wind speed time series at Huanghua station (0000 UTC 23 November 2004) are compared in Figure 7 for EXP1, EXP2, EXP5, and ECMWF reanalysis. It was found that the 10 m winds during the cold-air outbreak are reasonably well captured. The three numerical models produce similar results. When contrasting the three-way coupling with the two-way runs, the three-way model yields slightly lower wind speeds. This is attributed to the increased surface roughness computed via Equation (1) after SWAN was integrated into the WRF–ROMS framework. The results do suggest that wave height has notable effects on wind field and that the higher the wave height is, the larger the effect is. Through the comparison of the results between the two-way model and the WRF-only model, it can be seen that the wind velocities obtained by one-way model are systematically smaller, indicating that sea surface temperature significantly influences wind simulation.

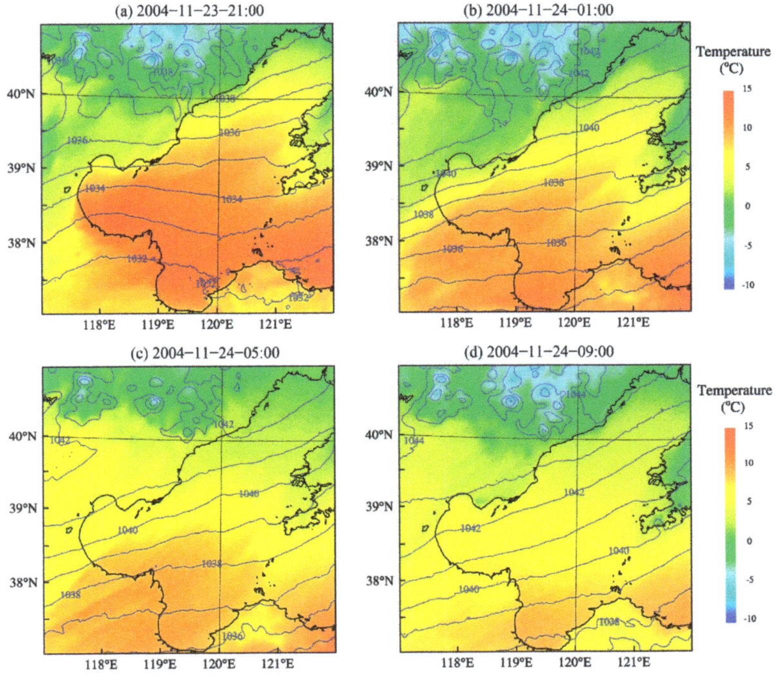

Figure 6. Sea level temperature and pressure snapshots every 4 h from the 2100 UTC 23 November 2004 as simulated by EXP1. The lines represent the pressure (hPa) and the colors represent the temperature (°C).

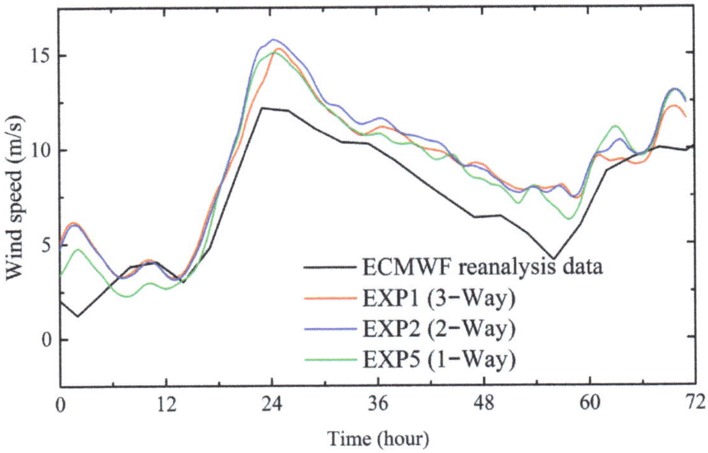

Figure 7. Comparison of wind speed from the 0000 UTC 23 November 2004 at station Huanghua. The red, blue, green, and black lines represent the EXP1, EXP2, EXP5, and ECMWF reanalysis data, respectively.

Unfortunately, in situ wind observations are unavailable, preventing direct validation. The simulated winds tend to exceed the reanalysis dataset. Previous work has shown that

WRF overestimates weak winds, whereas reanalysis data underestimates extremes [21]. The extreme value of predicted wind speed is around 16 m/s and the value extracted from ECMWF database is 12 m/s, and the difference is about 25%. It also indicates that the reanalysis data may significantly underestimate the extreme winds and the storm surge simulated based on reanalysis data may be underestimated.

Figure 8 compares 1-hourly wind vectors from the three-way model and ECMWF reanalysis data at Station Huanghua during the cold wave. As noted, the simulated winds are noticeably stronger than the winds extracted from reanalysis data. In addition, during the first several hours, the wind directions were west and the wind speeds were small. After that, the winds veered rapidly to the northeast and intensified. By ~0100 UTC 24 November, wind speeds peaked near 16 m/s. Subsequently, winds weakened and gradually shifted northwestward. The modeled wind directions exhibit a westward bias relative to reanalysis. Furthermore, after noon on 25 November, winds swung from north to west, with westerlies persisting ~20 h. These wind-field characteristics directly influence water-level simulations, as discussed in the next section.

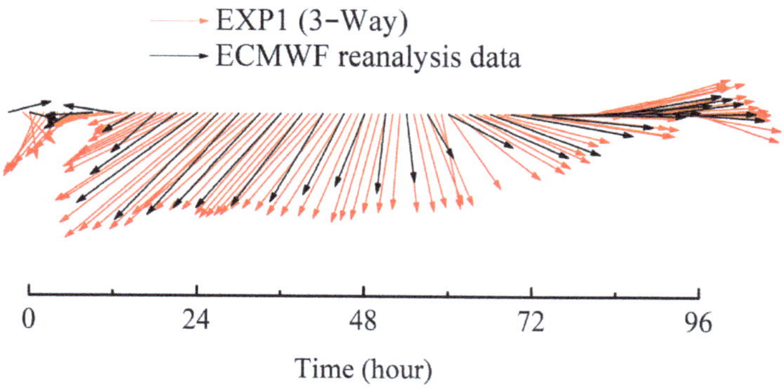

Figure 8. Time series of wind vector from the 0000 UTC 23 November 2004 at station Huanghua. The arrows represent direction and intensity of the 10 m wind.

4.2. Ocean Results

Storm surge denotes the anomalous sea level variation induced by severe meteorological disturbances. The measured and simulated water levels at Station Huanghua are compared in Figure 9. Before cold-air outbreak, sea level fluctuation primarily reflected the astronomical tide. During cold-air outbreak, the high water level within each tidal cycle rose progressively under strong winds, diverging from the astronomical tide. Maximum water elevation reached approximately 2.2 m during the high-tide event occurring at 0500 UTC on 24 November 2004. The surge peaked about four hours after the wind speed maximized, indicating a lag tied to tidal timing. Once the wind weakened and shifted direction, the elevated water levels receded rapidly. Additionally, after reaching peak water level, the high water level within a tidal cycle quickly decreased as the wind direction changed and intensity became weaker. In addition, the water level measured at Station Huanghua also showed a largely negative surge about 57 h after the peak surge arrived and each of the models can capture this negative surge accurately. The minimum water level reached nearly −2.8 m and the driving force is the sustained west wind, which lasted about 20 h (Figure 8). In Bohai Bay, extratropical storm surge is clearly governed by wind direction; offshore winds lower sea level, while onshore winds raise it.

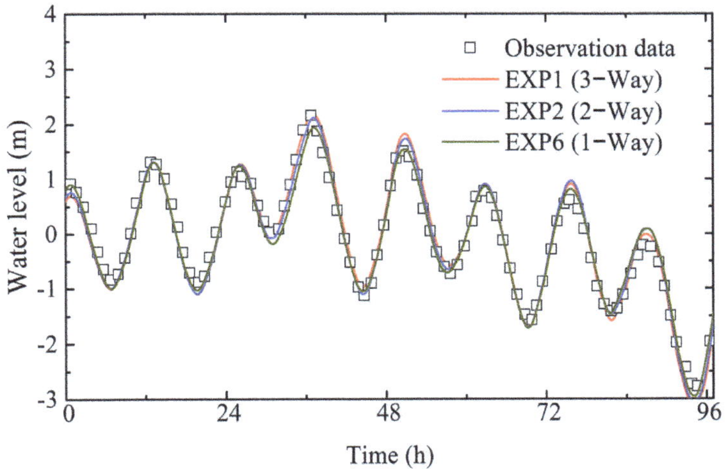

Figure 9. Comparison of water level from the 1600 UTC 22 November 2004 at station Huanghua. The red, blue, and dark yellow solid lines represent the EXP1, EXP2, and EXP6, respectively. The squares represent observed water levels at Stations Huanghua.

In the one-way case, good agreement is achieved except at extreme values, where simulated peaks are slightly lower than observations. A lag of approximately 0.8 h precedes the surge maximum, attributable to bulk-flux forcing by ERA-Interim, whose winds veer more southward than predicted (Figure 8). The significant improvements are obtained for peak values when atmosphere–ocean coupling between WRF and ROMS is activated (EXP1 and EXP2), demonstrating WRF's ability to reproduce the cold-air outbreak's strong wind field. Peak timing was well captured, and only marginal improvements were observed when waves were included, suggesting a limited impact of wind waves on Bohai Bay water levels for this event. Xia et al. [46] developed a coupled SWAN–ADCIRC model to investigate wind waves and storm tides during extratropical surges in the Bohai Sea, finding similarly small wave effects on water level.

For validation, Root Mean Square Error (RMSE) quantifies prediction error and Normalized Root Mean Square Error (NRMSE) normalizes it across scales, which are introduced here:

$$\text{RMSE} = \sqrt{\frac{\sum_{n=1}^{N}\left(\widehat{y}_n - y_n\right)^2}{N}} \tag{10}$$

$$\text{NRMSE} = \frac{\text{RMSE}}{y_{\max} - y_{\min}} \tag{11}$$

where N represents the number of observations. $\widehat{y}_n$ represents model simulation outputs, while y_n corresponds to actual observed measurements. $y_{\max}$ and $y_{\min}$ correspond to the upper and lower bounds of the dataset, representing the highest and lowest observed values, respectively.

Table 3 summarizes RMSE, NRMSE, and bias for all experiments. All configurations achieve NRMSE values below 5%. The best agreement is achieved by two-way coupling model (EXP2) and the model can preferably reproduce the dynamic change in the water level. The model obtains the smallest NRMSE (4.0%) and is slightly negatively biased around 18 cm. Although the three-way coupling model can accurately predict the maximum water level during the period of storm surge, the model obtains the highest RMSE (18 cm) and NRMSE (4.8%), alongside a positive bias (+0.025 m), indicating a tendency to overestimate water levels. The one-way ROMS run (EXP6) maintains a low global NRMSE (4.1%) but generally underestimates the extreme surge elevation (Figure 9).

Table 3. Statistics between experiments and observations of the water level from 1600 UTC 22 November 2004 to 1600 UTC 25 November 2004.

Experiment	Water Level (m)	
	RMSE (NRMSE)	Bias
EXP1	0.18 (4.8%)	0.025
EXP2	0.15 (4.0%)	−0.018
EXP6	0.15 (4.1%)	−0.064

As noted in the introduction, Bohai Bay is highly susceptible to rapid and pronounced sea-level fluctuations under extreme weather. Figure 10 illustrates the spatial distribution of total sea level obtained from the ocean model. Due to the strong eastward wind, vast volumes of seawater surged into Bohai Bay and the water level in this area increased sharply during storm surge, causing a sharp rise in water level (Figure 10a). Peak elevations exceeding 2 m occurred along the bay's western shore, whereas the northern Yellow Sea coast experienced a ~2 m depression. During the negative surge phase, westerly winds drove offshore currents from Bohai Bay, resulting in water accumulation along the northern Yellow Sea coast (Figure 10b). Wind strength and direction emerge as the primary controls on surge magnitude, with northeastern winds identified as the chief driver of extratropical storm surges in Bohai Bay, particularly along its western coasts.

Wind stress directly governs surface current velocities, as shown by the 12 h average outputs in Figure 11. Under initial surge conditions, mean surface speeds remained below 0.2 m/s in a chaotic flow regime (Figure 11a). Pre-peak surge conditions exhibited intensified inflow currents exceeding 0.5 m/s, particularly along the northeast coast, where velocities exceeded those in southern areas (Figure 11b). Post-peak, velocities declined rapidly, leaving residual higher flows concentrated nearshore (Figure 11c,d). Ahead of the pronounced negative surge, currents reversed out of Bohai Bay, again exceeding 0.5 m/s and driving increased southward flow (Figure 11e,f). Notably, strong westward currents appear during the early cold wave (Figure 11b), whereas eastward currents dominate during the negative surge (Figure 11f). These patterns underscore that average surface-current direction in Bohai Bay during a storm surge is firmly dictated by the prevailing wind direction (Figure 8).

4.3. Wave Results

One of direct consequence of sustained winds is the transfer of energy from the atmosphere to the ocean, producing surface waves. Owing to the sea-level rise induced by strong onshore wind, water depths in the nearshore zone increased during the extratropical storm surge. Consequently, bottom-boundary-layer energy dissipation diminished with greater depth, promoting storm waves of increased height and extended period. As wind-induced waves occupy a larger area, they influence the surge's hydrodynamic [47]. Moreover, wave growth modifies sea surface roughness, amplifying wind stress effects on the surge.

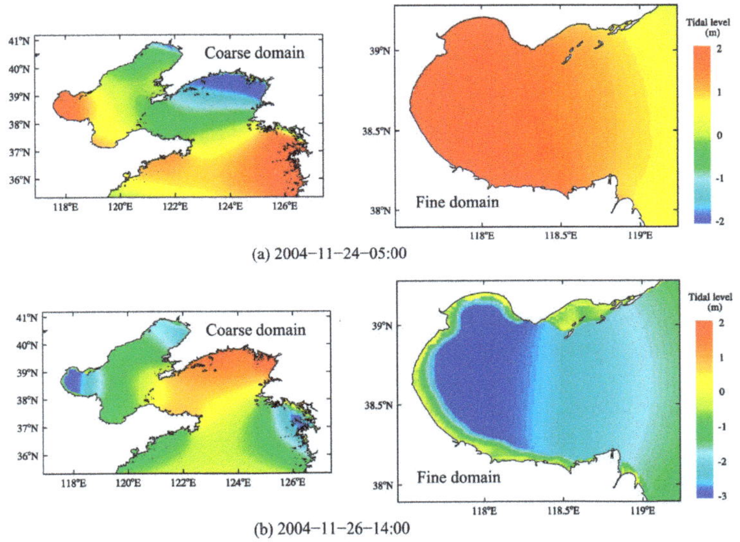

Figure 10. Oceanic snapshots for peak of water level during extratropical storm surges as simulated by EXP1. The color fields represent the water level.

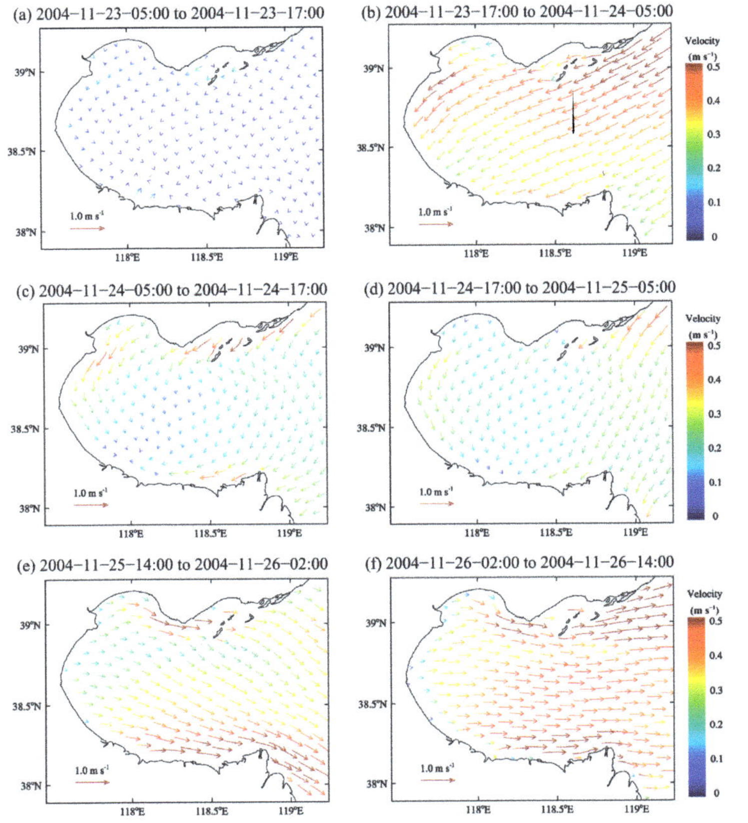

Figure 11. Average speed of the surface current from the 0500 UTC 23 November 2004 as simulated by EXP1. The arrows and colors represent the surface current direction and intensity.

The spatial distribution of significant wave heights (H_s) during storm surge for the fully coupled run (EXP1) is shown in Figure 12. The wave heights in the south were markedly stronger than those in the north. Due to the accumulation and propagation of energy and also because of the increase in alongshore water depths resulting from the high landward winds, which were beneficial to wave growth, the wave heights along the coasts of Bohai Bay increased. Compared with Figure 5, the significant wave heights rose with wind speed. At 0100 UTC on 24 November 2004, the maximum wave height exceeds 4 m in the central and eastern Bohai Bay. With the decrease in the wind speed, the wave height decreased correspondingly. The changes in wind velocities could lead to the changes in the significant wave heights and the results revealed a good correlation between them. In addition, the distribution of wave height is not entirely consistent with that of the wind speed, which may result from the difference in the seafloor topography.

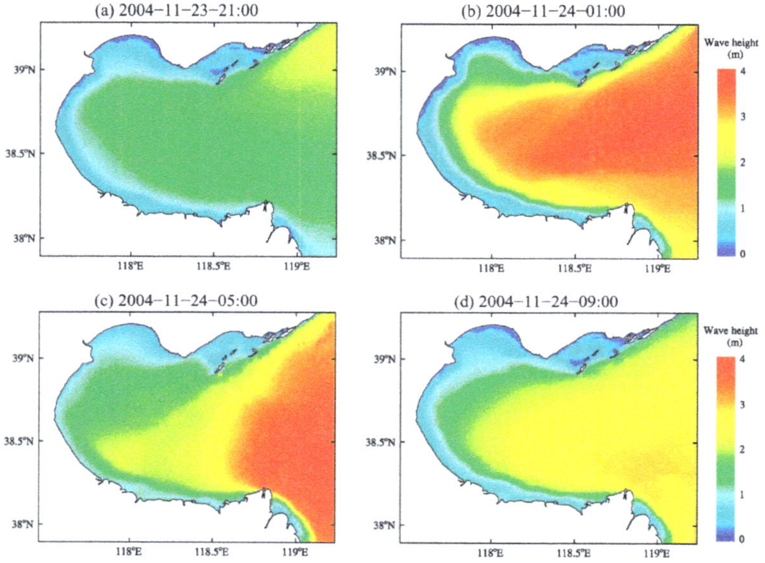

Figure 12. Surface wave snapshots every 4 h from the 2100 UTC 23 November 2004 as simulated by EXP1. The color fields represent the significant wave height.

To assess the influence of tides and storm surges on wind waves, the simulated results were compared with observations. Figure 13 compares modeled results and measured significant wave heights at Station Huanghua. Among the three configurations, the three-way coupling model achieved the closest agreement. It accurately captured the variability of H_s during the surge, resolving both high and low wave events. The maximum H_s peaked at 2.6 m at 0100 UTC on 24 November 2004, fell to 2.1 m at 0330 UTC on 24 November 2004, and then rose again to 2.5 m at 0630 UTC on 24 November 2004. Finally, the significant wave height dropped to 1.4 m at 1100 UTC on 24 November 2004. The three-way model can successfully depict this varied process, while the one- and two-way models fail. This demonstrates that the ocean–wave interaction is crucial to surface wave simulation and the three-way coupled model can well reflect the developing and changing process of storm wave. In addition, according to the growth process of wind waves (Figure 13), it can be seen that the duration of wave growth is short in the southwest coast of the Bohai Bay and the big waves are generated in a short period of time by strong landward winds.

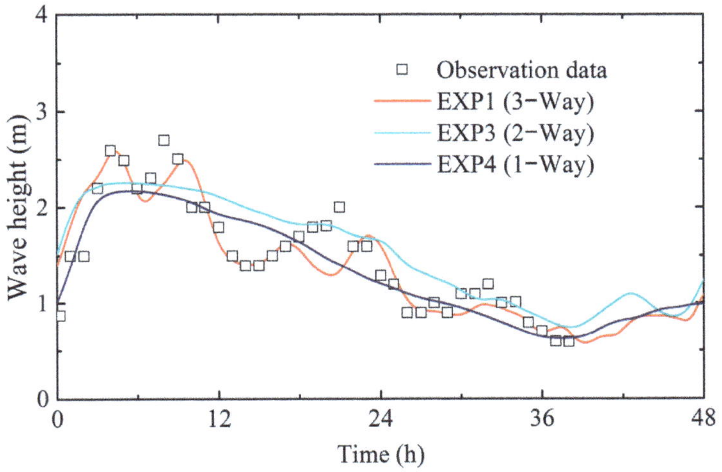

Figure 13. Time series of significant wave height from the 2100 UTC 23 November 2004 at Station Huanghua, showing 3-way (red solid line), 2-way (cyan solid line), 1-way (blue solid line), and in situ H_s (black box).

Comparing the two-way and SWAN-only runs (Figure 13) reveals that the one-way model consistently underestimates wave heights. This underestimation stems from the reanalysis wind speeds being lower than those in the coupled simulations (Figure 7). Relative to the two-way run, the three-way model's wave heights vary with tidal currents, producing more realistic patterns. Yin et al. [48] introduced a joint numerical of wave and tide-surge motion to examine the influence of interaction effects in the Bohai Sea. They reported an immediate 8% change in wave heights due to wave–tide-surge coupling. In this study, the mean interaction effect on wave height is approximately 10%, highlighting the importance of ocean–wave coupling for accurate storm wave simulation.

Table 4 summarizes statistics for significant wave height. All models reproduce wave heights effectively, with NRMSE values between 9.8% and 12.1%. The three-way coupling model (EXP1) achieves a low RMSE (26 cm, 9.6%) and an almost zero bias (1 cm). The one-way uncoupled model (EXP4) also shows globally low NRMSE (9.8%) but fails to capture H_s variability, resulting in slightly lower wave heights during the surge. These shortcomings persist in the two-way coupled model (EXP3), which yields the highest NRMSE of 12.1%. This model exhibits a consistent positive bias of ~0.17 m, indicating a systematic tendency of the two-way approach to overpredict wave heights.

Table 4. Statistics between experiments and observations of the significant wave height from 2100 UTC 23 November 2004 to 1100 UTC 25 November 2004.

Experiment	Significant Wave Height (m)	
	RMSE (NRMSE)	**Bias**
EXP1	0.26 (9.6%)	−0.01
EXP3	0.33 (12.1%)	0.17
EXP4	0.27 (9.8%)	−0.06

Figure 14 illustrates the spatial distribution of mean wave period from the 2100 UTC 24 November 2004 as simulated by EXP1. The mean period values in Bohai Bay were below 5 s, with longer periods in the south than that in the north and shorter intervals near the coast. In contrast to simulated results of winds (Figure 5), it is obvious that the east–northeast (ENE) winds were more helpful to the growth of wind wave than the northeast

(NE) winds for the central Bohai Bay. We can see that the significant wave heights were close to 4 m and the mean wave periods were over 5 s under ENE wind condition (Figure 12b or Figure 14b). Conversely, under NE winds, H_s stayed below 3 m and mean periods dropped below 4 s (Figure 12c or Figure 14c). Notably, the highest H_s and mean periods occurred just outside Bohai Bay's mouth.

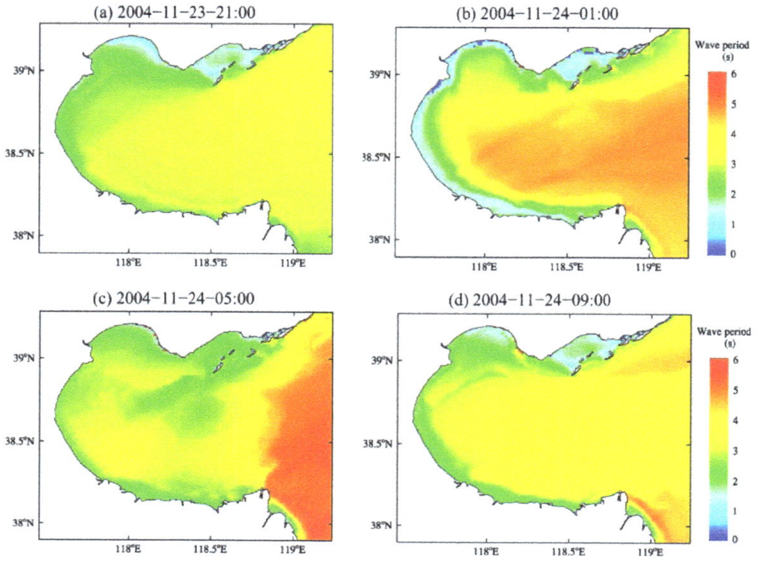

Figure 14. Surface wave snapshots every 4 h from the 2100 UTC 23 November 2004 as simulated by EXP1. The color fields represent the mean wave period.

Figure 15 presents mean wave period time series at Huanghua Station. The one- and two-way models overpredicted mean periods by approximately 30% compared to the three-way simulation. Observed values ranged from 1.5 to 4.5 s, displaying significant tidal-driven fluctuations. This reinforces that ocean–wave interaction must be incorporated for accurate surface-wave modeling during storm surges.

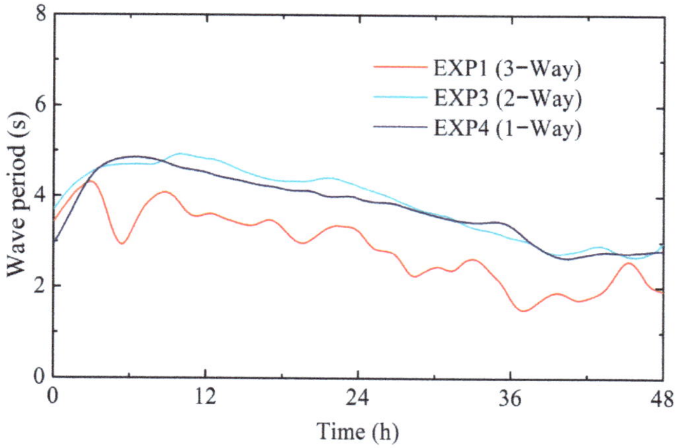

Figure 15. Time series of mean wave period from the 2100 UTC 23 November 2004 at Station Huanghua. The red, cyan, and blue solid lines represent the wind stress at Station Huanghua as simulated by EXP1, EXP3, and EXP4, respectively.

5. Summary and Conclusions

A three-way coupled atmosphere–ocean–wave model has been developed for Bohai Bay, integrating the WRF atmospheric model, ROMS oceanic model, and SWAN wave model via the MCT and SCRIP coupling libraries. The coupled model was applied to simulate an extratropical storm surge happened in November 2004. The dynamic process of the storm surge is analyzed. The coupling effects on the simulation results are investigated with six numerical experiments. Through comparative analysis, all model components are validated independently. The key conclusions are as follows:

- Minor differences in 10 m wind speed were observed between coupled and uncoupled runs, with sea surface temperature exerting a significant influence on wind simulation.
- During the cold-air outbreak, peak water-level timing was well reproduced when WRF–ROMS coupling was enabled, with the two-way configuration achieving the closest match. Although the three-way model accurately predicted maximum surge heights, it tended to overestimate water levels.
- A time lag of ~0.75 h was noted in the one-way ROMS run before the surge peak, and this configuration generally underestimated storm-surge magnitude. Inclusion of SWAN yielded small further improvements, indicating a limited impact of waves on water-level simulations in Bohai Bay for this event.
- Wind and tidal currents strongly modulated wave growth and, in turn, wave dynamics enhanced the effective wind stress driving the surge. The three-way coupling demonstrated the greatest improvement in wave prediction.

The coupled framework establishes a robust basis for integrating sediment transport modules. Future work will focus on sediment entrainment and transport patterns during storm surges using this multi-field coupling approach.

Author Contributions: Conceptualization: Y.L.; Methodology: Y.L., X.L. and J.L.; Investigation: J.L. and G.X.; Writing—original draft: Y.L., X.L. and J.L.; Writing—review and editing: Y.L., X.L. and G.X. All authors have read and agreed to the published version of the manuscript.

Funding: This research was funded by Qinhuangdao Science and Technology Plan Project (Nos. 202401A160 and 202401A092), Science Research Project of Hebei Education Department (No. QN2024047), Project of Hebei University of Environmental Engineering (No. GCZ202301), and Department of Water Resources of Hebei Province under the project "Development of a Tiered Early Warning System and Emergency Measures for Cyanobacterial Bloom Prevention and Control in Taolinkou Reservoir" (No. 2023-61).

Data Availability Statement: The original contributions presented in the study are included in the article; for any further inquiries, please email directly the corresponding author.

Conflicts of Interest: The authors declare that the research was conducted in the absence of any commercial or financial relationships that could be construed as a potential conflict of interest.

References

1. Li, Y.; Jiang, X.Y.; Tian, L.Z.; Li, J.F.; Wang, F. Storm surges on Bohai Sea. *China Min. Mag.* **2019**, *28*, 510–516.
2. Li, Y.; Wen, M.Z.; Yu, H.; Yang, P.; Wang, F.C.; Wang, F. Changes of coastline and tidal flat and its implication for ecological protection under human activities: Take China's Bohai Bay as an example. *China Geol.* **2024**, *7*, 26–35. [CrossRef]
3. Li, Y.Y.; Feng, H.; Vigouroux, G.; Yuan, D.K.; Zhang, G.Y.; Ma, X.D.; Lei, K. Storm surges in the Bohai Sea: The role of waves and tides. *Water* **2020**, *12*, 1509. [CrossRef]
4. Wang, F.; Liu, X.Z.; Li, Y.; Yu, H.; Wen, M.Z.; Hu, Y.Z. Risk assessment of coastal flooding disaster by storm surge based on Elevation-Area method and hydrodynamic models: Taking Bohai Bay as an example. *China Geol.* **2024**, *7*, 494–504. [CrossRef]
5. Warner, J.C.; Armstrong, B.; He, R.; Zambon, J.B. Development of a coupled ocean-atmosphere-wave-sediment transport (COAWST) modeling system. *Ocean Model.* **2010**, *35*, 230–244. [CrossRef]

6. Li, D.M.; Fu, Q.J.; Xie, Y.Y.; Bai, L. Numerical model of storm surge and inundation in Bohai Bay. *Trans. Tianjin Univ.* **2011**, *17*, 57–61. [CrossRef]
7. Antony, C.; Testut, L.; Unnikrishnan, A.S. Observing storm surges in the Bay of Bengal from satellite altimetry. *Estuar. Coast. Shelf Sci.* **2014**, *151*, 131–140. [CrossRef]
8. Liu, Q.; Ruan, C.; Zhong, S.; Li, J.; Yin, Z.; Lian, X. Risk assessment of storm surge disaster based on numerical models and remote sensing. *Int. J. Appl. Earth Obs. Geoinf.* **2018**, *68*, 20–30. [CrossRef]
9. Chao, W.T.; Young, C.C. Accurate Storm Surge Prediction with a Parametric Cyclone and Neural Network Hybrid Model. *Water* **2022**, *14*, 96. [CrossRef]
10. Agulles, M.; Marcos, M.; Amores, A.; Toomey, T. Storm surge modelling along European coastlines: The effect of the spatio-temporal resolution of the atmospheric forcing. *Ocean Model.* **2024**, *192*, 102432. [CrossRef]
11. Elahi, M.W.E.; Wang, X.H.; Ritchie, E.A. Cyclone-induced storm surge flooding in the Ganges-Brahmaputra-Meghna delta under different mean-sea level rise scenarios. *Ocean Dyn.* **2025**, *75*, 23. [CrossRef]
12. Zhao, P.; Jiang, W.S. A numerical study of the effects of coastal geometry in the Bohai Sea on storm surges induced by cold-air outbreaks. *J. Ocean Univ. China (Ocean. Coast. Sea Res.)* **2011**, *10*, 9–15. [CrossRef]
13. Mo, D.; Hou, Y.J.; Li, J.; Liu, Y.H. Study on the storm surges induced by cold waves in the Northern East China Sea. *J. Mar. Syst.* **2016**, *160*, 26–39. [CrossRef]
14. Xu, J.L.; Zhang, Y.H.; Cao, A.Z.; Liu, Q.; Lv, X.Q. Effects of tide-surge interactions on storm surges along the coast of the Bohai Sea, Yellow Sea, and East China Sea. *Sci. China Earth Sci.* **2016**, *59*, 1308–1316. [CrossRef]
15. Ding, Y.M.; Ding, L. A numerical simulation of extratropical storm surge and hydrodynamic response in the Bohai Sea. *Discret. Dyn. Nat. Soc.* **2014**, *1*, 282085. [CrossRef]
16. Ding, Y.M.; Wei, H. Modeling the impact of land reclamation on storm surges in Bohai Sea, China. *Nat. Hazards* **2017**, *85*, 1–15. [CrossRef]
17. Li, Y.; Tian, L.Z.; Pei, Y.D.; Wang, F.; Wang, H. Numerical simulation of storm surge inundation in the west zone of Bohai Bay. *Geol. Bull. China* **2016**, *35*, 1638–1645.
18. Zambon, J.B.; He, R.; Warner, J.C. Investigation of hurricane Ivan using the coupled ocean–atmosphere–wave–sediment transport (COAWST) model. *Ocean Dyn.* **2014**, *64*, 1535–1554. [CrossRef]
19. Olabarrieta, M.; Warner, J.C.; Armstrong, B.; He, R.; Zambon, J.B. Ocean–atmosphere dynamics during Hurricane Ida and Nor'Ida: An application of the coupled ocean–atmosphere–wave–sediment transport (COAWST) modeling system. *Ocean Model.* **2012**, *43–44*, 112–137. [CrossRef]
20. Wu, Z.; Chen, J.; Jiang, C.; Liu, X.; Xie, Z. Numerical investigation of typhoon kai-tak (1213) using a mesoscale coupled wrf-roms model—Part II: Wave effects. *Ocean Eng.* **2020**, *196*, 106805. [CrossRef]
21. Li, Y.; Chen, X.; Tian, L.Z.; Li, J.F. Numerical simulation of the storm surge in Bohai Bay based on a coupled ocean-atmosphere model. *J. Shanghai Jiaotong Univ.* **2017**, *51*, 1512–1519.
22. Mo, D.; Hu, P.; Li, J.; Hou, Y.; Li, S. Effect of wave-dependent mechanisms on storm surge and current simulation during three extreme weather systems. *J. Phys. Oceanogr.* **2024**, *54*, 25. [CrossRef]
23. Li, Y.; Chen, X.; Jiang, X.; Li, J.; Tian, L. Numerical simulations and comparative analysis for two types of storm surges in the Bohai Sea using a coupled atmosphere-ocean model. *Acta Oceanol. Sin.* **2019**, *38*, 35–47. [CrossRef]
24. Ghanavati, M.; Young, I.R.; Kirezci, E.; Liu, J. The impact of long-term changes in ocean waves and storm surge on coastal shorelline change: A case study of Bass Strait and south-east Australia. *Nat. Hazards Earth Syst. Sci.* **2024**, *24*, 2175–2190. [CrossRef]
25. Kwun, J.H.; Kin, Y.K.; Seo, J.W.; Jeong, J.H.; You, S.H. Sensitivity of MM5 and WRF mesoscale model predictions of surface winds in a typhoon to planetary boundary layer parameterizations. *Nat. Hazards* **2009**, *51*, 63–77. [CrossRef]
26. Myksvoll, M.S.; Sandvik, A.D.; Skarðhamar, J.; Sundby, S. Importance of high resolution wind forcing on eddy activity and particle dispersion in a Norwegian fjord. *Estuar. Coast. Shelf Sci.* **2012**, *113*, 293–304. [CrossRef]
27. Shiferaw, A.; Tadesse, T.; Rowe, C.; Oglesby, R. Weather research and forecasting model (wrf) sensitivity to choice of parameterization options over ethiopia. *Atmosphere* **2024**, *15*, 974. [CrossRef]
28. Drennan, W.M.; Taylor, P.K.; Yelland, M.J. Parameterizing the sea surface roughness. *J. Phys. Oceanogr.* **2005**, *35*, 835–848. [CrossRef]
29. Li, X.; Shen, D.; Zheng, G.; Zhou, L.; Liu, A. Comprehensive satellite observations and a numerical study of a wintertime shallow sea smoke event in the yellow sea. *J. Atmos. Sci.* **2022**, *79*, 3163–3179. [CrossRef]
30. García-Sánchez, G.; Mancho, A.M.; Ramos, A.G.; Coca, J.; Jiménez Madrid, J.A. Dynamical systems for remote validation of very high-resolution ocean models. *Nonlinear Dyn.* **2024**, *112*, 8653–8673. [CrossRef]
31. Haidvogel, D.B.; Arango, H.; Budgell, W.P.; Cornuelle, B.D.; Curchitser, E.; Di Lorenzo, E.; Fennel, K.; Geyer, W.R.; Hermann, A.J.; Lanerolle, L.; et al. Ocean forecasting in terrain-following coordinates: Formulation and skill assessment of the regional ocean modeling system. *J. Comput. Phys.* **2008**, *227*, 3595–3624. [CrossRef]

32. Hong, X.; Ji, Y.; Yan, J.; Cheng, J.; Wu, S.; Li, Q. Study on the wave-dissipation effect of oyster reefs based on the swan numerical model. *Water* **2023**, *15*, 2884. [CrossRef]
33. Zhang, W.; Zhao, H.; Chen, G.; Yang, J. Assessing the performance of swan model for wave simulations in the bay of bengal. *Ocean Eng.* **2023**, *285*, 115295. [CrossRef]
34. Warner, J.C.; Perlin, N.; Skyllingstad, E. Using the model coupling toolkit to couple earth system models. *Environ. Model. Softw.* **2008**, *23*, 1240–1249. [CrossRef]
35. Larson, J.; Jacob, R.; Ong, E. The model coupling toolkit: A new fortran90 toolkit for building multiphysics parallel coupled models. *Int. J. High Perform. Comput. Appl.* **2004**, *19*, 277–292. [CrossRef]
36. Jones, P.W. *A User Guide for Scrip: A Spherical Coordinate Remapping and Interpolation Package*, Version 1.4; Los Alamos National Laboratory: Los Alamos, NM, USA, 1998.
37. Lin, Y.L.; Farley, R.D.; Orville, H.D. Bulk parameterization of the snow field in a cloud model. *J. Clim. Appl. Meteorol.* **1983**, *22*, 1065–1092. [CrossRef]
38. Chen, F.; Dudhia, J. Coupling an advanced land surface-hydrology model with the Penn State-NCAR MM5 modeling system. Part I: Model implementation and sensitivity. *Mon. Weather Rev.* **2001**, *129*, 569–585. [CrossRef]
39. Mlawer, E.J.; Taubman, S.J.; Brown, P.D.; Iacono, M.J.; Clough, S.A. Radiative transfer for inhomogeneous atmosphere: RRTM, a validated correlated-k model for the longwave. *J. Geophys. Res.* **1997**, *102*, 16663–16682. [CrossRef]
40. Dudhia, J. Numerical study of convection observed during the winter monsoon experiment using a mesoscale two-dimensional model. *J. Atmos. Sci.* **1989**, *46*, 3077–3107. [CrossRef]
41. Hong, S.Y.; Noh, Y.; Dudhia, J. A new vertical diffusion package with an explicit treatment of entrainment processes. *Mon. Weather Rev.* **2006**, *134*, 2318–2341. [CrossRef]
42. Kain, J.S.; Fritsch, J.M. Convective parameterization in mesoscale models: The Kain-Fritsch scheme. *Meteorol. Monogr.* **1993**, *24*, 165–170.
43. Skamarock, W.C.; Klemp, J.B.; Dudhia, J.; Gill, D.O.; Barker, D.M.; Duda, M.G.; Huang, X.Y.; Wang, W.; Powers, J.G. *A Description of the Advanced Research WRF Version 3*, NCAR Technical Note, NCAR/TN-475+STR; National Center for Atmospheric Research: Boulder, CO, USA, 2008.
44. Choi, B.H.; Kim, K.O.; Eum, H.M. Digital bathymetric and topographic data for neighboring seas of Korea. *J. Korean Soc. Coast. Ocean Eng.* **2002**, *14*, 41–50.
45. Komen, G.J.; Hasselmann, S.; Hasselmann, K. On the existence of a fully developed wind-sea spectrum. *J. Phys. Oceanogr.* **1984**, *14*, 1271–1285. [CrossRef]
46. Xia, B.; Zhang, Q.H.; Jiang, C.B. A coupled wave-current numerical model on unstructured grids. *Oceanol. Et Limnol. Sin.* **2013**, *44*, 1451–1456.
47. Peng, Z.; Wang, P. Surface Wave Effects on Storm Surge: A Case Study of Typhoon Doksuri (2023). *J. Mar. Sci. Eng.* **2025**, *13*, 478. [CrossRef]
48. Yin, B.S.; Wang, T.; Hou, Y.J.; Cheng, M.H.; Fan, S.T.; Su, J.Z.; El-Sabh, M.I. Numerical study of the influence of waves and tide-surge interaction on waves in Bohai Sea. *Oceanol. Limnol. Sin.* **2001**, *32*, 109–116.

water

MDPI

Article

Monitoring Regional Terrestrial Water Storage Variations Using GNSS Data

Dejian Wu [1], Jian Qin [2,3,*] and Hao Chen [4,*]

1 Guangzhou Salvage, Guangzhou 510290, China; dejianwu@163.com
2 Tianjin Water Transport Engineering Research Institute Co., Ltd., Tianjin 300456, China
3 Tianjin Key Laboratory of Water Transport Engineering Surveying and Mapping Technology for Enterprises, Tianjin 300456, China
4 School of Civil and Environmental Engineering, Nanyang Technological University, 50 Nanyang Ave, Singapore 639798, Singapore
* Correspondence: qin_jian1@126.com (J.Q.); chen.hao@ntu.edu.sg (H.C.)

Abstract

Accurately monitoring terrestrial water storage (TWS) variations is essential due to global climate change and growing water demands. This study investigates TWS changes in Oregon, USA, using Global Navigation Satellite System (GNSS) data from the Nevada Geodetic Laboratory, Gravity Recovery and Climate Experiment (GRACE) level-3 mascon data from the Jet Propulsion Laboratory (JPL), and Noah model data from the Global Land Data Assimilation System (GLDAS) data. The results show that the GNSS inversion offers superior spatial resolution, clearly capturing a water storage gradient from 300 mm in the Cascades to 20 mm in the basin and accurately distinguishing between mountainous and basin areas. However, the GRACE data exhibit blurred spatial variability, with the equivalent water height amplitude ranging from approximately 100 mm to 145 mm across the study area, making it difficult to resolve terrestrial water storage gradients. Moreover, GLDAS exhibits limitations in mountainous regions. The GNSS can provide continuous dynamic monitoring, with results aligning well with seasonal trends seen in GRACE and GLDAS data, although with a 1–2 months phase lag compared to the precipitation data, reflecting hydrological complexity. Future work may incorporate geological constraints, region-specific elastic models, and regularization strategies to improve monitoring accuracy. This study demonstrates the strong potential of GNSS technology for monitoring TWS dynamics and supporting environmental assessment, disaster warning, and water resource management.

Keywords: GNSS; terrestrial water storage; GRACE; GLDAS; water resource management

1. Introduction

Terrestrial water storage includes surface water such as rivers and lakes, soil water, groundwater, etc. [1]. Against the backdrop of ongoing global climate warming, frequent occurrences of high temperatures and droughts have led to insufficient supplies of surface water and groundwater [2,3]. Meanwhile, the increasing demand for groundwater resources driven by agricultural and industrial development, as well as population growth, has resulted in the overexploitation of groundwater, further aggravating the water resource crisis [4–6]. Therefore, accurately quantifying changes in terrestrial water storage is of great significance for monitoring the health and security of both human and natural ecosystems, and can serve as a basis for local government to formulate water resource management policies [7,8].

The Gravity Recovery and Climate Experiment (GRACE), jointly developed by the United States and Germany, was successfully launched in 2002, thus inaugurating a new era of gravitational field measurement [7]. Since then, the method of estimating terrestrial water storage changes using GRACE gravity satellites has been widely applied in related research worldwide [1,9]. However, a major limitation of GRACE data lies in its spatial resolution of 300–500 km and monthly-scale temporal resolution, which restricts its ability to monitor finer-scale mass changes [10,11]. The follow-up satellite GRACE Follow-On (GFO) led to nearly a year of data gaps during its transition with the GRACE mission, necessitating alternative methods to fill this data void.

The Global Navigation Satellite System (GNSS) plays a critical role in geodetic and geodynamic research [12,13]. Across most regions of the globe, GNSS records primarily reflect the elastic response of the Earth's crust to surface loads such as snow and precipitation. During autumn and winter, increased surface loading from rainfall and snow causes downward elastic deformation of the crust. During summer, as snow melts and water evaporates, the reduced surface load leads to an upward elastic rebound of the crust [14,15]. Therefore, surface deformation caused by hydrological loading monitored via GNSS can be converted into changes in terrestrial water storage [16–19].

Recent advances in hydrogeodesy have demonstrated that GNSS-based observations can serve as a powerful complementary tool in conjunction with GRACE/GRACE-FO in resolving terrestrial water storage (TWS) variations, especially at regional scales where GRACE's coarse resolution is insufficient. Several studies have validated that GNSS vertical displacements, after correcting for non-hydrological signals (e.g., atmospheric, oceanic, thermal effects), can capture seasonal and interannual variations in hydrological loading, offering independent estimates of TWS anomalies [17,18]. For instance, time series decomposition and machine learning techniques such as PCA and VMD-BiLSTM have been applied to GNSS data to extract TWS signals in regions like the Amazon, yielding data in good agreement with that of GRACE, while providing finer spatial detail [20]. Moreover, GNSS-based methods have been successfully applied to quantify groundwater loss in areas like California and the North China Plain [21,22], highlighting their capability for drought monitoring and groundwater management. Sensitivity analyses have shown that the choice of Earth structure models significantly affects TWS estimates from GNSS data—especially at spatial scales <10 km—necessitating careful selection and validation using auxiliary datasets like GRACE [23]. In parallel, GNSS-reflectometry (GNSS-R) has emerged as a valuable approach for retrieving surface parameters like soil moisture, leveraging platforms ranging from ground-based to satellite-borne missions such as CYGNSS, with enhanced accuracy through deep learning integration [24,25]. These developments underscore the growing role of GNSS technologies in water cycle research, enabling higher-resolution and more timely assessments of terrestrial water dynamics.

In areas with a dense distribution of GNSS stations, retrieving water storage changes from GNSS data offers significant advantages, providing monitoring products of terrestrial water storage changes at a higher spatial resolution than that of GRACE [17,18]. Numerous studies have confirmed that GNSS can serve as an independent data source for monitoring terrestrial water storage changes, with related methods achieving significant results [17,18,20,23]. For example, research using GNSS data estimated that groundwater depletion in California's Central Valley occurred at a rate of 2.2 ± 0.7 km^3/year between 2006 and 2021, with a cumulative loss of 45 ± 21 km^3 of water during the severe drought period from October 2011 to October 2015 [22].

Given the coarse spatial resolution of GRACE/GFO data, which limits their ability to resolve the spatial distribution of terrestrial water storage in small regions, we focus on using densely distributed GNSS data to estimate the spatial distribution of regional terrestrial water storage changes, analyze their spatial variation characteristics, and compare them with multi-source satellite data.

2. Data Processing and Methods

2.1. GNSS Data and Processing Strategies

The GNSS data used in this study are provided by the Nevada Geodetic Laboratory (NGL), with daily resolution and alignment with the IGS14 frame [12]. The raw data were processed using the GIPSY/OASIS software (Version 6.0) developed by the Jet Propulsion Laboratory (JPL). JPL first determines the orbits of 24 satellites and the positions of 80 ground stations. During processing, corrections were applied for ionospheric delays, tropospheric delays, and antenna phase center deviations. Solid Earth tides, polar tides, and ocean tides were corrected according to the standards of the International Earth Rotation and Reference Systems Service (IERS), but non-tidal loading effects, such as atmospheric, oceanic, and hydrological loading, were not considered [26,27]. To accurately extract surface deformations caused by terrestrial water storage changes, we processed the GNSS data through three steps: (a) using the earthquake occurrence times provided by NGL, the TSAnalyzer tool (v3.2.5.0) was employed to correct step jumps in the GNSS time series caused by antenna replacements or seismic events [28]; (b) a 3σ (three-sigma) method was used to detect and remove gross errors in the time series; (c) environmental loading products released by the German Geoscience Research Center (GFZ) were applied to correct non-tidal atmospheric and oceanic loading displacements in the GNSS observation time series, ensuring that the remaining signals exclusively reflect deformations caused by terrestrial water storage changes.

For each GNSS time series y(t), we model it using the following formula, which includes an intercept term a, a rate term b, and seasonal annual and semi-annual sinusoidal terms, as follows:

$$y(t) = a + b \cdot t + \Sigma\ (k = 1 \text{ to } 2)\ [C_k \cdot \cos(2\pi k \cdot t) + S_k \cdot \sin(2\pi k \cdot t)] \quad (1)$$

where time t is expressed in integer years, and the annual amplitude, *Amp*, is calculated as

$$\text{Amp} = \text{sqrt}\ (C1^2 + S1^2) \quad (2)$$

2.2. GRACE Data

In this study, the GRACE data product used is the JPL level-3 mascon (mass concentration) product, which divides the Earth into numerous discrete blocks and assumes uniform mass distribution within each block. The mass change of each block is expressed in equivalent water height. This product has been corrected for glacial isostatic adjustment (GIA), non-tidal atmospheric and oceanic effects, ocean tides, and high-frequency solid tide signals [29], with low-degree terms replaced [30–32], and it can be used directly. Compared with traditional spherical harmonic solutions, the mascon product is less affected by gravitational signal leakage outside specific regions and applies spatial and temporal regularization constraints to reduce correlated errors (i.e., north–south striping artifacts) [26].

2.3. Hydrological Model Data

The changes in terrestrial water storage can also be quantitatively estimated using hydrological models. This study uses the Noah hydrological model, released by the

Global Land Data Assimilation System (GLDAS). The model integrates both ground-based and satellite observational data products, including components such as surface water, soil moisture, canopy water, and snow water equivalent. It generates optimal fields of surface states and fluxes through advanced techniques in surface modeling and data assimilation. The GLDAS hydrological model can not only be used to study the evolution of terrestrial hydrological loading but can also be applied in climate simulation, meteorological forecasting, and assessing water resource sustainability. The GLDAS monthly surface hydrological model used here has a spatial resolution of $0.25° \times 0.25°$. Data for the GLDAS shallow surface hydrological model can be downloaded at https://disc.gsfc.nasa.gov/datasets?keywords=GLDAS (accessed on 20 May 2025).

2.4. *TRMM_3B43 Rainfall Data*

Terrestrial water storage is closely related to precipitation, and we also collected TRMM_3B43 rainfall data for empirical validation. The TRMM (Tropical Rainfall Measuring Mission) rainfall data are global precipitation datasets obtained from a joint satellite remote sensing mission conducted by NASA (National Aeronautics and Space Administration) and JAXA (Japan Aerospace Exploration Agency). Launched in 1997, the primary objective of the TRMM satellite mission was to observe precipitation in tropical regions and provide critical data for studying global precipitation patterns and monitoring climate change. The TRMM_3B43 rainfall data are collected by instruments such as the satellite's microwave imager (TMI) and radar system (PR), covering regions near the equator (approximately $\pm 50°$ latitude). The satellite scans the Earth's surface every few hours to generate precipitation data. This data product is released in a gridded format, with a spatial resolution of $0.25° \times 0.25°$ and a monthly temporal resolution spanning from 1998 to 2019, capable of providing large-scale, long-term precipitation information.

2.5. *Method for Inverting Terrestrial Water Storage by GNSS*

There is a relatively high density distribution of GNSS stations in the state of Oregon, allowing us to invert mass changes with higher spatial resolution. The study area (latitude: 41.5°–46.5° N, longitude: 125°–116° W) is subdivided into $0.25° \times 0.25°$ grids. Using the vertical displacements from all GNSS stations within the study area as observations, the well-known Green's function is used to relate surface displacements to changes in surface mass [33]. For point load, the resulting vertical displacement observed by GNSS can be expressed as

$$u = G \cdot m + \delta\_error \tag{3}$$

where u is the vertical displacement (m), m is the load mass (kg), G is the Green's function related to the angular distance between the point load and the GNSS station, and δ_error is the observation error. Typically, we define an objective function for minimizing errors to solve the problem, which is expressed as

$$\min\{||G \cdot m - u||^2\} \tag{4}$$

However, since the number of unknown parameters for mass changes far exceeds the number of GNSS observations, the Green's function coefficient matrix G is rank-deficient and cannot be directly solved. Therefore, additional constraints are required to invert the mass change parameters. Here, the Laplace operator L is used as an additional constraint to smooth the solution, and the Tikhonov regularization method is employed for solving. The objective function then becomes

$$\min\{||G \cdot m - u||^2 + k^2 \cdot ||L(m)||^2\} \tag{5}$$

after considering the Laplace constraint, the solution for mass, m, is

$$m = (G^t \cdot P \cdot G + k \cdot L^t \cdot L)^{-1} \cdot G^t \cdot P \cdot u \quad (6)$$

where P is the weight matrix of GNSS observations, and u and k are unknown non-negative parameters representing the roughness factor, which quantifies the difference between adjacent pixel values. There are multiple methods to determine k, including principal component estimation [34], Bayesian information criterion, and the L-curve method [35]. Here, we use generalized cross validation (GCV) [36] to determine the roughness factor, which operates by solving for the variable that minimizes the GCV function. The Laplace operator L has a template form, as follows:

$$L_1 = [\ [0, 1, 0], [1, -4, 1], [0, 1, 0]\ ] \quad (7)$$

However, the L_1 cannot smooth the edges effectively. To solve this problem, this paper introduces two templates, namely, the L_2 template and the L_3 template:

$$L_2 = [1; -2; 1], L_3 = [1, -2, 1] \quad (8)$$

3. Results and Analysis of Land Water Storage Inversion by GNSS

Using the GNSS vertical position time series, we inverted three types of terrestrial water storage changes in Oregon: (1) checkerboard test simulations, (2) terrestrial water storage changes over time, and (3) the spatial distribution of multi-year average seasonal terrestrial water storage.

3.1. Simulation Test

Since the inversion of terrestrial water storage using GNSS is influenced by the locations of the GNSS stations, we conducted a checkerboard test to verify the sensitivity and robustness of our inversion method. Figure 1 show the input model and inversion results of the checkerboard test, respectively, to evaluate the spatial resolution and robustness of the inversion method. Figure 1a shows the synthetic input model (alternating positive and negative TWS anomalies), and Figure 1b shows the inversion results based on simulated GNSS station distribution. The test demonstrates that with the dense distribution of GNSS stations in Oregon, the input model for the region can be well recovered, confirming the sensitivity of the method. However, due to the constraints of the Laplace operator, the inversion results are smoother than those of the true input model.

3.2. Land Water Storage Changes over Time

To investigate terrestrial water storage changes at higher temporal resolutions, we used detrended GNSS vertical time series to invert the spatiotemporal water load changes in Oregon. Terrestrial water storage changes in Oregon from 2015 to 2020 were calculated using Equation (4). With the aid of the high temporal resolution of GNSS, we obtained daily terrestrial water storage changes (Figure 2). The long-term trend reveals strong interannual periodic variations in Oregon. We also show the time-varying regional average equivalent water height (EWH) time series derived using two independent techniques: GRACE and the GLDAS hydrological model. These three EWH time series exhibit good consistency. Additionally, we present the average precipitation change curve for the state, clearly demonstrating a strong correlation between the three terrestrial water storage solutions and precipitation. The TWS typically reaches its maximum, around 150 mm equivalent water height, between January and April each year, and its minimum, around −150 mm equivalent water height, between July and October.

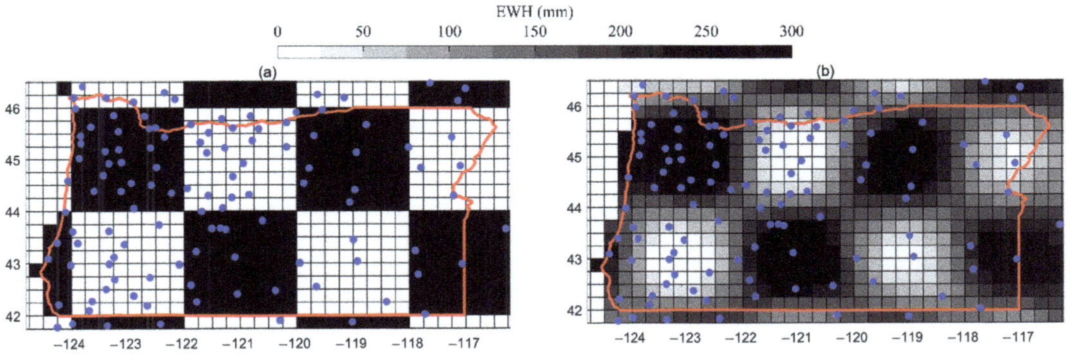

Figure 1. Checkerboard test in Oregon: (**a**) synthetic input model (alternating positive and negative TWS anomalies); (**b**) inversion results based on simulated GNSS station distribution. The red line represents the boundary of the study region.

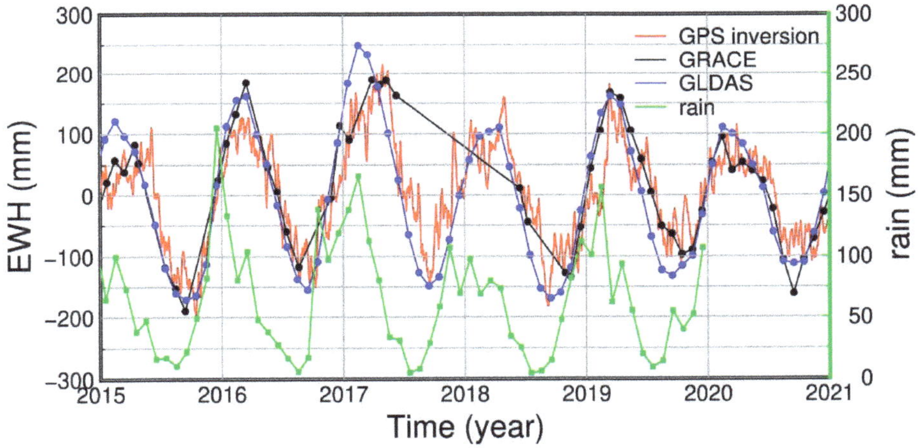

Figure 2. Changes in terrestrial water storage.

To further investigate the spatiotemporal patterns of terrestrial water storage (TWS) in the region, we also present maps of monthly mean TWS distributions. As shown in Figures 3–8, TWS in Oregon reaches its annual minimum during summer, with values as low as −300 mm equivalent water height. This is primarily due to high temperatures causing snowmelt and soil moisture evaporation. In winter, TWS reaches its annual maximum of up to a 300 mm equivalent water height, driven by rainfall and snowfall, which aligns with the known hydrological conditions of the region. Moreover, both GRACE and GLDAS data show that the TWS variations inverted from GNSS are in good agreement with observational and model values. This indicates that the inversion accuracy meets the requirements for TWS monitoring in the region, and that GNSS vertical displacements effectively capture the elastic response of the solid Earth to seasonal terrestrial water variations.

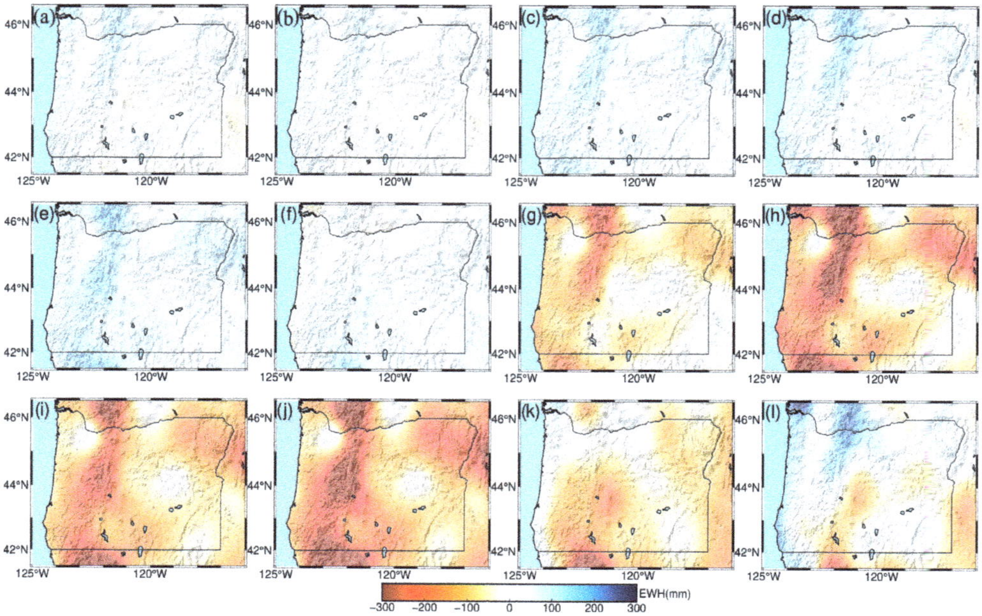

Figure 3. Monthly spatial and temporal terrestrial water storage changes in Oregon for the year 2015. (**a**) January; (**b**) February; (**c**) March; (**d**) April; (**e**) May; (**f**) June; (**g**) July; (**h**) August; (**i**) September; (**j**) October; (**k**) November; (**l**) December.

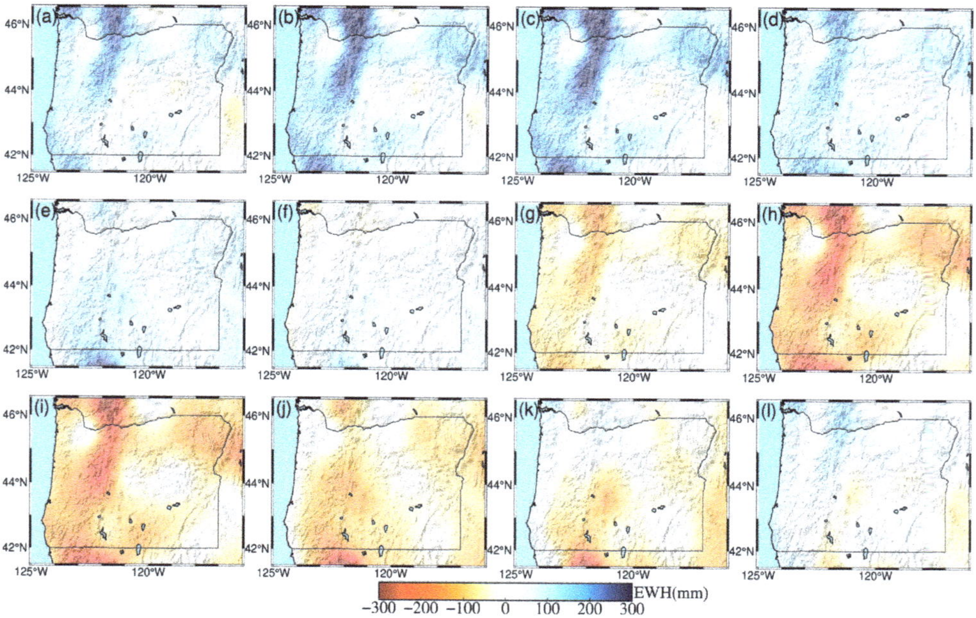

Figure 4. Monthly spatial and temporal terrestrial water storage changes in Oregon for the year 2016. (**a**) January; (**b**) February; (**c**) March; (**d**) April; (**e**) May; (**f**) June; (**g**) July; (**h**) August; (**i**) September; (**j**) October; (**k**) November; (**l**) December.

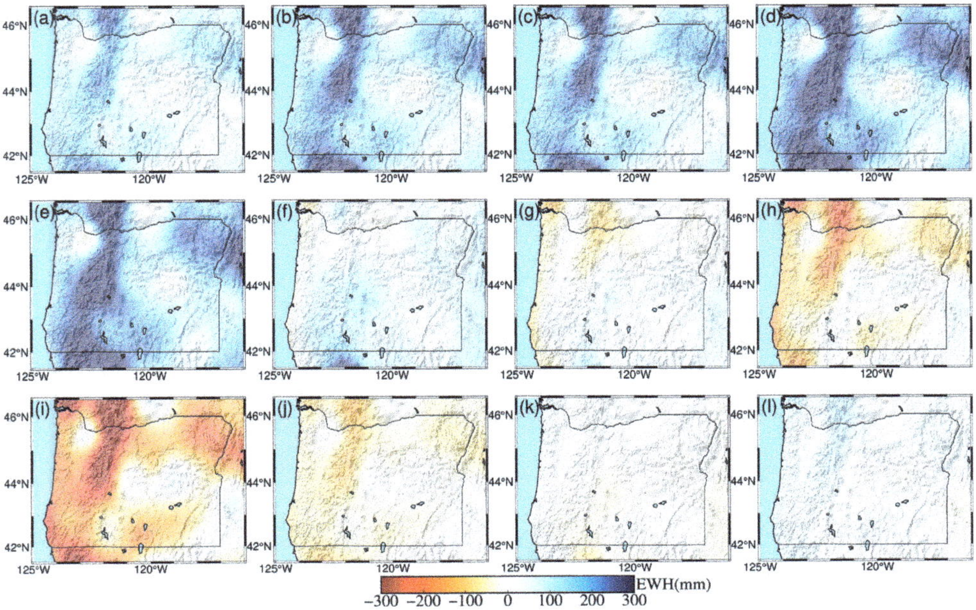

Figure 5. Monthly spatial and temporal terrestrial water storage changes in Oregon for the year 2017. (**a**) January; (**b**) February; (**c**) March; (**d**) April; (**e**) May; (**f**) June; (**g**) July; (**h**) August; (**i**) September; (**j**) October; (**k**) November; (**l**) December.

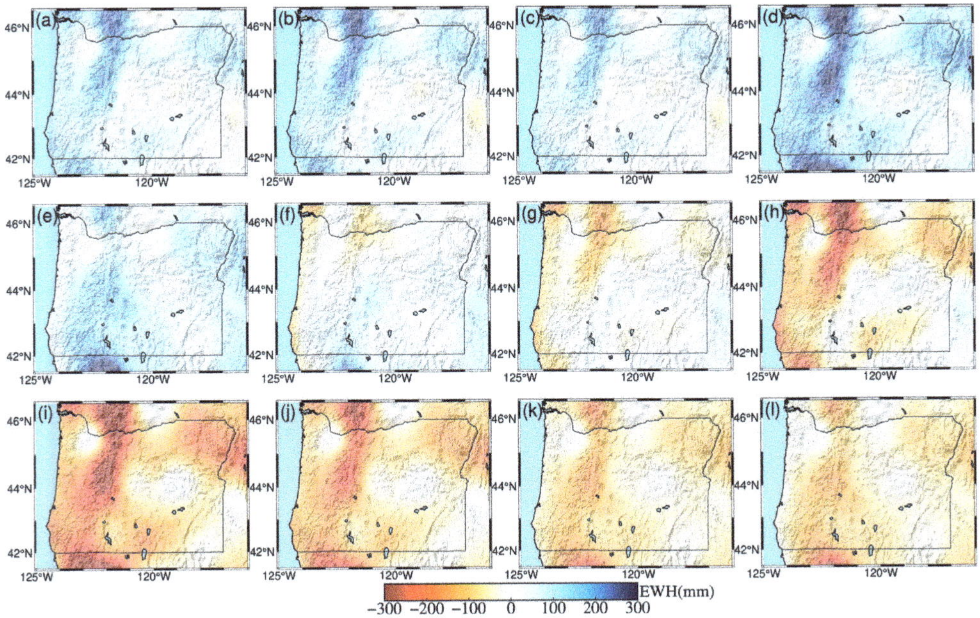

Figure 6. Monthly spatial and temporal terrestrial water storage changes in Oregon for the year 2018. (**a**) January; (**b**) February; (**c**) March; (**d**) April; (**e**) May; (**f**) June; (**g**) July; (**h**) August; (**i**) September; (**j**) October; (**k**) November; (**l**) December.

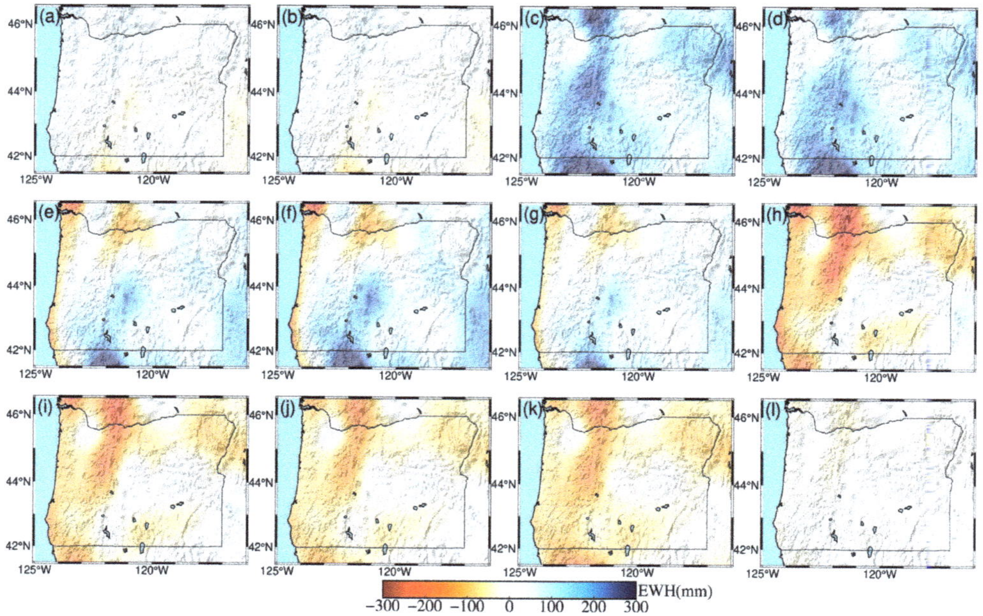

Figure 7. Monthly spatial and temporal terrestrial water storage changes in Oregon for the year 2019. (**a**) January; (**b**) February; (**c**) March; (**d**) April; (**e**) May; (**f**) June; (**g**) July; (**h**) August; (**i**) September; (**j**) October; (**k**) November; (**l**) December.

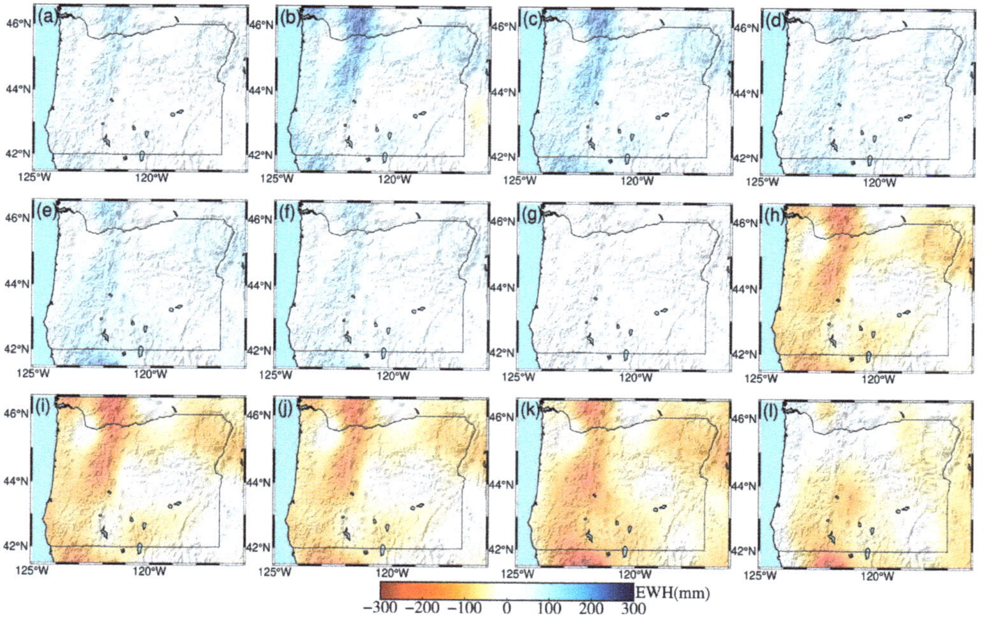

Figure 8. Monthly spatial and temporal terrestrial water storage changes in Oregon for the year 2020. (**a**) January; (**b**) February; (**c**) March; (**d**) April; (**e**) May; (**f**) June; (**g**) July; (**h**) August; (**i**) September; (**j**) October; (**k**) November; (**l**) December.

3.3. Spatial Distribution of Seasonal Terrestrial Water Storage Amplitude

To investigate the spatial distribution of terrestrial water storage, we present the annual amplitude of terrestrial water storage inverted using three independent techniques—GNSS, GRACE, and GLDAS—for Oregon, as shown in Figure 9a–c. The annual amplitude at each grid point is derived by fitting the time series using Equations (1) and (2), which extract the seasonal (annual) harmonic component. Figure 9a reveals that the annual mean equivalent water height (EWH) amplitude of terrestrial water storage (TWS) estimated by GNSS in Oregon reaches up to 300 mm, with the maximum observed in the Cascade Mountains near the northern border with Washington State. The amplitude gradually decreases from the mountains toward the adjacent basins, ranging from 20 to 100 mm, and also diminishes from north to south along the mountain range. Figure 9b illustrates the spatial distribution of long-term average seasonal TWS variations derived from GRACE data. Due to GRACE's coarse spatial resolution (~300 km), the spatial variation across the study area is not obvious, and the amplitude is minimal. The amplitude in the southern part is approximately 100 mm, slightly lower than the 145 mm observed in the north, with a difference of less than 50 mm. Furthermore, the annual amplitude in the Cascade Mountains is about 150 mm, which is significantly smaller than the 300 mm annual TWS amplitude derived from the GNSS inversion. Figure 9c shows the spatial distribution of terrestrial water storage from the GLDAS hydrological model, which is primarily divided into the eastern and western parts. The average amplitude in the western region is approximately 250 mm, which is greater than the 50–100 mm observed in the eastern region. This is consistent with the GNSS results, as the Cascade Range in the east exhibits pronounced seasonal variations. However, west of the mountains, GLDAS's spatial features are coarser compared to those of GNSS, and the maximum amplitude of seasonal changes calculated by the GLDAS hydrological model is 250 mm, which is less than that of the annual amplitude, about 300 mm, inverted by GNSS. Collectively, these results demonstrate that GNSS can resolve the spatial distribution of terrestrial water storage in this region with higher spatial resolution.

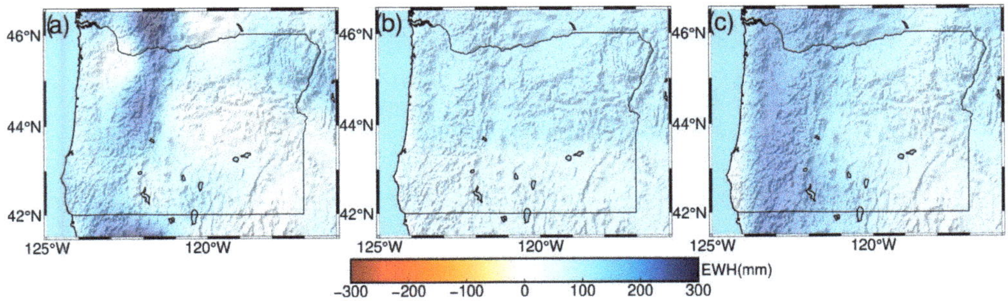

Figure 9. Spatial distribution of seasonal amplitude of terrestrial water storage changes inferred by (**a**) GNSS; (**b**) GRACE; (**c**) GLDAS.

4. Discussion

This study investigates terrestrial water storage (TWS) changes in Oregon, USA, by inverting GNSS vertical position time series data and performing a comparative analysis using GRACE satellite data and the GLDAS hydrological model. The study explores in depth the coupling mechanisms between the surface hydrological system and the elastic response of the crust, summarizes key findings and limitations, and provides guidance for future research.

The checkerboard test shows that the GNSS-inverted TWS results are smoother than those of the true input model, mainly due to the influence of relatively dense GNSS station distribution and the constraints imposed by the Laplacian operator.

When examining the spatial annual amplitude variations in TWS derived from GNSS vertical deformation, the GNSS results display clear gradient features, consistent with the accumulation and release patterns of TWS driven by precipitation and snowmelt in mountainous regions. In contrast, GRACE shows little spatial variability, while GLDAS, although capable of indicating gradient changes, exhibits coarse spatial characteristics in regions with weak seasonal variations. This can be attributed to GRACE's relatively coarse spatial resolution and potential gravity field errors, which limit its ability to provide accurate and detailed data for regional hydrological processes or fine-scale TWS changes. Additionally, GRACE data exhibits data discontinuities, such as the data void between 2017 and 2018 following the end of the GRACE mission and before GRACE-FO became operational.

GLDAS standard products typically display spatial resolutions of 0.25° to 1° (approximately 25–100 km), making it challenging to capture hydrological process details in regions with complex terrain (e.g., mountains, small catchments) or high land-use heterogeneity (e.g., urban areas, farmlands). For example, local precipitation, soil moisture, and runoff in mountainous regions may show significant spatial variability due to topographic differences, but coarse-resolution models tend to average out these differences, leading to simulation errors. Furthermore, the model simulates multilayer soil moisture (e.g., 0–10 cm, 10–40 cm), which may present vertical scale mismatch issues during direct assimilation.

Although GNSS excels in high-resolution monitoring, its inversion accuracy is highly dependent on station density and data quality. In areas with sparse stations or near the boundaries of the study area, inversion performance can deteriorate significantly due to insufficient observations. Moreover, most mainstream crustal elastic models are based on the assumption of a homogeneous elastic half-space and fail to adequately account for local geological heterogeneity, lithological contrasts, or complex 3D crustal structures. Therefore, such simplifications may introduce systematic errors into the inversion results.

In terms of regional time series inversion, although the GNSS-derived TWS results exhibit strong agreement with GRACE and GLDAS data, they show a 1–2 month phase lag compared with the data for precipitation series. This phenomenon highlights the buffering and regulating properties of the hydrological system, indicating that TWS changes are not solely driven by precipitation but are also modulated by multiple hydrological processes such as evapotranspiration, soil infiltration, and groundwater recharge. This finding not only reflects the complexity of hydrological processes but also suggests higher requirements for parameter optimization and process simulation in current hydrological models.

Based on the above discussion, future research could be further advanced in the following aspects: (1) introducing geological structural constraints or regionally varying elastic models to enhance the physical realism of crustal response modeling; (2) developing regularization strategies adapted to station spatial distribution characteristics to improve inversion stability and boundary fidelity in sparse station regions. These efforts would further improve the spatiotemporal accuracy and interpretability of TWS monitoring and help fill the observational gaps between the GRACE and GFO missions.

5. Conclusions

This study employed GNSS data to invert terrestrial water storage (TWS) changes in Oregon, USA, with the following key conclusions:

(1) The TWS in Oregon exhibits obvious interannual periodic variations. The equivalent water height (EWH) reaches peak values between January and April each year (maximum of approximately 150 mm), while troughs occur between July and October

(minimum of approximately −150 mm), showing strong consistency with regional precipitation cycles. During summer (June–August), high temperatures lead to snowmelt and soil moisture evaporation, causing TWS to reach its annual minimum, with EWH in the Cascade Mountains dropping to as low as −300 mm. In winter (December–February), rainfall and snowfall increase TWS to peak levels (up to 300 mm), with spatial patterns consistent with the region's climate and topographic features.

(2) The daily-scale dynamic variations in TWS inverted from GNSS data show good agreement with GRACE satellite observations and GLDAS hydrological model results, confirming the reliability of the inversion. The multi-year average seasonal amplitude indicates that GNSS-inverted EWH amplitude reaches a maximum of 300 mm in the northern Cascade Mountains, decreasing toward surrounding basins.

(3) Compared with traditional methods, the GNSS inversion approach offers distinct advantages in regards to both spatial and temporal resolution. With the aid of a dense GNSS station distribution, it enables the resolution of fine-scale local variations that are difficult to capture using conventional methods (e.g., amplitude differences across mountain slopes). The spatial details of the inversion results (such as the 300 mm amplitude peak) are significantly superior to those of GRACE (spatial resolution ≈300 km, capturing only regional averages), while the daily monitoring capability of the time series greatly outperforms the monthly resolution of GRACE and the limited dynamic response of the GLDAS model.

Author Contributions: Conceptualization, H.C. and J.Q.; methodology, J.Q.; software, D.W.; validation, D.W., J.Q., and H.C.; formal analysis, D.W.; investigation, D.W.; resources, J.Q.; data curation, D.W.; writing—original draft preparation, D.W.; writing—review and editing, H.C. and J.Q.; visualization, D.W.; supervision, J.Q.; project administration, H.C.; funding acquisition, J.Q. All authors have read and agreed to the published version of the manuscript.

Funding: This research received no external funding.

Data Availability Statement: Data used during the study will be made available upon reasonable request to the corresponding author.

Acknowledgments: The authors would like to gratefully acknowledge that the study presented in this paper is supported by Guangzhou Salvage in China, Nanyang Technological University in Singapore, and the Tianjin Research Institute for Water Transport Engineering Co., Ltd., in China.

Conflicts of Interest: Author Jian Qin was employed by Tianjin Water Transport Engineering Research Institute Co., Ltd. Author Dejian Wu was employed by Guangzhou Salvage, Guangzhou. The remaining authors declare that the research was conducted in the absence of any commercial or financial relationships that could be construed as a potential conflict of interest.

Abbreviations

The following abbreviations are used in this manuscript:

GRACE	Gravity Recovery and Climate Experiment
GLDAS	Global Land Data Assimilation System
GNSS	Global Navigation Satellite System
NGL	Nevada Geodetic Laboratory
JPL	Jet Propulsion Laboratory
TWS	Terrestrial Water Storage
EWH	Equivalent Water Height

References

1. Rodell, M.; Velicogna, I.; Famiglietti, J.S. Satellite-based estimates of groundwater depletion in India. *Nature* **2009**, *460*, 999–1002. [CrossRef] [PubMed]
2. Famiglietti, J.S. The global groundwater crisis. *Nat. Clim. Change* **2014**, *4*, 945–948. [CrossRef]
3. Trenberth, K.E.; Dai, A.; Van Der Schrier, G.; Jones, P.D.; Barichivich, J.; Briffa, K.R.; Sheffield, J. Global warming and changes in drought. *Nat. Clim. Change* **2014**, *4*, 17–22. [CrossRef]
4. Wada, Y.; Van Beek, L.P.H.; Van Kempen, C.M.; Reckman, J.W.T.M.; Vasak, S.; Bierkens, M.F.P. Global depletion of groundwater resources. *Geophys. Res. Lett.* **2010**, *37*, L20402. [CrossRef]
5. Konikow, L.F. Contribution of global groundwater depletion since 1900 to sea-level rise. *Geophys. Res. Lett.* **2011**, *38*. [CrossRef]
6. Scanlon, B.R.; Faunt, C.C.; Longuevergne, L.; Reedy, R.C.; Alley, W.M.; McGuire, V.L.; McMahon, P.B. Groundwater depletion and sustainability of irrigation in the US High Plains and Central Valley. *Proc. Natl. Acad. Sci. USA* **2012**, *109*, 9320–9325. [CrossRef] [PubMed]
7. Tapley, B.D.; Bettadpur, S.; Ries, J.C.; Thompson, P.F.; Watkins, M.M. GRACE measurements of mass variability in the Earth system. *Science* **2004**, *305*, 503–505. [CrossRef] [PubMed]
8. Long, D.; Longuevergne, L.; Scanlon, B.R. Deriving groundwater storage changes in China from GRACE with an improved approach. *Remote Sens. Environ.* **2013**, *129*, 79–91.
9. Wahr, J.; Molenaar, M.; Bryan, F. Time variability of the Earth's gravity field. *JGR Solid Earth* **1998**, *103*, 30205–30229.
10. Long, D.; Chen, X.; Scanlon, B.R.; Wada, Y.; Hong, Y.; Singh, V.P.; Chen, Y.; Wang, C.; Han, Z. Have GRACE satellites overestimated groundwater depletion in the Northwest India Aquifer? *Sci. Rep.* **2015**, *5*, 12843. [CrossRef] [PubMed]
11. Razeghi, M.; Han, S.-C.; McClusky, S.; Sauber, J. A Joint Analysis of GPS Displacement and GRACE Geopotential Data for Simultaneous Estimation of Geocenter Motion and Gravitational Field. *J. Geophys. Res. Solid Earth* **2019**, *124*, 11. [CrossRef]
12. Blewitt, G.; Hammond, W.; Kreemer, C. Harnessing the GPS data explosion for science. *Eos* **2018**, *99*, e2020943118. [CrossRef]
13. Calais, E.; Dong, L.; Wang, M.; Shen, Z.; Vergnolle, M. Continental deformation in Asia from a combined GPS solution. *Geophys. Res. Lett.* **2006**, *33*, L08303. [CrossRef]
14. Van Dam, T.; Wahr, J.; Milly, P.C.D.; Shmakin, A.B.; Blewitt, G.; Lavallée, D.; Larson, K.M. Crustal displacements due to continental water loading. *Geophys. Res. Lett.* **2001**, *28*, 651–654. [CrossRef]
15. Tregoning, P.; Watson, C.; Ramillien, G. Effects of snow and ice mass variations on seasonal deformation in the western Himalaya. *GRL* **2009**, *36*. [CrossRef]
16. Argus, D.F.; Landerer, F.W.; Wiese, D.N.; Martens, H.R.; Fu, Y.; Famiglietti, J.S.; Thomas, B.F.; Farr, T.G.; Moore, A.W.; Watkins, M.M. Sustained water loss in California's mountain ranges. *JGR Solid Earth* **2017**, *44*, 6247–6255.
17. Fu, Y.; Chen, Y.W.; Hsu, Y.J.; Shum, C.K.; Chiang, H.W. Inversion of terrestrial water storage from GNSS in Taiwan. *JGR Solid Earth* **2021**, *126*, e2020JB021157.
18. Jiang, Y.; Li, X.; Zhang, H.; Wang, J. Terrestrial water storage from GNSS vertical displacement: Sichuan case. *Water* **2021**, *13*, 961.
19. Swarr, M.A.; Moore, A.W.; Lavallée, D.; Blewitt, G. Sensitivity of GNSS-Derived Estimates of Terrestrial Water Storage to Assumed Earth Structure. *JGR Solid Earth* **2024**, *129*. [CrossRef]
20. Liu, Y.; Xu, K.; Guo, Z.; Li, S.; Zhu, Y. GNSS and GRACE fusion using PCA and VMD-BiLSTM. *Sci. Total Environ.* **2021**, *776*, 145758.
21. Tangdamrongsub, N.; Blewitt, G.; Liang, C.; Shen, Z. High-resolution TWS from GPS and GRACE in North China Plain. *JGR Solid Earth* **2017**, *122*, 6940–6959.
22. Li, J.; Zhang, X.; Wang, C.; Shen, Z. Monitoring drought-induced groundwater loss using GPS. *JGR Solid Earth* **2021**, *126*, e2021JB022489.
23. Jiang, Z.; Hsu, Y.J.; Yuan, L.; Feng, W.; Yang, X.; Tang, M. GNSS-derived TWS error sensitivity. *ETH Res. Collect.* **2022**, *26*, 1–7.
24. Alemohammad, S.H.; Madani, K.; Zhao, X.; Zadeh, M.S.; Kaymakci, E.; Mahmoodian, S. Review of GNSS-R soil moisture retrieval. *Remote Sens.* **2021**, *13*, 1350.
25. Wang, W.; Li, X.; Zhang, Y.; Chen, Z.; Liu, H.; Zhao, S. GNSS-R and meteorological fusion for soil moisture. *Environ. Syst. Res.* **2024**, *13*, 2.
26. Carlson, G.; Werth, S.; Shirzaei, M. Joint Inversion of GNSS and GRACE for Terrestrial Water Storage Change in California. *J. Geophys. Res. Solid Earth* **2022**, *127*, e2021JB023135. [CrossRef] [PubMed]
27. Blewitt, G.; Kreemer, C.; Hammond, W.C.; Goldfarb, J.M. Terrestrial reference frame NA12 for crustal deformation studies in North America. *J. Geodyn.* **2013**, *72*, 11–24. [CrossRef]
28. Wu, D.; Yan, H.; Shen, Y. TSAnalyzer, A GNSS time series analysis software. *GPS Solut.* **2017**, *21*, 1389–1394. [CrossRef]
29. Peltier, W.R.; Argus, D.; Drummond, R. Space geodesy constrains ice age terminal deglaciation: The global ICE-6G_C (VM5a) model. *J. Geophys. Res. Solid Earth* **2015**, *120*, 450–487. [CrossRef]
30. Sun, Y.; Riva, R.; Ditmar, P. Optimizing estimates of annual variations and trends in geocenter motion and J2 from a combination of GRACE data and geophysical models. *J. Geophys. Res. Solid Earth* **2016**, *121*, 8352–8370. [CrossRef]

31. Swenson, S.; Chambers, D.; Wahr, J. Estimating geocenter variations from a combination of GRACE and ocean model output. *J. Geophys. Res. Solid Earth* **2008**, *113*, JB005338. [CrossRef]
32. Loomis, B.D.; Rachlin, K.E.; Wiese, D.N.; Landerer, F.W.; Luthcke, S.B. Replacing GRACE/GRACE-FO with satellite laser ranging: Impacts on Antarctic Ice Sheet mass change. *Geophys. Res. Lett.* **2020**, *47*, e2019GL085488. [CrossRef]
33. Farrell, W. Deformation of the Earth by surface loads. *Rev. Geophys.* **1972**, *10*, 761–797. [CrossRef]
34. Xu, P.L.; Rummel, R. A Simulation Study of Smoothness Methods in Recovery of Regional Gravity Fields. *Geophys. J. Int.* **1994**, *117*, 472–486. [CrossRef]
35. Hansen, P.C. Analysis of Discrete Ill-Posed Problems by Means of the L-Curve. *Siam Rev.* **1992**, *34*, 561–580. [CrossRef]
36. Golub, G.H.; Heath, M.; Wahba, G. Generalized cross-validation as a method for choosing a good ridge parameter. *Technometrics* **1979**, *21*, 215–223. [CrossRef]

water

MDPI

Article

Morphodynamic Interactions Between Sandbar, Beach Profile, and Dune Under Variable Hydrodynamic and Morphological Conditions

Alirio Sequeira [1,*], Carlos Coelho [1] and Márcia Lima [1,2]

1 CERIS & Department of Civil Engineering, University of Aveiro, Campus Universitário de Santiago, 3810-193 Aveiro, Portugal; ccoelho@ua.pt (C.C.); marcia.lima@ulusofona.pt (M.L.)
2 FCNET—Faculty of Natural Sciences, Engineering and Technology, Porto University Center (CUP), Lusofona University, 4000-098 Porto, Portugal
* Correspondence: aliriosequeira@ua.pt

Abstract

Coastal areas are increasingly vulnerable to erosion, a process that can lead to severe consequences such as flooding and land loss. This study investigates strategies for preventing and mitigating coastal erosion, with a particular focus on nature-based solutions, notably artificial sand nourishment. Artificial nourishment has proven to be an effective method for erosion control. However, its success depends on factors such as the placement location, sediment volume, and frequency of operations. To optimize these interventions, simulations were conducted using both a numerical model (CS-Model) and a physical flume model, based on the same cross-section beach/dune profile, to compare cross-shore nourishment performance across different scenarios. The numerical modeling approach is presented first, including a description of the reference prototype-scale scenario. This is followed by an overview of the physical modeling, detailing the experimental 2D cross-section flume setup and tested scenarios. These scenarios simulate nourishment interventions with variations in beach profile, aiming to assess the influence of water level, berm width, bar volume, and bar geometry. The results from both numerical and physical simulations are presented, focusing on the cross-shore morphological response of the beach profile under wave action, particularly the effects on profile shape, water level, bar volume, and the position and depth of the bar crest. The main conclusion highlights that a wider initial berm leads to greater wave energy dissipation, thereby contributing to the mitigation of dune erosion.

Keywords: artificial nourishment; coastal erosion; coastal protection; CS-Model; flume tests

1. Introduction

Sandbar–beach–dune systems are highly dynamic features exposed to waves and currents, particularly during extreme events. Their evolution is mainly governed by the complex exchange of sediments through hydrodynamic processes. In particular, during storms and surges, the action of waves can cause severe erosion, with widespread morphological changes, namely in the bar volumes, beach berm width, and/or slope. In fact, coastal erosion poses a significant threat to sandy shorelines worldwide, driven by a complex combination of natural and anthropogenic factors. Key processes include wave action, storm surges, sea-level rise, and sediment transport dynamics, all of which interact with the beach and dune morphology. Understanding how specific physical parameters—such

https://doi.org/10.3390/w17142112

as water level, berm width, and sandbar characteristics—affect beach profiles is essential for designing effective mitigation and adaptation strategies.

This phenomenon can lead to serious consequences for coastal ecosystems, infrastructure, and human activities. Addressing erosion challenges requires modeling tools capable of simulating and predicting the evolution of beach profiles under changing conditions. Both numerical and physical models play a vital role in this context: numerical models allow for efficient scenario testing, while physical modeling enables detailed observation and validation of sediment transport and morphological responses. In fact, coastal erosion has significant impacts on both natural ecosystems and coastal infrastructure, with direct consequences for biodiversity, tourism, and the safety of coastal populations [1]. Recent studies emphasize the importance of integrated modeling and the adoption of hybrid solutions to mitigate these effects. Understanding erosion dynamics and implementing effective mitigation measures requires the use of modeling tools capable of simulating and predicting the complex interactions between different components of the coastal system, including the beach, dune, and sandbars.

In the field of coastal simulations, the use of numerical and physical models has been widely recognized as fundamental to understanding these processes. Numerical models, such as the CS-Model [2], offer the advantage of enabling rapid and comprehensive simulations of erosional processes, with the flexibility to explore a wide range of scenarios and parameters [3]. These models often simplify physical conditions to make computations feasible, but they may lose accuracy when representing highly dynamic and variable phenomena. To overcome these limitations, small-scale physical modeling conducted in laboratory environments allows for direct observation and measurement of the interactions between waves, sediment transport, and the morphological characteristics of the beach and dune [4]. This experimental approach provides valuable data to validate and refine numerical models, ensuring that simulations represent more accurately real coastal system conditions.

Various numerical models have been employed to simulate coastal erosion processes. For instance, the XBeach model [5] is widely used for simulating wave dynamics and sediment transport in coastal environments, while MIKE21 [6] has been applied to predict the effects of erosion and sediment transport in dynamic coastal areas. While these models provide robust results, the selection of the CS-Model for this study is based on its capability to efficiently and directly model interactions between the beach, dune, and sandbar, which are cross-shore sediment transport processes that represent key aspects of coastal erosion dynamics [7,8]. The CS-Model, with its ability to integrate variables such as bar volume and berm width, offers an effective framework for assessing coastal erosion mitigation and dune protection.

Although the CS-Model is a powerful tool for numerical modeling, it is crucial to balance the simulation results with physical laboratory experiments. The interaction between environmental conditions and erosional processes can vary significantly between models and real-world scenarios. Physical modeling permits a more detailed analysis, allowing direct observation of interactions and morphological changes over time [7]. Although physical models allow for controlled and intuitive investigation of hydrodynamic and sediment transport processes, they also involve inherent simplifications, particularly related to scaling laws, sediment homogeneity, and boundary effects. As noted by [1], physical models may not fully replicate natural variability, and therefore their results should be interpreted with caution. Thus, the combination of these two complementary methods (numerical and physical) offers a more robust and complete assessment of solutions for coastal erosion. Several recent studies have explored the use of physical and numerical models to simulate coastal erosion processes, with particular emphasis on beach–dune systems and their response to extreme events. D'Alessandro et al. [1] conducted a comprehensive review on the morphodynamics of these systems, highlighting the role of flume experiments and computational modeling in understanding the interactions among waves, sediments, and coastal vegetation. Complementarily, the study by [9] presents large-scale experimental results on wave–dune interactions, while Larson et al. [10] developed an analytical model that estimates dune erosion based on successive wave impacts. These studies underscore the importance of integrating physical, analytical, and numerical models for accurate prediction and effective mitigation of coastal erosion

Building on this integrated approach, this study aims to analyze coastal erosion dynamics through numerical and physical simulations, using the CS-Model in conjunction with a physical laboratory model. Rather than confirming general principles already established in the literature, the objective is to quantify the specific morphological effects of each parameter, namely water level, berm width, bar volume, and bar shape, under controlled conditions. A key goal is to compare and validate the CS-Model outputs against physical flume results, providing empirical support for the use of numerical models in the assessment of coastal protection strategies, particularly in the context of bar–beach–dune interactions when looking at cross-shore sediment transport processes. As previously stated, this study focuses on cross-shore dynamics, although the significance of the longshore component in the overall sediment budget is duly acknowledged. Integrated models, such as the one proposed by [11], which account for both sediment transport vectors, represent a natural progression for future research efforts.

2. Materials and Methods

To meet the objectives of this study, a combined methodology was adopted, integrating both numerical and physical modeling techniques. The numerical component relies on the CS-Model, described in Section 2.1, while the physical experiments were conducted in a laboratory flume, as detailed in Section 2.2. A reference scenario common to both models was defined (Section 2.3), followed by the definition of test scenarios used to assess morphological changes under different conditions (Section 2.4).

2.1. CS-Model Description

The CS-Model is a cross-shore numerical model developed to simulate sand transport and morphological evolution over medium-term timescales (some months to few years). While a detailed description is available in [1], this section provides a summary of its structure and key processes. The model simulates sand movement along the cross-shore profile and the resulting profile response over periods of up to approximately ten years, considering the main medium-term cross-shore processes: dune erosion and overtopping, wind-driven sand transport, and sediment exchanges between the bar and the beach berm. Each of these processes is represented by a specific module integrated into the CS-Model, containing algorithms based on physical principles and validated by laboratory and field data [2].

To model the profile response over time, sand volume conservation equations are used and solved alongside sediment transport equations to describe the evolution of the main morphological features of the cross-shore profile. These morphological parameters represent the cross-shore profile and include dune height (S), the location of the dune toe on the landward and seaward sides (Y_L and Y_S, respectively), the location of the berm crest (Y_B), and the volume of the bar (V_B), as illustrated in Figure 1. The angles βL and βS correspond to the slopes of the dune face on the landward and seaward sides, respectively, while βF represents the beach slope (these parameters are considered constant throughout the simulations). D_B and D_C denote berm height and closure depth, respectively, and MSL represents Mean Sea Level.

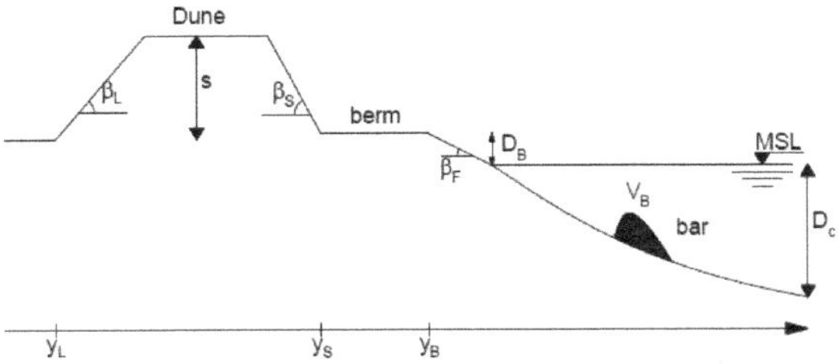

Figure 1. Cross-shore profile scheme in the CS-Model [8].

Sediment movement along the profile, which alters its geometry, is commonly attributed to wave and wind action, and is influenced by water levels. These changes, which represent the profile's response, are determined geometrically to maintain mass balance, although the fundamental parameters vary over time.

The CS-Model consists of several integrated modules: two modules for calculating aerial sediment exchange processes (dune erosion and overtopping, as well as aeolian sediment transport) and one module for calculating underwater material exchange (material exchange between the bar and the berm, based on bar theory). A brief explanation of dune erosion and overtopping processes is presented.

The determination of dune erosion is carried out through an analytical model proposed by [10], which is based on the studies of [12,13] on dune erosion. In this model, the eroded volume of the dune is considered to be proportional to the wave impact force hitting its face.

To illustrate the evolution of the profile, a storm impact is considered as an example. If the waves, along with the water level, generate sufficient runup (R), meaning that the wave runup exceeds the base level of the dune, the dune will lose volume (ΔV_D) and supply sand to the beach berm (Equation (1)). As a consequence of this erosion, the dune's base shifts inland, and Y_S decreases, assuming a constant dune slope.

$$\Delta V_D = 4C_S(R - Z_D)^2 \frac{\Delta t}{T} \tag{1}$$

In Equation (1), Δt represents the time step of the simulation, Z_D indicates the vertical distance between the dune toe level and the water level at each time interval (as illustrated in Figure 2), T denotes the wave period, and C_S is an empirical impact coefficient, which is based on the analytical dune erosion model proposed by [10], in which the eroded volume is related to the wave runup height and its interaction with the dune face. The coefficient was calibrated using experimental and field data. The smaller the value of Z_D, the higher the risk of dune erosion. Additionally, a smaller value of Z_D increases the likelihood that waves will reach the upper part of the profile, resulting in $R > Z_D + S$. In this scenario, it is assumed that the wave impact is reduced due to additional overtopping flow over the dune (as expressed in Equation (2)). If the wave runup exceeds the dune crest (i.e., overtopping occurs), the wave energy is assumed to be distributed across the entire dune height. In this case, the eroded volume is no longer proportional to the square of the impact height but instead scales linearly with the product $(R - Z_D)\cdot S$.

$$\Delta V_D = 4C_S(R - Z_D)S \frac{\Delta t}{T} \tag{2}$$

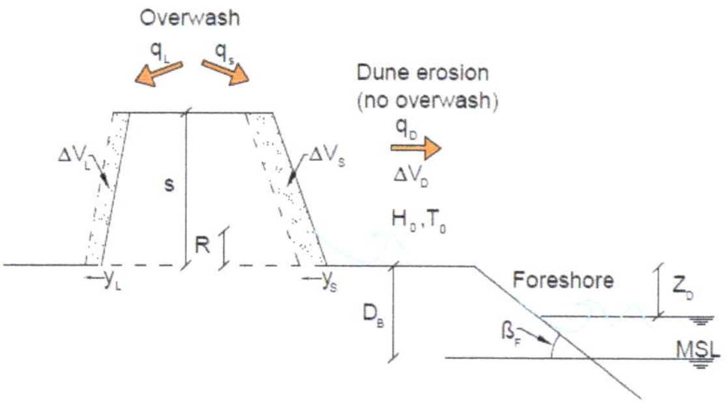

Figure 2. Diagram of dune erosion and overtopping processes, adapted from [8].

During an overtopping event, the total volume of sediment eroded from the dune due to wave impact (ΔV_D) is redistributed, with a portion (ΔV_L) deposited on the landward side of the dune and the remainder (ΔV_S) transported seaward, resulting in a reduction in Y_L (i.e., inland displacement). In this case, the slope of the dune's landward face, represented by β_L, is assumed to remain constant. The remaining material is displaced toward the ocean (ΔV_S). The distribution of ΔV_D between ΔV_L and ΔV_S, the amounts of the eroded dune volume directed landward and seaward, respectively, is determined by the coefficient α (Equation (3)), as expressed by $\Delta V_L = \Delta V_D \alpha/(1 + \alpha)$ and $\Delta V_S = \Delta V_D/(1 + \alpha)$.

$$\alpha = \left(\frac{R - Z_D}{S} - 1 \right) \frac{1}{A} \tag{3}$$

The empirical coefficient A, approximately established as 3 by [14] through a comparative analysis with field data, plays a crucial role. When the variation in sediment volume (ΔV_D) exceeds the initial volume (V_D), it is interpreted as an indication that the dune is undergoing erosive processes, as described by [14].

2.2. Experimental Setup and Flume Characteristics

The hydraulic flume of the Civil Engineering Department at the University of Aveiro, Portugal, has a usable length of 10 m, with a cross-sectional area of 0.40 m × 0.50 m (Figure 3). Based on the limitations of the flume identified in previous studies [15], a geometric scale of 1/40 was adopted. This scale, derived from Froude similarity, was used to construct a model of the cross-shore beach and dune profile.

(**a**) Cross-sectional scheme (**b**) View of the flume

Figure 3. Hydraulic flume of the Civil Engineering Department (University of Aveiro, Portugal).

A manual wave generator was installed at the upstream end of the flume to produce controlled wave conditions for the experiments. The device operates through pendulum motion and consists of acrylic plates connected to a system of steel shafts. The setup is illustrated in Figure 4.

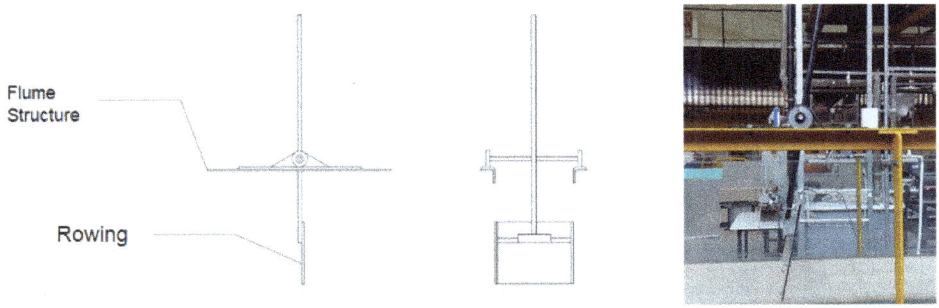

(**a**) Cross-sectional and front view schemes (**b**) View

Figure 4. Wave generation mechanism.

The water supply for the flume is provided by a pumping system. In the context of this study, the flume was kept in a horizontal position and, after adjusting the water level to the desired height, the pumping system was turned off, initiating the tests in the flume.

To evaluate and quantify the characteristics of the waves generated during the tests, an ultrasonic sensor was used (Figure 5). This sensor operates by emitting ultrasonic pulses at intervals of 40 ms, measuring the distance between the sensor and the water surface. The sensor was installed 4.30 m from the wave paddle, in the area near the beach profile.

(**a**) Installed location (**b**) General view of the flume

Figure 5. Ultrasonic sensor for water level measurement.

2.3. Reference Scenario

In this study, the reference scenario (Scenario C0) corresponds to a base beach–dune profile configuration used as a benchmark for both numerical and physical simulations. The term prototype refers to the real-world scale, while the physical model refers to the scaled-down experimental setup (1:40) used in the flume. The CS-Model was applied to both the prototype scale and the scale of the physical model, allowing comparisons across modeling approaches and scales.

Several preliminary tests were carried out with the objective of defining a profile vulnerable to wave attack but without overtopping, so that different reinforcement measures for the profile and their respective performance could later be evaluated. By varying the geometric parameters of the profile and the wave climate, some preliminary conclusions were drawn. In this regard, the profile represented in Figure 6 was considered for the reference scenario with the following values (prototype): 4 m (Y_L) for the location of the dune toe on the landward side, dune height of 1.5 m, and the position of the dune toe on the seaward side at 16 m (Y_S). This position coincides with the berm crest (Y_B), meaning that the reference profile does not include a berm.

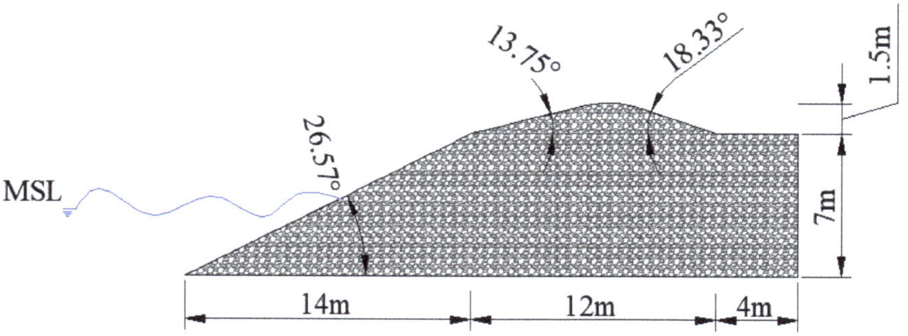

Figure 6. Profile of the reference scenario at the prototype scale.

For the wave climate, waves perpendicular to the beach were considered, with a wave height (H_0) of 2 m. The wave period (T) was calculated as a function of H_0 using the following expression [16], resulting in a wave period of 9.34 s:

$$T = 1.2H_0 + 6.92 \tag{4}$$

The beach profile in the physical model was designed to simulate sediment redistribution under wave forcing, with a double-slope configuration and without an initial bar. As highlighted in [9], simplified representations in flume tests are necessary to control variables and ensure repeatability. Therefore, the reference scenario profile assumes a zero volume for the bar. Subsequently, scenarios considering the presence of an initial bar volume are presented. It should be noted that the CS-Model does not simulate the geometry of the bar nor specify its exact position, considering only an available volume of sediments for bar–berm exchanges.

Considering the sediment grain size, as no variations are available in the CS-Model, homogeneous sand was also used in the flume. A representative sample of the sand employed in the physical tests was collected for the purpose of determining the median grain diameter (D_{50}), a critical parameter for the subsequent numerical simulations. Based on the particle size distribution analysis and the construction of the corresponding grading curve, the sand was characterized by a D_{50} value of 0.40 mm. In both the physical and numerical flume-scale models, the same sediment sample characterized by D_{50} = 0.40 mm was used to maintain direct comparability. For the prototype-scale numerical simulations, this same value was retained.

2.4. Scenarios Definition

The developed tests aim to characterize the behavior of the beach and dune profile morphology under different conditions of water level, berm width, bar volume, and bar shape. For this reason, values were established for different parameters, which varied independently, while keeping all other variables equal to the reference scenario. The reference scenario corresponds to Scenario C0.

Three scenarios were defined with water surface levels varying between 12 cm, 11 cm, and 10 cm (denoted as Pi, representing the water level in each scenario, as shown in Figure 7a), corresponding to scenarios C0, C1, and C2, respectively. These levels allow for testing a beach profile without a berm for different water surface heights. The selected water levels were also chosen to ensure the required wave height could be generated effectively, while maintaining sufficient depth and preventing wave overtopping during experiments.

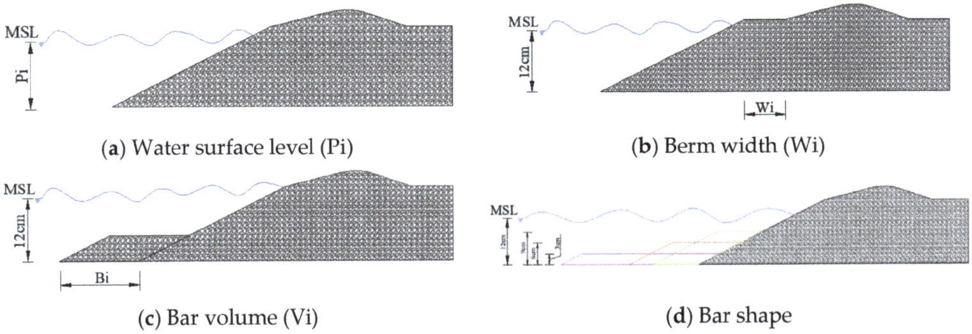

Figure 7. Schemes of the evaluated scenarios.

To evaluate the berm width, a volume of sand was added to the reference profile, creating a berm with increasing width, with Y_B values of 50 cm, 62.5 cm, 75 cm, and 87.5 cm, corresponding to scenarios C3, C4, C5, and C6, as shown in Figure 7b. The study of the variation in berm width (Wi) allows the simulation of artificial sand nourishment, enabling inferences about the morphology of profiles under the same wave climate.

For the study of the bar volume's effect on beach morphology, three scenarios were defined with volumes of 35 cm^3/cm, 115.30 cm^3/cm, and 196 cm^3/cm, corresponding to scenarios C7, C8, and C9, respectively, as shown in Figure 7c. The volume of 35 cm^3/cm corresponds to the equilibrium volume of the bar obtained in the CS-Model simulation at the model scale. The volume of 115.30 cm^3/cm represents the equilibrium volume of the bar in the reference scenario from the laboratory flume experiment. This volume corresponds to the final bar volume obtained in the CS-Model simulation of the reference scenario (C0) after morphological evolution reached a near-equilibrium condition, characterized by negligible changes in bar geometry over time. However, for the CS-Model simulation of the prototype, the bar volume was 409 cm^3/cm, which was considered too large to be represented in the laboratory flume. For this reason, a volume of 196 cm^3/cm was considered for scenario C9. For these scenarios, the same bar height of 6 cm was considered, with the bar width (Bi) varying according to each of the mentioned volumes.

Three scenarios were considered to evaluate the bar shape, all with the same bar volume. Thus, scenario C8, defined previously, was used as the reference, with a bar volume of 115.30 cm^3/cm. From this, two additional scenarios were defined: C10, with a bar height of 9 cm; and finally, scenario C11, with a submerged bar height of 3 cm (Figure 7d). In these two former scenarios, the width was adjusted to maintain the volume of 115.30 cm^3/cm.

The parameters common to all scenarios are presented in Table 1. Table 2 summarizes the parameters considered to evaluate the performance of the beach and dune profile as a function of the water level (scenarios C0 to C2), berm width (scenarios C0 and C3 to C6), berm volume (scenarios C0 and C7 to C9), and berm shape (scenarios C8, C10, and C11).

Table 1. Beach profile parameters, common to all scenarios.

H_0 (cm)	T (s)	Y_L (cm)	Y_S (cm)	β_L (°)	β_S (°)	β_F (°)	S (cm)
5	1.48	10	40	18.33	13.75	26.56	3.75

Table 2. Defined scenario and beach profile parameters at the physical model scale.

Scenarios	Y_B (cm)	D_B (cm)	D_C (cm)	V_{B0} (cm^3/cm)
C0	40.00	5.50	12.00	0.00
C1 C2	40.00	6.50 7.50	11.00 10.00	0.00
C3 C4 C5 C6	50.00 62.50 75.00 87.50	5.50	12.00	0.00
C7 C8 C9	40.00	5.50	12.00	35.00 115.30 196.00
C10 * C11	40.00	5.50	12.00	115.30

Note: * The shape of the bar varies as shown in Figure 7d.

3. Results

This section presents the results obtained from the simulations performed for each of the defined scenarios, followed by an interpretation of the morphological responses observed. At the end of each laboratory experiment (which lasted 6 min), the beach and dune profiles were recorded to assess the morphological changes.

3.1. CS-Model

The CS-Model was used to simulate the same scenarios at both the physical model scale and the prototype scale (1:40 geometric scale) to enable comparison with flume experiments and real-scale behavior. The outcomes are organized by test type, focusing on water level, berm width, bar volume, and profile shape. To facilitate comparison between the physical model and the CS-Model, all values presented here refer to the physical model scale (1:40).

3.1.1. Water Level

Figure 8 shows the final berm position in the CS-Model (i.e., the base of the dune slope and the top of the beach slope) as a function of the water level. At the model scale, all three scenarios (C0 to C2) reached the same value for the final berm position of 38.35 cm, while at the prototype scale, all scenarios resulted in the same value of 20.75 cm. In this context, the term 'final' refers to the state of the system at the end of the simulation period.

It was observed that the numerical model is not sensitive to changes in the water level. In the CS-Model simulations at the prototype scale, a greater retreat of the dune was noted.

Figure 9 shows the time evolution of the berm position, allowing for a comparison between the numerical simulation results. It can be seen that the time evolution of the berm position in the CS-Model is independent of the tested scenario. At the model scale, there is a nonlinear reduction from 40 cm to 38.35 cm, while at the prototype scale, there is also a nonlinear decrease of the berm position, from 40 cm to 20.75 cm.

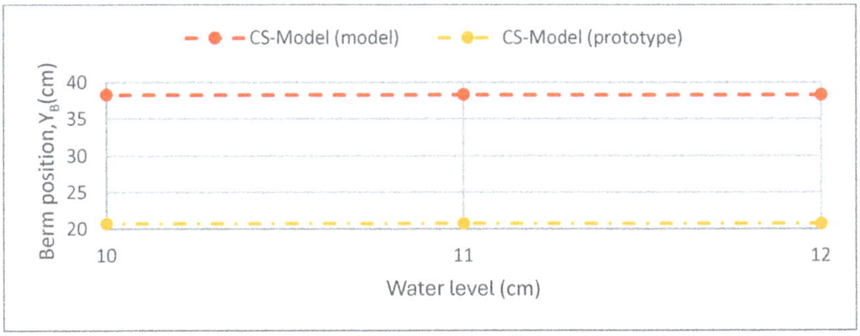

Figure 8. Final berm position as a function of the water level (scenarios C0 to C2).

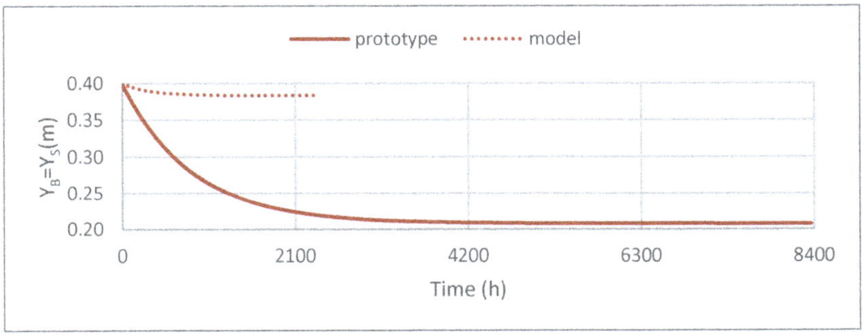

Figure 9. Evolution of the berm position over time, for scenarios C0 to C2.

3.1.2. Berm Width

For the study of berm width, four scenarios (C3 to C6) were defined with initial berm widths of 10 cm, 22.5 cm, 35 cm, and 47.5 cm. The results from the numerical modeling are shown in Figure 10, which presents the final berm width as a function of the initial berm width. The CS-Model simulation at the model scale for scenarios C0 and C3 to C6 resulted in final berm width values ranging from 0 cm to 45.50 cm. In contrast, the CS-Model at the prototype scale yielded values between 0 cm and 24.13 cm.

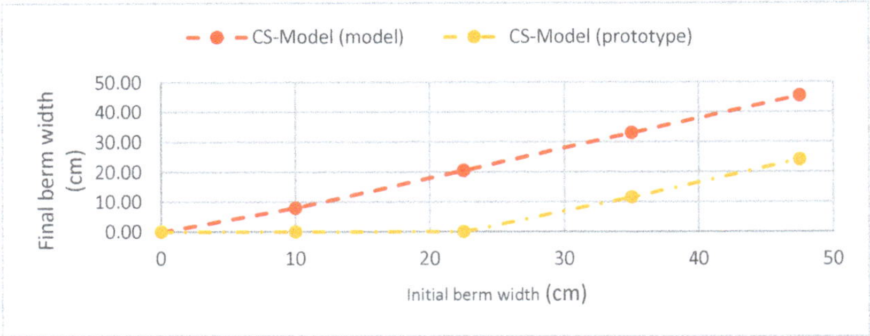

Figure 10. Berm width as a function of the initial width (CS-Model).

Figure 11 shows the time evolution of the berm width for each scenario (C0 and C3 to C6), based on the calculations performed by the numerical model. In the CS-Model, a nonlinear decrease in berm width over time is observed, tending to converge toward equilibrium. At the model scale, berm width decreased from 10 cm to 8 cm, from 22.50 cm to 20.50 cm, from 35 cm to 33 cm, and from 47.50 cm to 45.55 cm for scenarios C3 to C6, respectively. At the prototype scale, the berm width also decreased from 10 cm to 0 cm, from 22.50 cm to 0 cm, from 35 cm to 11.63 cm, and from 47.50 cm to 24.13 cm, according to scenarios C3 to C6, respectively. Therefore, the variations at the prototype scale are more significant. The results from the CS-Model at the model scale show similar differences from the beginning to the end of the simulation across all scenarios, with the berm width becoming null in scenarios C3 and C4.

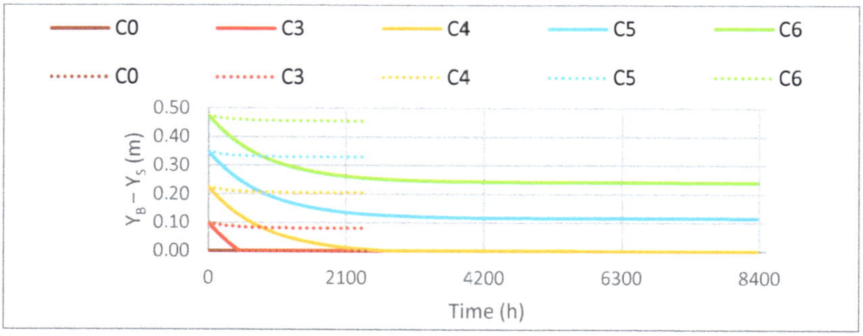

Figure 11. Evolution of the berm width over time in scenarios C0 and C3 to C6 (CS-Model).

3.1.3. Volume of the Bar

In the study of the bar volume, three scenarios were considered (C7, C8, and C9). The results obtained from the CS-Model at both model and prototype scales are shown in Figure 12, relating the berm position to bar volumes. In the CS-Model, at the model scale, the final position of the berm is 40 cm, 45 cm, and 49.2 cm for scenarios C7 to C9, emphasizing that in these simulations, the beach berm increased in width. For the results of the CS-Model at the prototype scale, the position of the beach berm was 22.40 cm, 26.18 cm, and 29.98 cm, for the respective scenarios C7 to C9. The CS-Model simulations show a linear variation in berm position as a function of bar volume, regardless of the scale (model or prototype).

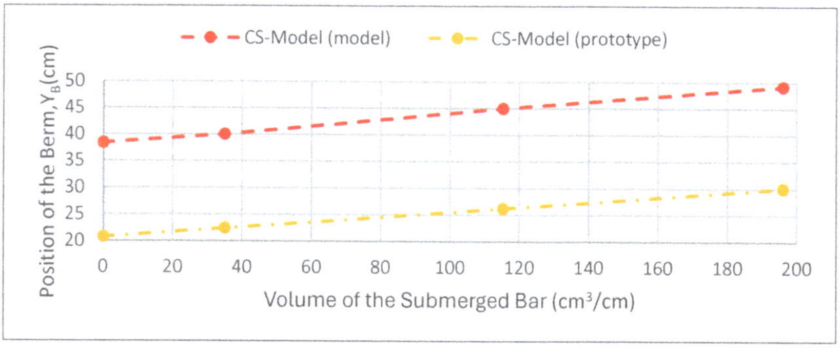

Figure 12. Evaluation of the volume of the bar in the physical model.

In Figure 13a, the evolution of the berm width over time at the model scale is presented. It is observed that the berm width increases over time, with increasing initial bar volumes, according to scenarios C7 to C9. At the prototype scale (Figure 13b), the wave climate caused a gradual recession of the berm position, resulting in final values of 22.40 cm, 26.18 cm, and 30 cm for scenarios C7, C8, and C9, respectively.

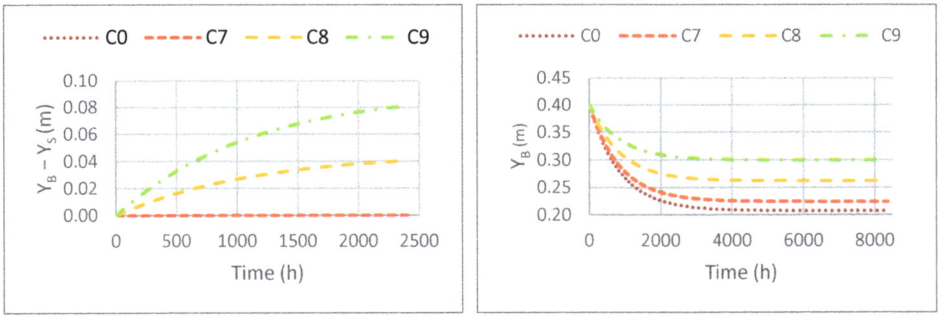

(**a**) Berm width at the model scale (**b**) Berm position at the prototype scale

Figure 13. Evolution of the berm width and position over time in the CS-Model, for scenarios C0 and C7 to C9.

It is observed that the CS-Model, at the model scale, created a berm for scenarios C8 and C9, while in scenario C7, there was virtually no change in the morphology, because the volume corresponding to the smallest bar in these scenarios is equal to the equilibrium volume of the bar. It was noted that at the prototype scale, the equilibrium volume of the bar is higher than the volume considered in the three scenarios. Therefore, in this situation, the trend is for the beach and/or dune to lose material to the bar until the equilibrium volume of the bar is reached. For this reason, in the CS-Model simulations at the prototype scale, there was always a recession in the dune.

3.1.4. Bar Shape

To analyze the shape of the bar, three scenarios were considered (C8, C10, and C11). The results from scenario C8 were presented in the previous section and serve as a reference for the simulations of the other bar shapes. In the numerical simulations using the CS-Model, it is not possible to reproduce the exact shape of the bar, because the model defines the bar only through its volume, VB. Thus, the results and values of the CS-Model for both the model scale and the prototype scale are presented in Figure 14. For the CS-Model at the model scale, the final berm position is 45 cm, implying a gain in material for the beach berm. However, for the CS-Model at the prototype scale, the final berm position is 26.18 cm, resulting in a recession of the berm position.

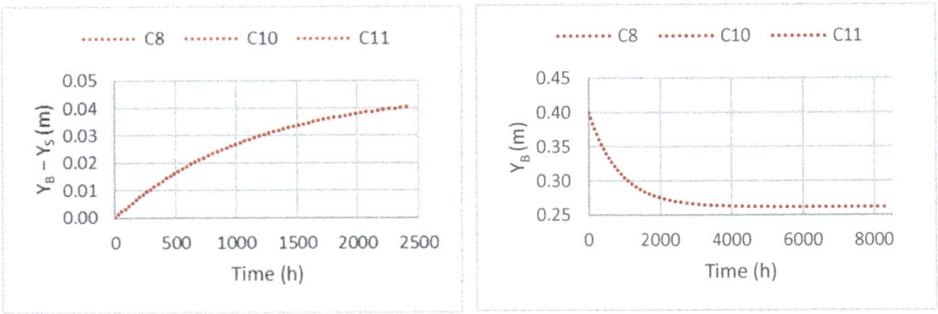

(**a**) Model (**b**) Prototype

Figure 14. Evolution of the berm width and position over time in the CS-Model for scenarios C8, C10, and C11.

Figure 14a presents the time evolution of the berm width, representing the result of the CS-Model at the model scale for the three scenarios (C8, C10, and C11), while Figure 14b presents the time evolution of the berm in the CS-Model at the prototype scale, with the berm position varying from 40 cm to 26.18 cm.

3.2. Flume Tests Results Assessment

Laboratory tests were carried out for the same scenarios defined in the numerical simulations, allowing for a comparison between physical and numerical results. The results are organized by test type—with particular focus on the profile shape—water level, as well as the bar volume, position, and crest depth.

Throughout the experiments, in order to record the evolution of the beach profile, a grid of vertical lines spaced 2 cm apart was drawn on the side wall of the flume. This setup facilitated the measurement of the sand surface elevation at the end of each time step using a ruler graduated in millimeters. No significant sediment loss occurred during the experiments; therefore, the discrepancies in volume measurements are attributed to the limitations of the measurement method.

3.2.1. Water Level

To assess the morphological effects observed in tests C0 to C2, Figure 15 presents the final beach and dune profiles recorded at the end of each laboratory experiment.

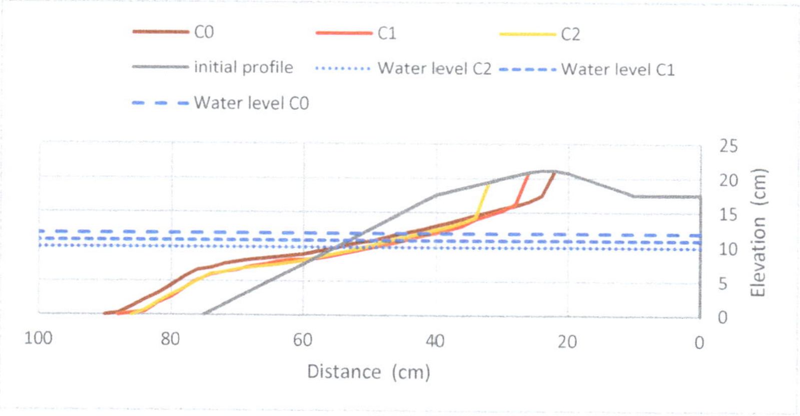

Figure 15. Final profile morphology for scenarios C0 to C2.

Throughout the tests, it was observed that the sand gradually migrated downward along the profile, with the erosion volume defined as the area below the initial profile. Simultaneously, a bar forms in the lower section of the profile (area above the initial configuration), considered as the deposition volume. The morphology of each test, at each one-minute time step, was compared to the reference scenario profile, enabling the quantification of the volumes corresponding to the erosion and accretion zones, as presented in Table 3.

Upon analyzing Table 3, it was observed that almost all the relative difference percentages are below 10%, a condition sufficient to ensure mass conservation in the profile during the laboratory experiments. In scenario C1, a slight decrease in the bar area was observed between the 5th and 6th minutes. This behavior is believed to be related to a reduced measurement precision, as all other results indicate a progressive evolution toward equilibrium.

Table 3. Erosion and deposition volume, as well as the absolute and relative differences in profile volumes, for scenarios C0 to C2 of the laboratory tests.

Test	Time (Minutes)	Volume		Differences	
		Erosion ($\times10^{-3}$ m^3/m)	Deposition ($\times10^{-3}$ m^3/m)	Absolute [1] ($\times10^{-3}$ m^3/m)	Relative [2] (%)
C0	1	5.09	6.35	1.27	24.88
	2	8.56	9.36	0.80	9.35
	3	9.56	10.67	1.11	11.61
	4	9.71	11.32	1.61	16.58
	5	10.12	11.31	1.19	11.76
	6	10.86	11.53	0.66	6.08
C1	1	6.59	6.33	0.26	4.00
	2	7.54	7.41	0.13	1.77
	3	8.61	7.93	0.68	7.93
	4	9.12	8.81	0.30	3.33
	5	9.61	9.00	0.61	6.36
	6	11.66	7.99	3.67	31.47
C2	1	5.63	5.41	0.22	3.95
	2	6.52	6.05	0.47	7.24
	3	6.93	6.97	0.04	0.54
	4	7.31	7.16	0.14	1.95
	5	8.15	7.40	0.74	9.11
	6	8.09	7.85	0.24	2.99

Notes: [1]: Difference between erosion volume and bar volume. [2]: Relative difference in relation to erosion volume (accretion in green and erosion in red).

The scenario with the lowest water level, corresponding to scenario C2, shows the smallest morphological variations. Scenario C1, in turn, presents the highest erosion volume, with 11.66 × 10^{-3} m^3/m, while scenario C0 reveals the highest deposition volume, approximately 11.53 × 10^{-3} m^3/m. In scenario C1, beach and dune erosion were more significant, while in scenario C0, dune erosion was greater than in the other scenarios. In all scenarios, the largest morphological variation occurred within the first 60 s. During the first 60 s, the erosion rates of 46.83%, 56.50%, and 69.54%, as well as the accretion rates of 55.14%, 79.14%, and 68.85%, respectively, were observed for scenarios C0, C1, and C2 when compared to the final situation of the test.

Figure 16 shows the variation over time in the depth and position of the bar crest in scenarios C0 to C2. The bar crest was identified as the inflection point of the bar shape, characterized by the change in the slope of the upper face of the submerged profile.

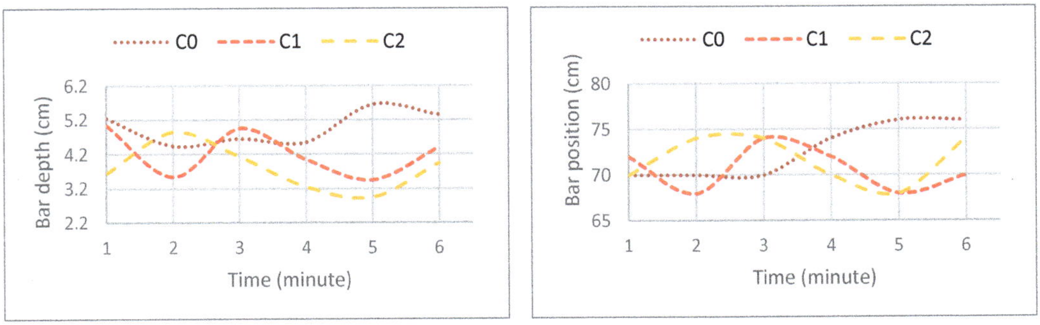

(**a**) Bar depth (**b**) Bar position

Figure 16. Variation in bar depth and position over time, for scenarios C0 to C2.

In this regard, Figure 17 shows the position of the final berm (base of the dune slope and top of the beach slope) as a function of the water level. Thus, the physical model results are situated between 23.97 cm and 33.25 cm.

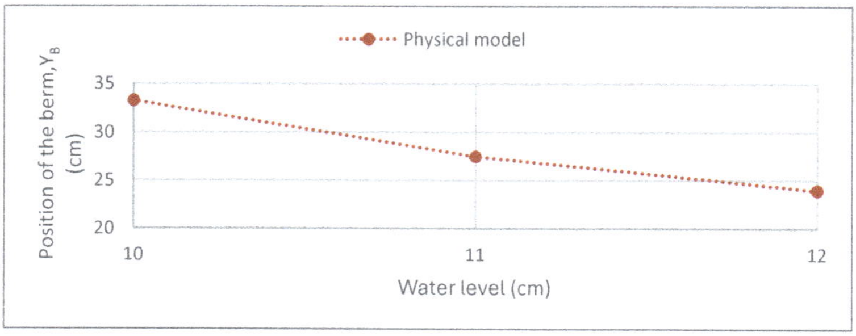

Figure 17. Final position of the berm, depending on the water level.

In Figure 18, it is observed that the temporal evolution of the berm position in the physical model is less pronounced as the water level decreases. It is also noted that the position of the seaward base of the berm is directly proportional to the water surface level, showing smaller retreats as the water level decreases.

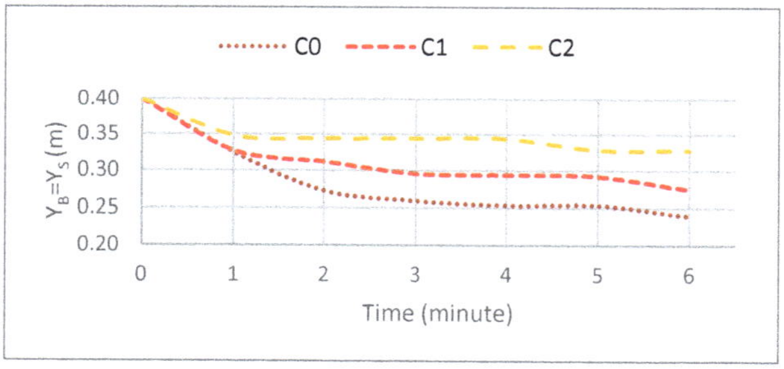

Figure 18. Evolution of the berm position, for scenarios C0 to C2, over time.

3.2.2. Berm Width

For the study of berm width, four scenarios (C3 to C6) were defined with the following berm widths: 10 cm; 22.5 cm; 35 cm; and 47.5 cm. Figure 19 represents the morphology of each laboratory test at the end of six minutes, thus allowing the observation of the effect of increasing berm width. It should be noted that the greater the initial berm width, the smaller the retreat of the dune toe position at the end of the test, and consequently, the smaller the erosion caused to the dune.

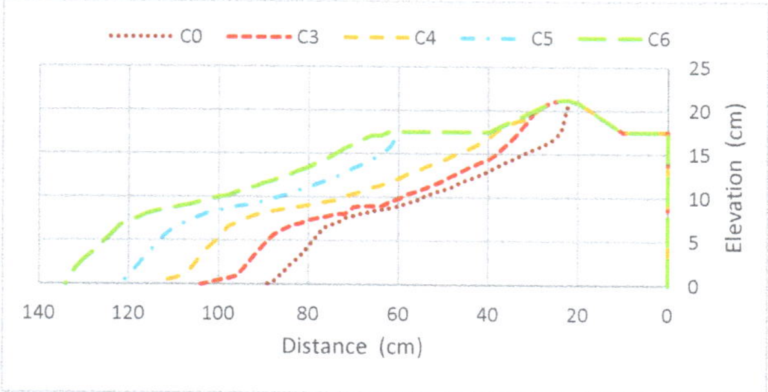

Figure 19. Profile morphology at the end of the laboratory tests, for scenarios C0 and C3 to C6.

Table 4 presents the erosion and deposition volume for each berm width scenario, minute by minute.

Table 4. Erosion and deposition volume, as well as the absolute and relative differences in profile volumes, for scenarios C3 to C6 from the laboratory tests.

Test	Time (Minutes)	Volume		Differences	
		Erosion ($\times 10^{-3}$ m^3/m)	Deposition ($\times 10^{-3}$ m^3/m)	Absolute [1] ($\times 10^{-3}$ m^3/m)	Relative [2] (%)
C3	1	7.35	8.13	0.79	10.69
	2	9.03	9.26	0.23	2.56
	3	9.73	9.59	0.13	1.38
	4	10.25	10.09	0.15	1.50
	5	11.19	10.31	0.87	7.81
	6	11.53	10.68	0.85	7.40
C4	1	11.66	9.72	1.94	16.63
	2	12.40	10.40	2.00	16.12
	3	12.66	10.94	1.72	13.58
	4	12.68	11.08	1.60	12.60
	5	13.25	11.09	2.16	16.29
	6	13.30	10.90	2.40	18.03
C5	1	6.63	7.64	1.01	15.15
	2	7.63	8.80	1.17	15.27
	3	8.37	9.15	0.79	9.38
	4	8.50	9.31	0.82	9.59
	5	9.02	10.07	1.06	11.70
	6	9.54	10.56	1.02	10.64
C6	1	7.27	7.12	0.15	2.06
	2	9.25	8.42	0.83	8.97
	3	9.89	10.13	0.24	2.43
	4	10.74	10.33	0.41	3.82
	5	11.41	10.54	0.87	7.62
	6	11.44	10.75	0.69	6.03

Notes: [1]: Difference between erosion volume and bar volume. [2]: Relative difference in relation to erosion volume (accretion in green and erosion in red).

In scenario C4, a slight decrease in the bar volume was observed in the last minute of the test. It is noteworthy that, regardless of the berm width, the final bar volume ranges from 10.56 to 10.90 $\times$ 10^{-3} m^3/m, indicating a trend that suggests the formation of the bar is not dependent on the berm width. It is also relevant to note that the greatest erosions occur in scenarios C3 and C4, because the dune is directly affected by wave action, while in scenarios C5 and C6, the berm is sufficiently wide, preventing dune erosion, and the final erosion volume is lower than in C3 and C4.

Figure 20 shows the variation over time in bar depth and position for scenarios C0 and C3 to C6. In Figure 20a, it is observed that the bar crest depth is independent of the amount of material used, displaying an oscillatory behavior over time, which tends to converge to the same value of 5.35 cm at the end of the test. In Figure 20b, it is observed that the bar position is directly proportional to the berm width, with larger berm widths corresponding to a further offshore bar crest position.

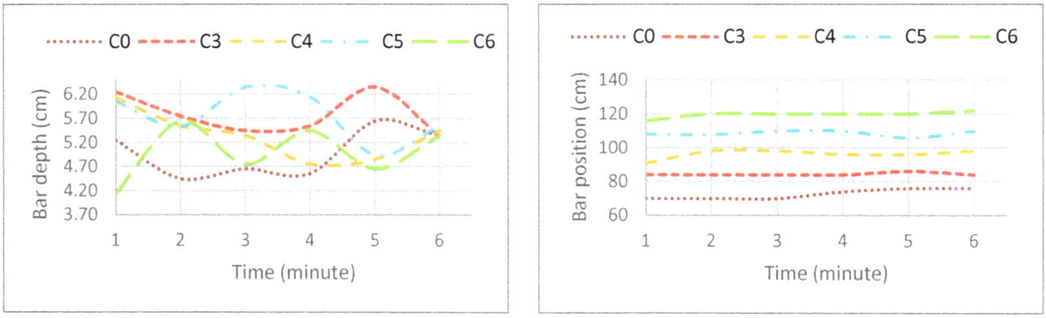

(**a**) Bar depth (**b**) Bar position

Figure 20. Variation in bar depth and position over time, for scenarios C0 and C3 to C6.

The results of the physical model are shown in Figure 21, where the final berm width is presented as a function of the initial berm width. The physical model shows final berm widths ranging from 0 cm to 22 cm.

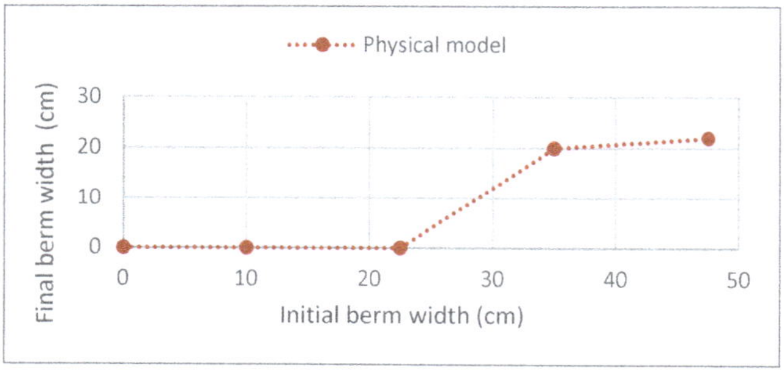

Figure 21. Berm width as a function of initial berm width (physical model).

Figure 22 presents the temporal evolution of berm width for each scenario (C0 and C3 to C6) of the laboratory tests.

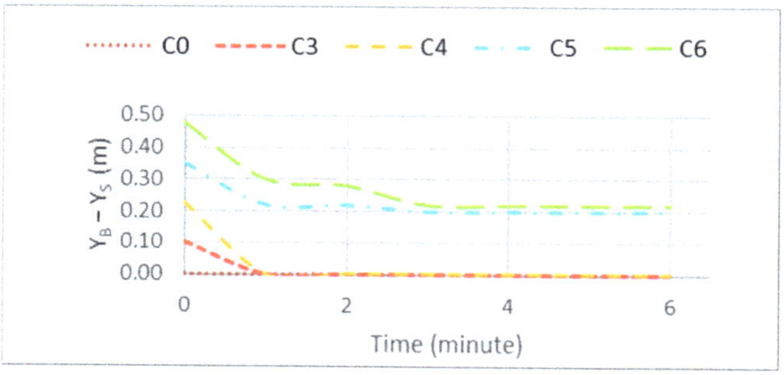

Figure 22. Evolution of berm width over time, for scenarios C0 and C3 to C6 (physical model).

Upon analyzing the laboratory tests, it was observed that the results for scenarios C3 and C4 are quite similar, with final berm position (Y_B) values of 33.25 cm and 38.33 cm, respectively. Both of these scenarios resulted in dune erosion, with scenario C3 showing a higher erosion rate. Therefore, it can be stated that increasing the berm width was beneficial for better dune shape preservation. In scenarios C5 and C6, the dune was preserved as the erosion did not affect the dune, with wave energy dissipating along the berm. Additionally, in these two scenarios, a residual berm width was maintained. At the end of the tests for scenarios C5 and C6, the berm position was 60 cm and 62 cm, resulting in final berm widths of 20 cm and 22 cm, respectively.

3.2.3. Volume of the Bar

In the study of the bar volume, three scenarios were considered: C7, C8, and C9. After obtaining the morphological evolution of the conducted tests, it was possible to represent their morphology at the end of the test (Figure 23).

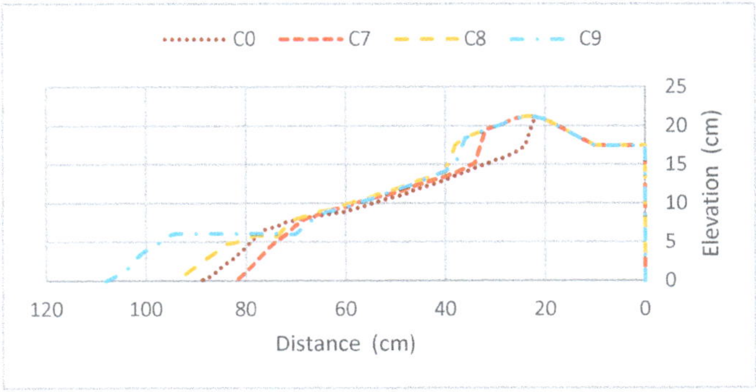

Figure 23. Profile morphology at the end of the test, for scenarios C0 and C7 to C9.

Based on the morphological evolution recorded throughout the tests, it was possible to determine the erosion and deposition volumes resulting from sediment dynamics for each minute of the tests performed (Table 5). It is observed that adding an initial volume to the bar improved dune preservation compared to the reference scenario, C0. However, in scenarios C7, C8, and C9, erosion on the dune still occurred. The values obtained for the dune toe position at the end of their respective tests were 32.92 cm, 37.90 cm, and 36.57 cm, respectively. Therefore, the smallest bar volume, scenario C7, corresponds to the greatest dune erosion. However, despite scenario C8 having a smaller bar volume than scenario C9, it resulted in better dune preservation than scenario C9.

In scenarios C7 and C9, a slight decrease in the erosion volume was observed in the last minute of the test, showing some stability in the behavior, which tends to converge towards equilibrium. In Figure 24, it can be seen that the initial bar volume caused the initial position and depth of the crest to differ, but they tend to evolve towards the same final result, with the exception of just one scenario, C0, which corresponds to the reference scenario.

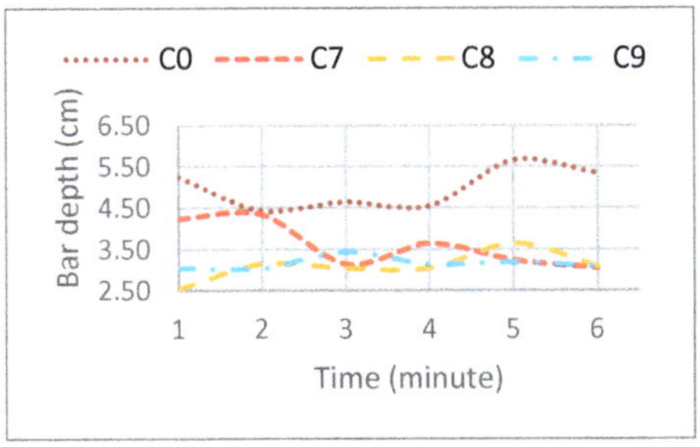

(**a**) Bar depth

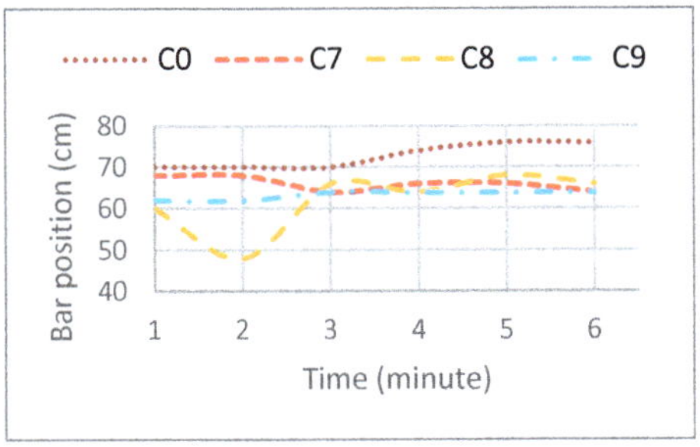

(**b**) Bar position

Figure 24. Variation in depth and position of the bar over time for scenarios C0 and C7 to C9.

Table 5. Erosion and deposition volume, as well as the absolute and relative differences in the volumes on the profile, for scenarios C7 to C9 of the laboratory tests.

Test	Time (Minutes)	Volume		Differences	
		Erosion ($\times 10^{-3}$ m^3/m)	Deposition ($\times 10^{-3}$ m^3/m)	Absolute [1] ($\times 10^{-3}$ m^3/m)	Relative [2] (%)
C7	1	4.02	3.22	0.80	19.89
	2	4.68	3.56	1.12	23.92
	3	5.18	4.02	1.16	22.39
	4	5.22	3.88	1.34	25.66
	5	5.50	4.42	1.08	19.63
	6	5.49	4.61	0.88	16.03
C8	1	1.90	2.03	0.13	7.00
	2	2.53	2.74	0.20	8.01
	3	2.83	3.46	0.62	22.00
	4	2.81	3.39	0.58	20.46
	5	2.94	3.85	0.91	30.78
	6	3.28	3.93	0.66	19.97
C9	1	2.55	2.23	0.32	12.72
	2	2.68	2.28	0.39	14.76
	3	2.99	2.58	0.41	13.87
	4	3.18	2.91	0.27	8.64
	5	3.32	3.03	0.29	8.87
	6	3.30	3.13	0.17	5.29

Notes: [1]: Difference between erosion volume and bar volume. [2]: Relative difference in relation to erosion volume (accretion in green and erosion in red).

The laboratory test results for C0 and C7 to C9 are presented in Figure 25, relating the berm position to the bar volumes.

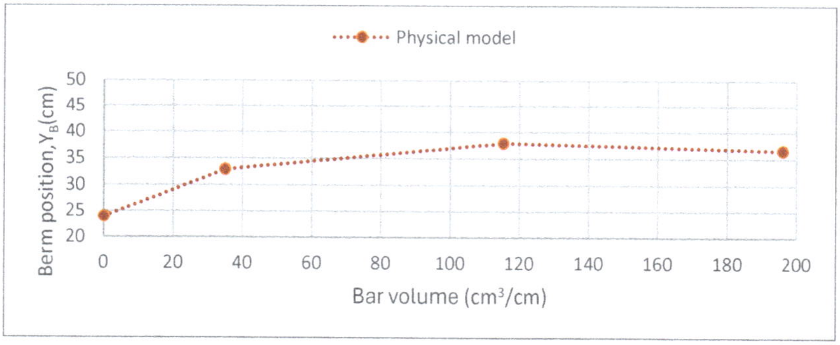

Figure 25. Evaluation of the bar volume in the laboratory model.

3.2.4. Bar Shape

To analyze the shape of the bar, three scenarios were considered: C8, C10 and C11. The results for scenario C8 were presented in the previous section and serve as a reference for the tests involving other bar shapes. Figure 26 depicts the morphology of each of these scenarios at the final stage of the laboratory tests, illustrating how the shape of the bar can influence the profile morphology.

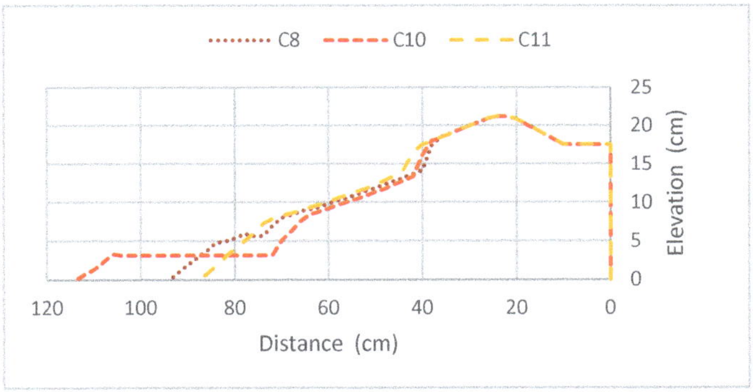

Figure 26. Profile morphology at the end of the test, for scenarios C8, C10, and C11.

Based on the morphology of each test, it was possible to determine the erosion and deposition volume over time (Table 6).

Table 6. Erosion and deposition volume, as well as the absolute and relative differences in the volumes on the profile, for scenarios C10 and C11 of the laboratory tests.

Test	Time (Minutes)	Volume		Differences	
		Erosion ($\times10^{-3}$ m^3/m)	Deposition ($\times10^{-3}$ m^3/m)	Absolute [1] ($\times10^{-3}$ m^3/m)	Relative [2] (%)
C10	1	1.12	1.59	0.47	41.56
	2	1.67	2.31	0.65	38.75
	3	2.43	3.19	0.77	31.54
	4	2.49	3.54	1.05	41.95
	5	2.47	3.70	1.22	49.52
	6	2.76	3.88	1.13	40.84
C11	1	0.65	0.95	0.30	45.69
	2	1.06	1.15	0.09	8.93
	3	1.06	1.25	0.19	17.63
	4	1.03	1.29	0.27	26.00
	5	0.97	1.28	0.31	31.64
	6	0.89	1.32	0.44	49.27

Notes: [1]: Difference between erosion volume and bar volume. [2]: Relative difference in relation to erosion volume (accretion in green).

In scenario C11, the morphological changes over time were reduced, suggesting that the initial bar shape is closer to the equilibrium, thereby minimizing sediment dynamics. Similarly to scenario C8, scenarios C10 and C11 showed improved dune conservation compared to the reference scenario, C0. The final values of the berm position (Y_B) were 37.9 cm, 38.53 cm, and 40 cm for scenarios C8, C10, and C11, respectively. It was observed that scenario C10, despite having a lower bar height, exhibited a lower erosion rate than scenario C8. However, in both scenarios (C8 and C10), erosion slightly affected the dune. Therefore, among these scenarios, only scenario C11 ensures dune conservation, as erosion did not impact the dune at any point. Figure 27 further confirms that scenario C11 exhibits reduced sediment dynamics, maintaining the bar crest position and corresponding depth constant over time.

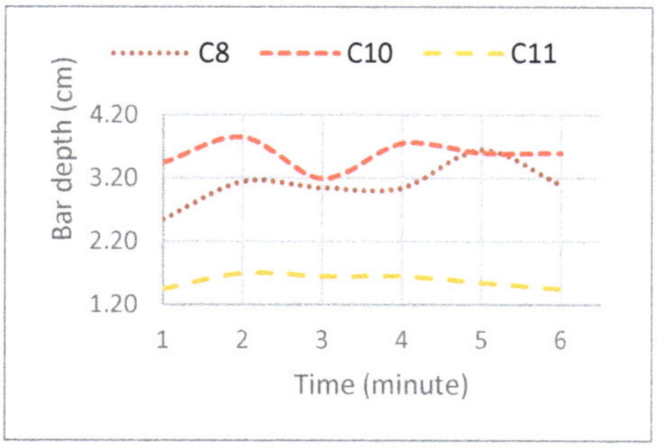

(a) Bar depth

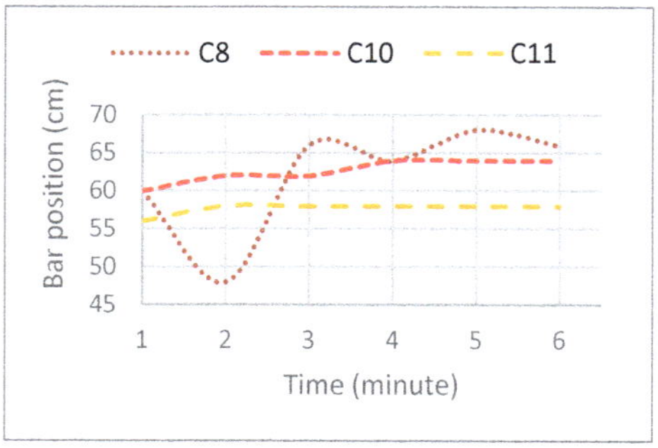

(b) Bar position

Figure 27. Variation in bar depth and position over time, for scenarios C8, C10, and C11.

4. Discussion

This section presents a critical analysis of the results from the numerical simulations and laboratory tests. A comparison of the performance of the CS-Model at both prototype and model scales is discussed, aiming to better understand the coastal erosion processes, particularly the interactions among the beach profile, dune system, and bar.

The study demonstrated that the CS-Model is not sensitive to variations in water level, as the results were identical for different water levels in scenarios C0 to C2. This behavior may be attributed to the simplification of boundary conditions in the numerical model, which does not account for dynamic variations in water levels, unlike sandy beaches associated with coastal dunes models that integrate tides and other factors [3]. Moreover, dune retreat was more pronounced in the CS-Model at the prototype scale, indicating an overestimation of erosion. This is likely due to the lack of variability in erosive processes, such as the interaction between water and sediments [3]. According to [17], an increase in the water surface level amplifies the interaction between waves and the beach profile, accelerating erosion. The results of this study confirm this trend, with scenario C1, which features a higher water level, causing greater erosion in the upper part of the beach.

It was observed that a greater berm width acts as a buffer, reducing dune and beach erosion, especially in scenarios C5 and C6. However, the simulations at model and prototype scales showed differences in berm width variations. These discrepancies can be explained by the CS-Model's limitation in simulating the dynamics of water and sediment transport in the berm interaction zone [18]. The relationship between berm width and the reduction in dune erosion is well documented, as shown by [19]. This study supports these findings, demonstrating that a wider berm results in the dissipation of wave energy, preventing dune erosion. In scenarios C5 and C6, berm width was crucial for dune protection, which is consistent with the results of other authors [8,18].

In scenarios C7 to C9, at the model scale, an increase in the bar volume resulted in an expansion of the berm width, indicating material deposition on the beach. However, in the prototype, continuous dune erosion was observed, suggesting that the equilibrium volume of the bar in the prototype exceeds the volume used in the model-scale simulations. This highlights the inadequacy of model-scale simulations in accurately replicating real coastal system conditions [20]. Previous studies, such as those by [21], suggest that the volume of the bar can mitigate dune erosion, albeit with diminishing returns, as the volume increases. The laboratory tests confirm that, although the increase in bar volume, as observed in scenarios C8 and C9, reduces dune erosion, this relationship is not linear. The analysis suggests that an optimized adjustment of bar volume could lead to more effective and cost-efficient solutions.

Sediment properties and scale effects may have impact on physical–numerical discrepancies. These issues have been previously highlighted by [9,10], who emphasized the challenges in reconciling laboratory-scale and prototype-scale behaviors in beach and dune morphodynamic modeling. The CS-Model only uses a single sediment grain size to characterize all the beach profile, from the submerged part to the dune. As discussed in [22], geometric similarity alone does not ensure dynamic similarity in sediment transport processes, particularly given the nonlinear dependence of critical shear stress and transport rates on grain size. While geometric and hydraulic scaling were respected, the sediment grain size (D_{50} = 0.40 mm) was held constant across all model scenarios due to the limitations of the CS-Model. This approach simplifies the setup but introduces potential scale effects. As demonstrated by [22], the sediment transport response is highly sensitive to grain size, especially in shallow flows where incipient motion thresholds and bedform evolution are strongly grain-dependent. Furthermore, sediment-related scale effects do not follow a linear law, making direct comparison between physical and prototype results subject to distortion. Therefore, morphodynamic results obtained from prototype-scale simulations should be interpreted cautiously, recognizing that sediment transport rates and erosion/deposition patterns may be over- or underestimated due to the fixed sediment size assumption. Despite this limitation, relative patterns remain valid, particularly when used for qualitative assessments or scenario comparisons.

The CS-Model does not account for the shape of the bar, relying solely on its volume. This limitation restricts the model's ability to accurately simulate the morphological interactions between waves, currents, and sediments, which are crucial for beach evolution and dune protection [4,23]. In the physical model, the shape of the bar, in terms of height and width, also plays a key role in wave energy dissipation. Borsje et al. [21] highlights the importance of these factors, which is confirmed in this study. The results indicate that the difference between the water level and the height of the bar affects energy dissipation, with the width of the bar becoming more significant in dune protection than its height, when this difference reaches 6 cm.

One limitation of the current modeling approach is the assumption of constant dune face slopes (βL and βS) throughout the simulation. In reality, dune slopes may evolve due to variations in sediment characteristics (grain size, for instance), sediment transport, and erosional processes. Main conclusions of the developed work apply only under the tested conditions and sediment types and should not be generalized to full-scale applications of the CS-Model. Future works can explore models that allow for dynamic adjustment of dune slopes to better represent morphological responses under varying wave conditions.

5. Conclusions

This study addressed a relevant gap in coastal engineering research by systematically comparing the effectiveness of different artificial nourishment strategies under controlled laboratory and numerical conditions. While many previous studies have investigated berm widening and bar reinforcement individually, few have directly compared them using a consistent experimental framework at multiple scales. The scenarios explored here were designed to provide new insights into how variations in berm width and bar volume and shape, as well as influence beach morphological changes and dune erosion, supporting more effective and targeted coastal protection strategies.

Simulations of the CS-Model at the model scale resulted in dune retreat values lower than those obtained in simulations at the prototype scale, as evaluated by the berm position. In evaluating the effect of the bar volume in the CS-Model simulations, it is important to note the equilibrium bar volume. At the model scale, as the bar volume increased, the position of the beach berm increased due to the fact that the 35 cm^3/cm volume is greater than the equilibrium bar volume. Whenever the initial bar volume exceeds the equilibrium bar volume, the bar volume decreases, and this material moves toward the beach, increasing the beach width. On the other hand, in the three scenarios of the bar volume study at the prototype scale, dune erosion was always observed. The equilibrium volume of the bar in the prototype is approximately 409 cm^3/cm, while the volumes considered in the studied scenarios were lower. In this situation, an increase in the bar volume was observed, leading to the material moving towards the bar, resulting in beach and dune erosion. The CS-Model only considers the bar volume, making this parameter independent of the shape and position of the bar. Thus, for the same bar volume, the result obtained is always the same.

In all the laboratory tests, the highest profile variation rates, resulting in accretion on the bar and erosion in the upper part of the profile, occurred during the first 60 s, with changes exceeding 40% of the total morphological variation recorded at the end of the test. In evaluating the effect of the water surface level in the physical model, a higher rate of dune erosion variation was recorded for scenario C0, while scenario C2 exhibited a lower rate of dune erosion variation. Thus, it can be concluded that the lower the water surface height, the less likely the wave climate will cause dune erosion. Regarding the berm width's influence, it was found that the wider the berm, the lower the chances of dune erosion occurring, as the beach width contributes to wave energy dissipation. When the width is sufficient in dissipating wave energy, no dune erosion occurs, and only a reduction in berm width is observed, as confirmed in scenarios C5 and C6. For the evaluation of the bar volume in the laboratory, the scenarios suggest that increasing the bar volume, although beneficial in reducing dune erosion, does not necessarily guarantee the most effective solution when applied in isolation or without optimization of other parameters, as seen in the results from scenarios C8 and C9. The relationship between bar volumes and dune erosion reduction is not linear, which suggests that studying the bar volume for dune erosion mitigation could lead to a more optimized artificial nourishment project, requiring fewer financial resources. During the laboratory tests, it was observed that for the same bar volume, when the difference between the water surface level and the bar height is at least 3 cm (scenario C10), the height of the bar becomes more crucial in dissipating wave energy. If this difference exceeds 6 cm (scenario C8), the width of the bar becomes more important than the height of the bar, leading to improved dune erosion mitigation. Therefore, scenario C11, with a 3 cm bar height, yielded better results in dune shape preservation than scenario C8, which had a height of 6 cm.

In conclusion, the laboratory tests showed that artificial nourishment involving bar volume was more effective at mitigating dune erosion than those focused on increasing berm width. Notably, only scenario C11 fully preserved the dune. These findings may offer valuable preliminary guidance for future artificial nourishment projects, contributing to more effective and sustainable coastal management.

Author Contributions: Conceptualization, C.C. and M.L.; Methodology, C.C. and M.L.; Software, A.S.; Validation, A.S. and C.C.; Formal analysis, A.S.; Writing—original draft, A.S.; Writing—review and editing, C.C. and M.L.; Supervision, C.C. and M.L. All authors have read and agreed to the published version of the manuscript.

Funding: This research received no external funding.

Data Availability Statement: The original contributions presented in this study are included in the article. Further inquiries can be directed to the corresponding author.

Conflicts of Interest: The authors declare no conflicts of interest.

References

1. D'Alessandro, F.; Tomasicchio, G.R.; Frega, F.; Leone, E.; Francone, A.; Pantusa, D.; Barbaro, G.; Foti, G. Beach–Dune System Morphodynamics. *J. Mar. Sci. Eng.* **2022**, *10*, 627. [CrossRef]
2. Larson, M.; Palalane, J.; Fredriksson, C.; Hanson, H. Simulating cross-shore material exchange at decadal scale. Theory and model component validation. *Coast. Eng.* **2016**, *116*, 57–66. [CrossRef]
3. Marinho, B.; Coelho, C.; Larson, M.; Hanson, H. Simulating cross-shore evolution towards equilibrium of different beach nourishment schemes. *Proc. Coast. Dyn.* **2017**, *2017*, 15.
4. Masselink, G.; Hughes, M.; Knight, J. *Introduction to Coastal Processes and Geomorphology*; Routledge: London, UK, 2014.
5. Roelvink, D.; Reniers, A.; van Dongeren, A.; van Thiel de Vries, J.; McCall, R.; Lescinski, J. Modelling storm impacts on beaches, dunes and barrier islands. *Coast. Eng.* **2009**, *56*, 1133–1152. [CrossRef]
6. DHI MIKE 21. *User Manual*; DHI Water & Environment: Singapore, 2012.
7. Guimarães, A.; Coelho, C.; Veloso-Gomes, F.; Silva, P.A. 3D Physical Modeling of an Artificial Beach Nourishment: Laboratory Procedures and Nourishment Performance. *J. Mar. Sci. Eng.* **2021**, *9*, 613. [CrossRef]
8. Marinho, B. Artificial Nourishments as a Coastal Defense Solution: Monitoring and Modelling Approaches. Ph.D. Thesis, Universidade de Aveiro, Aveiro, Portugal, 2018.
9. D'Alessandro, F.; Tomasicchio, G.R. Wave–dune interaction and beach resilience in large-scale physical model tests. *Coast. Eng.* **2016**, *116*, 15–25. [CrossRef]
10. Larson, M.; Erikson, L.; Hanson, H. An analytical model to predict dune erosion due to wave impact. *Coast. Eng.* **2004**, *51*, 675–696. [CrossRef]
11. Vitousek, S.; Barnard, P.L.; Limber, P.; Erikson, L.; Cole, B. A model integrating longshore and cross-shore processes for predicting long-term shoreline response to climate change. *J. Geophys. Res. Earth Surf.* **2017**, *122*, 782–806. [CrossRef]
12. Fisher, J.S.; Overton, M.F.; Chisholm, T. Field Measurements of Dune Erosion. In *Coastal Engineering 1986*; American Society of Civil Engineers: Reston, VA, USA, 1987; pp. 1107–1115.
13. Nishi, R.; Kraus, N. Mechanism and calculation of sand dune erosion by storms. In Proceedings of the 25th Coastal Engineering Conference, American Society of Civil Engineers (ASCE), Orlando, FL, USA, 2–6 September 1996; pp. 3034–3047.
14. Larson, M.; Donnelly, C.; Jiménez, J.A.; Hanson, H. Analytical model of beach erosion and overwash during storms. *Proc. Inst. Civil. Eng.-Marit. Eng.* **2009**, *162*, 115–125. [CrossRef]
15. Vale, D.; Pires, R. Avaliação do Espraiamento e Galgamento em Estruturas Costeiras. Projeto de Licenciatura em Engenharia Civil. Não Publicado. Universidade de Aveiro, Aveiro, Portugal, 2022. (In Portuguese)
16. Coelho, C. Riscos de Exposição de Frentes Urbanas para Diferentes Intervenções de Defesa Costeira. Ph.D. Thesis, Universidade de Aveiro, Aveiro, Portugal, 2005. (In Portuguese)
17. van Rijn, L.C.; Tonnon, P.K.; Walstra, D.J.R. Numerical modelling of erosion and accretion of plane sloping beaches at different scales. *Coast. Eng.* **2011**, *58*, 637–655. [CrossRef]
18. Puleo, J.A.; Lanckriet, T.; Conley, D.; Foster, D. Sediment transport partitioning in the swash zone of a large-scale laboratory beach. *Coast. Eng.* **2016**, *113*, 73–87. [CrossRef]
19. Ferreira, A.M.; Coelho, C.; Silva, P.A. Medium-Term Effects of Dune Erosion and Longshore Sediment Transport on Beach–Dune Systems Evolution. *J. Mar. Sci. Eng.* **2024**, *12*, 1083. [CrossRef]
20. Van Rijn, L.C.; Tonnon, P.K.; Sánchez-Arcilla, A.; Cáceres, I.; Grüne, J. Scaling laws for beach and dune erosion processes. *Coast. Eng.* **2011**, *58*, 623–636. [CrossRef]
21. Borsje, B.W.; van Wesenbeeck, B.K.; Dekker, F.; Paalvast, P.; Bouma, T.J.; van Katwijk, M.M.; de Vries, M.B. How ecological engineering can serve in coastal protection. *Ecol. Eng.* **2011**, *37*, 113–122. [CrossRef]
22. Raquel, S. Avaliação Experimental e Numérica de Parâmetros Associados a Modelos de Evolução da Linha de Costa. Ph.D. Thesis, Universidade do Porto, Porto, Portugal, 2010. (In Portuguese)
23. Chun, I.; Lim, H.S.; Shim, J.S.; Park, K.S. Numerical Analysis of the Performance of Perforated Coastal Structures under Irregular Wave Conditions. *J. Coast. Res.* **2015**, *315*, 1159–1169. [CrossRef]

water

MDPI

Article

Study on Characteristics of the Water Diversion Ratio and Impact of the Diversion Dyke at South and North Ports of the Minjiang River During Wet Season

Cheng Chen [1], Weijia Yang [1], Zhihui Wang [2], Kailong Huangfu [2], Feng Cai [1,*], Haoyan Chen [1] and Youlin Chen [1]

[1] College of Civil Engineering, Fuzhou University, Fuzhou 350108, China; chencheng_1117@163.com (C.C.); ywj92500@163.com (W.Y.); c123456202503@163.com (H.C.); youlin13950298186@163.com (Y.C.)
[2] Fujian Port & Waterway Investigation and Design Institute Limited, Fuzhou 350002, China; zhwang928@163.com (Z.W.); kailonghuangfu@163.com (K.H.)
* Correspondence: caifeng@fzu.edu.cn

Abstract: The hydrodynamic processes in estuarine regions play a crucial role in the morphological and ecological stability of coastal zones. As a key hydrodynamic characteristic of bifurcated rivers, the water diversion ratio (WDR) influences flow distribution, sediment transport, and shoreline changes in estuaries. This study focuses on the lower Minjiang River and employs a MIKE 21-based two-dimensional hydrodynamic model to quantify the WDR variations between the South and North Ports on the scale of a tidal cycle during the wet season and to reveal the regulatory effects of diversion dyke length and angle. The results indicate that the WDR of the North Port exhibits significant variation with tidal stages. The WDR of the North Port increases with the length of the diversion dyke. The current 110 m-long dyke has little effect on regulating water flow between the North and South Ports, and its WDR remains unaffected by changes in angle. In contrast, a 450 m-long dyke is highly sensitive to angle variations. This study not only provides scientific support for channel regulation in the lower Minjiang River but also offers indirect insights into shoreline stability and ecological management under the combined influence of human activities and natural processes in estuarine environments.

Keywords: water diversion ratio; mathematical model; diversion dyke; Minjiang River; anthropogenic impacts

1. Introduction

In estuarine regions, the water diversion ratio (WDR), a critical metric for quantifying flow distribution within bifurcated channels, not only reflects inherent hydrodynamic processes but also directly influences sediment transport and deposition patterns, thereby potentially affecting estuarine morphological evolution and ecosystem health [1,2]. Estuarine topographies exhibit the dual impacts of tidal amplitude and river flow, while also incorporating representations of alluvial deposits [3]. The evolution of the riverbed, in turn, exerts a feedback effect on the development of the river [4].

Meandering river systems exhibiting mid-channel bar development constitute one of the most ubiquitous planform configurations observed in fluvial environments [5–7]. These dynamic systems exhibit distinct planimetric characteristics through central mid-channel bars that induce bifurcation [8]. Notably, these bifurcation points exhibit persistent flow asymmetry that governs differential sediment routing pathways, demonstrating significant temporal variations in hydraulic partitioning efficiency [9]. Understanding the decadal-scale evolutionary trajectories of such complex anastomosing systems requires comprehensive analysis of their response mechanisms to both autogenic processes and allogenic disturbances, particularly through the lens of coupled hydrodynamic forces and sediment dynamics [10].

The dynamic adjustment of the WDR significantly shapes hydro-sedimentary processes, as evidenced by multidisciplinary studies spanning urban water systems, fluvial morphology, and debris flow dynamics. In urbanized river basins, Song and Pang [11] demonstrated that shifts in the WDR in Nanjing's Qinhuai River Basin directly modulate pollutant distribution and hydraulic equilibrium. Their modeling revealed that unequal flow partitioning between bifurcated channels creates distinct water quality gradients, with pollutant dilution exhibiting nonlinear responses to upstream flow variations. These findings highlight the dual role of WDRs as both hydraulic regulators and environmental stressors in managed water systems. Fluvial systems exhibit heightened sensitivity to diversion ratio alterations. Alomari et al. [12] experimentally quantified scour depth and length in main channels under diversion flows, revealing reduced scour with smaller diversion angles and establishing empirical relationships linking depth/length to angles, bed width ratios, and discharges. Complementary research by Abdulhafedh et al. [13] demonstrated that engineering interventions, particularly entrance edge modifications, can attenuate scour depth by up to 22% at critical discharge conditions, underscoring the potential for structural optimization to buffer sedimentary impacts. Debris flow studies have further expanded this paradigm. Xu et al. [14] developed a two-phase flow model showing that WDRs in cross-channels critically depend on solid–liquid composition and geometric constraints, with bifurcation dynamics dictating flow path sedimentation risks. Similarly, flood management research by Zhang and Yu [15] in the Han River basin identified threshold behaviors in diversion efficiency, where initial flow partitioning disproportionately influences breach evolution stages and sediment transport capacity. Therefore, diversion ratio management serves dual environmental roles: mitigating pollution/scour and triggering sediment instability. Developing adaptive frameworks and coupled flow–sediment models is critical for balancing objectives under climate-driven hydrological shifts. In the process of sediment transport, the dynamic evolution of permeability is a critical factor, as both permeability and flow characteristics are directly influenced by changes in the pore structure of sediments. Recent studies have highlighted the importance of advanced measurement techniques in quantifying permeability variations under varying stress conditions [16,17].

As critical engineered systems for managing flow distribution in bifurcated rivers, diversion dykes have persistently attracted research attention in river engineering, particularly regarding their hydrodynamic performance and failure mechanisms under complex hydraulic loads. Recent studies have systematically explored the hydrodynamic behavior and engineering applications of flow diversion structures. Gao et al. [18] conducted 3D numerical simulations on non-submerged diversion dykes, demonstrating that blockage ratios critically govern momentum redistribution and turbulent kinetic energy. Their results highlighted that higher blockage ratios intensify flow velocity gradients, amplify vortex shedding frequency, and elevate bed shear stress around the dyke, while Froude numbers exhibited minimal influence on recirculation zone dynamics. In parallel, Ranjbar-Zahedani et al. [19] investigated an optimized submerged triangular diversion structure for pier scour mitigation. Through systematic experimentation, they identified that strategic positioning and sizing of the structure significantly reduced both maximum scour depth and total scour volume. Flow field analysis revealed the structure's effectiveness in attenuating downflow intensity and suppressing horseshoe vortex development, key mechanisms driving scour reduction. The research links theory to application, emphasizing tailored diversion structures for specific flow regimes. The stability and failure mechanisms of diversion dykes under extreme hydrological events (e.g., floods and tidal surges) have also drawn significant attention. Recent experimental studies have shown that optimizing the pile shape of combined dykes in flood scenarios can reduce the intensity of horseshoe vortices, lowering scour risks and protecting riverbanks [20]. Moreover, despite extensive research on spur dyke design, knowledge gaps remain regarding failure mechanisms under extreme conditions. Nandhini et al. [21] provided a comprehensive review of the flow characteristics, scour effects, and design standards of spur dykes under both normal and extreme conditions, offering valuable insights for research and engineering practice.

As the largest independent river discharging into the East China Sea in Fujian Province, southeastern China [22], the Minjiang River's upstream flow regime is primarily regulated by the Shuikou Dam, with a mandated minimum ecological discharge threshold of 308 m^3/s [23]. Investigating the water diversion ratio characteristics and the impact of the diversion dyke structure at the South–North Port bifurcation holds significant implications for advancing hydrological science, optimizing water resource management strategies, and informing sustainable engineering practices within estuarine environments.

However, existing research on the WDR of the South and North Ports mainly applies modern statistical methods to analyze current hydrological and topographical data, while numerical simulation studies remain relatively limited. To address this gap, this study develops a two-dimensional hydrodynamic model. Unlike previous research that emphasized long-term spatiotemporal analyses of WDR dynamics in the lower Minjiang River, this study focuses on the dynamic variation in the WDR during the wet season, particularly during spring and neap tidal cycles. Furthermore, it explores the influence of the diversion dyke's length and rotation angle on the WDR under the current channel morphology of the South and North Ports.

2. Study Area and Methodology

2.1. Study Area

The Minjiang River originates in Jiankou Town, Jianxian County, located at the border of Fujian and Jiangxi provinces. In its lower reaches, the river bifurcates at Huaiankou. The northern branch, known as the North Port, flows through the urban area of Fuzhou and reaches Mawei Port. The southern branch, referred to as the South Port or Wulong River, flows along the southern side of Nantai Island, joins the Dazhangxi River at Jiangkou, exits the Xiadou, and reaches Mawei. The South and North Ports encircle Nantai Island before converging at Luoxingta in Mawei. The South and North Ports differ significantly in channel morphology. The North Port is narrow and deep, functioning as a navigable waterway, with its narrowest section measuring only 130 m in width during the dry season. In contrast, the South Port is wide and shallow, with its widest section spanning up to 4600 m. The terrain along both banks alternates between plains and hills. Compared to the North Port, the South Port features more sandbars, shallows, and bifurcated channels, resulting in scattered water flow. As a result, it serves as the main channel for flood discharge and sediment transport. A schematic diagram of the lower reaches of the Minjiang River is shown in Figure 1.

Before 1975, the WDR between the South and North Ports remained relatively stable. During the dry season, the WDR was 3:7, while it reversed during the wet season. However, after 1975, large-scale sand extraction in the North Port caused a steady increase in its WDR. In 1998, the catastrophic "98·6" flood eroded the riverbed at the entrance of the South Port, leading to the opening of the South Port entrance. In recent years, the complete ban on sand extraction in the North Port, combined with development and channel regulation projects in the South Port, has significantly deepened the channel. Since 1998, the WDR between the South and North Ports has undergone structural changes. This has resulted in a trend of the South Port's WDR being significantly greater than that of the North Port. Currently, the trend requires urgent attention from relevant authorities. If left unregulated, it could cause issues with navigation safety, bridge integrity, and the ecological environment. Therefore, new research from a different perspective is needed to propose waterway regulation strategies that align with the current trend.

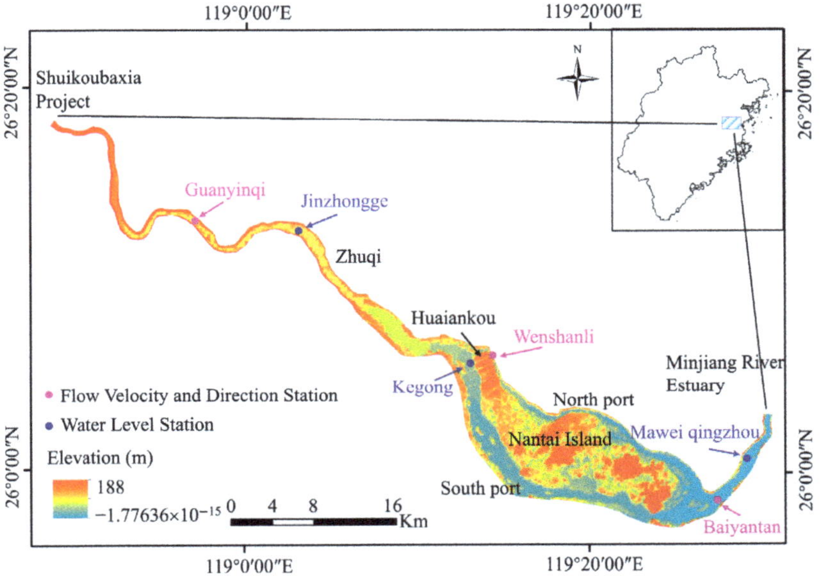

Figure 1. Schematic map of the study area and monitoring stations.

The arrangement of hydraulic structures significantly affects the WDR. In 1981, a 450 m-long diversion dyke was constructed on the left side of the South Port's diversion entrance in the Minjiang River. This structure played a crucial role in regulating the WDR between the South and North Ports. However, the regulatory effectiveness of the diversion dyke now faces several challenges due to changes in WDR trends. First, the diversion dyke, built in 1981, is no longer suited to current WDR conditions. Second, the spur dyke groups along the Zhuqi to Huaiankou river reach have been damaged by continuous water scouring over time. Notably, the diversion dyke at Huaiankou has experienced severe damage from repeated floods, significantly reducing its ability to control water flow. Therefore, exploring rehabilitation measures for the diversion dyke and assessing its impact on the WDR under the current morphological evolution of the South and North Ports holds significant engineering and academic value.

2.2. Model Establishment

This study utilizes the MIKE 21 two-dimensional hydrodynamic model developed by the DHI for the calculations. The computational principles of this model are based on the Reynolds-averaged Navier–Stokes equations and adhere to the Boussinesq assumption and the hydrostatic pressure assumption. The equations are solved using a finite volume method for spatial discretization. The continuity equation for water flow is expressed as follows:

$$\frac{\partial h}{\partial t} + \frac{\partial h\overline{u}}{\partial x} + \frac{\partial h\overline{v}}{\partial y} = hS \tag{1}$$

The momentum equations for water flow in the x and y directions are as follows:

$$\frac{\partial h\overline{u}}{\partial t}+\frac{\partial h\overline{u}^2}{\partial x}+\frac{\partial h\overline{v}\overline{u}}{\partial y}=f\overline{v}h-gh\frac{\partial \eta}{\partial x}-\frac{h\partial p_a}{\rho_0\partial x}-\frac{gh^2\partial \rho}{2\rho_0\partial x}+\frac{\tau_{sx}}{\rho_0}-\frac{\tau_{bx}}{\rho_0}-\frac{1}{\rho_0}\left(\frac{\partial S_{xx}}{\partial x}+\frac{\partial S_{xy}}{\partial y}\right)+\frac{\partial}{\partial x}(hT_{xx})+\frac{\partial}{\partial y}(hT_{xy})+hu_sS \quad (2)$$

$$\frac{\partial h\overline{v}}{\partial t}+\frac{\partial h\overline{u}\overline{v}}{\partial x}+\frac{\partial h\overline{v}^2}{\partial y}=-f\overline{u}h-gh\frac{\partial \eta}{\partial y}-\frac{h\partial p_a}{\rho_0\partial y}-\frac{gh^2\partial \rho}{2\rho_0\partial y}+\frac{\tau_{sy}}{\rho_0}-\frac{\tau_{by}}{\rho_0}-\frac{1}{\rho_0}\left(\frac{\partial S_{yx}}{\partial x}+\frac{\partial S_{yy}}{\partial y}\right)+\frac{\partial}{\partial x}(hT_{xy})+\frac{\partial}{\partial y}(hT_{yy})+hv_sS \quad (3)$$

where $\overline{u}$ and $\overline{v}$ are depth-averaged velocities; S is the magnitude of discharge due to point sources; t is time; x and y are Cartesian coordinates; η is surface elevation; d is still water depth; $h = \eta + d$ is total water depth; f is the Coriolis parameter; g is gravitational acceleration; p_a is atmospheric pressure; ρ is the density of water; ρ_0 is the reference density of water; τ_{sx} and τ_{sy} are the x and y components of the surface wind; τ_{bx} and τ_{by} are the x and y components of the bottom stresses; S_{xx}, S_{xy}, S_{yx}, and S_{yy} are components of the radiation stress tensor; u_s and v_s represent the velocity by which the water is discharged into the ambient water; and T_{xx}, T_{xy}, and T_{yy} are horizontal stress terms.

The study area encompasses the Minjiang River from the Shuikoubaxia project to the river's estuary, spanning approximately 117 km. The section from the Shuikoubaxia project to Huaiankou extends about 60 km, while the North Port and South Port measure approximately 32 km and 34 km, respectively. The computational domain is shown in Figure 1. The model simulates hydrodynamic processes during the wet season, covering the period from 25 May to 10 June 2017. A spring tide occurred on 26 May, followed by a neap tide on 2 June.

The model mesh was generated using the Gauss–Kruger projection system (CGCS2000 3 Degree GK CM 120E) and employed an unstructured triangular mesh to discretize the computational domain. To ensure the quality of the mesh and better focus on the changes in the WDR of the South and North Ports, mesh refinement was applied at the Huaiankou bifurcation and along the South and North Channels, with a minimum mesh size of 50 m in critical areas and a maximum size of 80 m in peripheral regions. The study area included 49,492 nodes and 94,760 elements.

Closed boundaries were applied to levees on both sides and the land boundaries of river islets. An average discharge of 1108.92 m^3/s, measured at Zhuqi Station during the 2017 wet season, was applied as the upstream boundary condition. The downstream boundary condition was defined by hourly tidal levels at the Minjiang River estuary, with tidal data simulated using the MIKE 21 tidal prediction function. The dynamic boundary condition adopted a "flood and dry" approach, setting drying depth to 0.005 m, flooding depth to 0.05 m, and wetting depth to 0.1 m. Table 1 summarizes the boundary conditions.

Table 1. Model boundary conditions.

Boundary Type	Location Description	Parameter Settings
Upstream Boundary	Shuikoubaxia project	Discharge of 1108.92 m^3/s, measured at Zhuqi Station during the 2017 wet season.
Downstream Boundary	Minjiang River estuary	Defined by hourly tidal levels at the Minjiang River estuary, simulated using the MIKE 21 tidal prediction function.
Closed Boundaries	Levees on both sides and land boundaries of river islets	Impermeable boundary.
Dynamic Boundary	Wetting and drying boundaries across the entire domain	Drying depth = 0.005 m; flooding depth = 0.05 m; wetting depth = 0.1 m.

The CFL number was set to 0.8, and the time step interval to 10 s. To maintain numerical stability, MIKE 21 ensures that the CFL condition is satisfied at all mesh points by setting a maximum and minimum time step. The computational time step (Δt) is automatically adjusted during model simulation. We set the maximum time step to 10 s and the minimum time step to 0.01 s. The Manning's roughness coefficient ranged from 0.014 to 0.017, increasing gradually from the Shuikoubaxia project to the estuary.

2.3. Model Validation

The model was validated using water levels, flow velocity and direction, and the WDR. For water level validation, data were collected from three stations: Jinzhongge, Kegong, and Maweiqingzhou. Flow velocity and direction validation used data from Guanyinqi, Wenshanli, and Baiyantan stations. The locations of these stations are shown in Figure 1.

The water level validation used hydrological measurement data from three water level stations collected from 12:00 on 17 November 2017 to 00:00 on 1 December 2017. Water levels were referenced to the zero point at Luoxingta. The validation results are shown in Figure 2. To quantify the agreement between the simulated and observed water levels, the mean absolute error (MAE) and mean squared error (MSE) were calculated, as presented in Table 2. The results indicate that the simulated water levels align well with the measured values. Although minor discrepancies occurred occasionally, the overall simulation results are reliable.

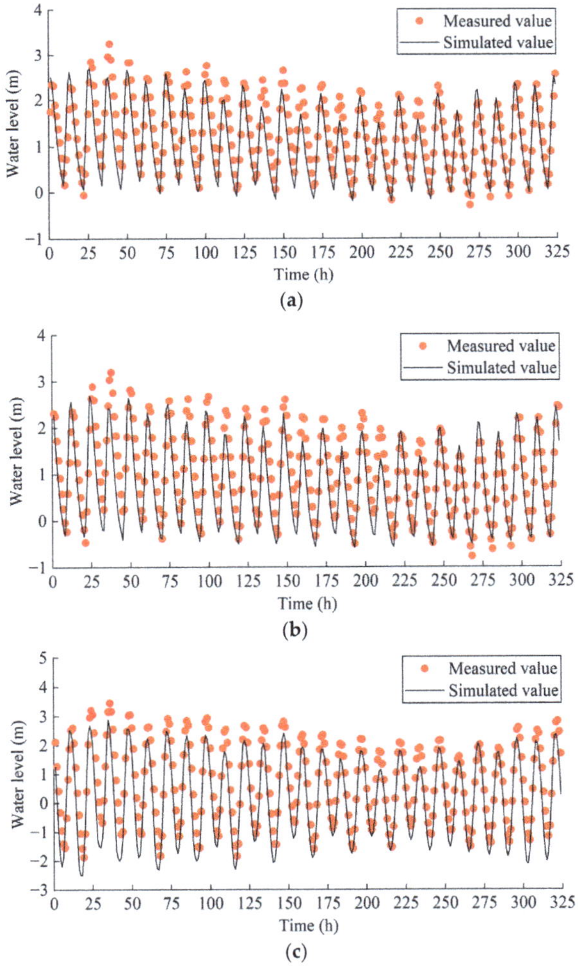

Figure 2. Validation of water levels at monitoring stations. (**a**) Jinzhongge. (**b**) Kegong. (**c**) Maweiqingzhou.

Table 2. The comparison between the simulated and observed values of water level.

Monitoring Stations	Mean Absolute Error (MAE)	Mean Squared Error (MSE)
Jinzhongge	0.37 m	0.20
Kegong	0.44 m	0.26
Maweiqingzhou	0.61 m	0.57

The flow velocity and direction validation utilized observed data from three monitoring stations, collected from 13:00 on 18 November 2017 to 16:00 on 19 November 2017. As shown in Figure 3, although the simulated values generally align with the measured values in terms of overall trend, phase errors may cause deviations in flow direction and velocity during rapid tidal flow reversals. The MAE of flow velocity at each monitoring station ranges from 0.13 to 0.37 m/s, with the MSE ranging from 0.04 to 0.23. The overall simulation performance for flow velocity is satisfactory. Additionally, the MAE of flow direction calculated by the model ranges from 36° to 51°. To further verify the accuracy and reliability of the model, the WDR between the South and North Ports during the wet season was examined. The validation data were based on the average WDR for the South and North Ports during the wet season of 2017. The simulated water diversion ratio for the North Port was 20.16%, while the measured ratio was 15.50%, with a relative error within 5%, demonstrating the model's reliability [24].

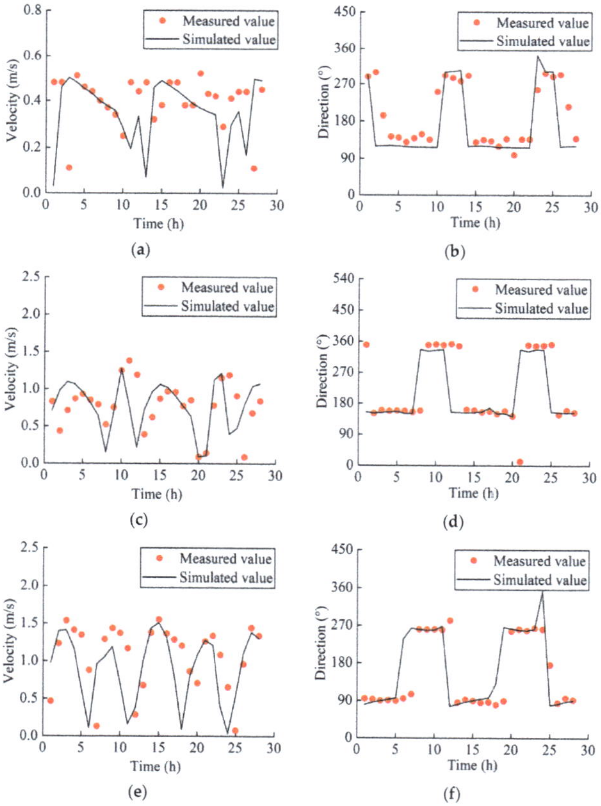

Figure 3. Validation of flow velocity and direction (flow direction is defined with 0° as true north and increases clockwise). (**a**,**b**) Guanyinqi monitoring station. (**c**,**d**) Wenshanli monitoring station. (**e**,**f**) Baiyantan monitoring station.

2.4. Calculation Method of the WDR

Yu et al. [25] redefined the tidal fluctuation WDR, which refers to the percentage of tidal inflow or outflow through all cross-sections between the South and North Ports of the Mingjiang River. This includes the shoals and main channels during the entire process of flood tide or ebb tide. The WDR is relative to the total tidal inflow or outflow measured simultaneously at the South and North Ports. Therefore, the WDR formula for the South and North Ports analyzed in this paper is

$$\mu_N = \frac{Q_N}{Q_N + Q_S}; \mu_S = \frac{Q_S}{Q_N + Q_S} \tag{4}$$

$$\mu_N{}' = \frac{|Q_N|}{|Q_N| + |Q_S|}; \mu_S{}' = \frac{|Q_S|}{|Q_N| + |Q_S|} \tag{5}$$

where μ_S is the WDR of the South Port; μ_N is the WDR of the North Port; Q_S and $|Q_S|$ are the flow at the Kegong cross-section in the South Port and its absolute value; Q_N and $|Q_N|$ are the flow at the Wenshanli cross-section in the North Port and its absolute value; and $\mu_S{}'$ and $\mu_N{}'$ are the WDR of the South and North Ports calculated using absolute flow values.

The average WDRs for the South and North Ports during different periods are calculated using the following equations:

$$\overline{\mu_S} = \frac{1}{T_2 - T_1} \int_{T_1}^{T_2} \mu_S dT \tag{6}$$

$$\overline{\mu_N} = \frac{1}{T_2 - T_1} \int_{T_1}^{T_2} \mu_N dT \tag{7}$$

where T_1 and T_2 are the start time and end time of the calculation period, respectively; $\overline{\mu_S}$ and $\overline{\mu_N}$ are the average WDRs of the South Port and North Port during the calculation period; and μ_S and μ_N are the instantaneous WDRs of the South Port and North Port at a specific time.

3. Results and Discussion

3.1. Spatiotemporal Dynamic Characteristics of the WDR Changes in the South and North Ports

This section analyzes the WDR simulation results for both the spring and neap tides during the wet season based on tidal cycle scales. The WDR for the South and North Ports on 26 May, calculated using Equation (4), is shown in Figure 4. During certain tidal phases, such as around high and low slack tides or one to two hours before or after these phases, the WDR between the South and North Ports may take negative values. For example, during the first ebb tide stage at 00:00, the WDR of the North Port is −53%; during the second ebb tide stage at 12:00, it is −41%; and during the second flood tide stage at 21:00, it is −11%. At other times, the WDR mainly ranges between 10% and 27%.

To explore the hydrodynamic mechanisms and dynamic characteristics of the WDR during these moments, the flow field and cross-sectional flow rates at 00:00 are analyzed as an example. Figure 5 shows the flow field near Huaiankou at 00:00 on 26 May. At this time, the flow pattern at the South and North Ports is complex. The flow rate at the Wenshanli section of the North Port is −475.51 m^3/s, with the flow direction towards the downstream Minjiang River estuary. The flow rate at the Kegong section of the South Port is 1371.56 m^3/s, with the flow direction towards the upstream Zhuqi. When calculated with Equation (4), the total flow of the South and North Ports is positive, but the flow at the North Port is negative, resulting in a negative WDR for the North Port.

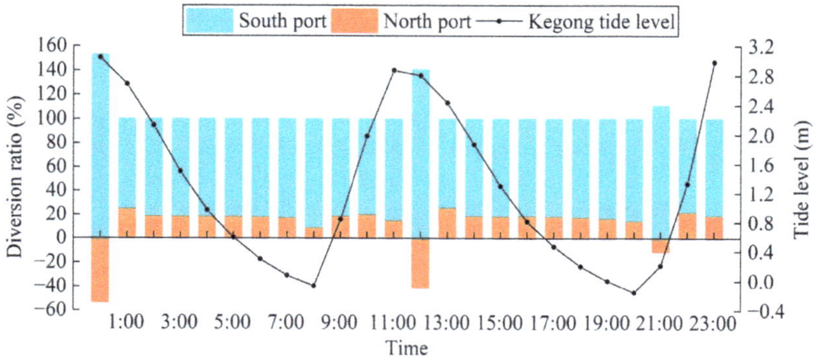

Figure 4. Hourly WDR of the South and North Ports during spring tides in the wet season calculated using Equation (4).

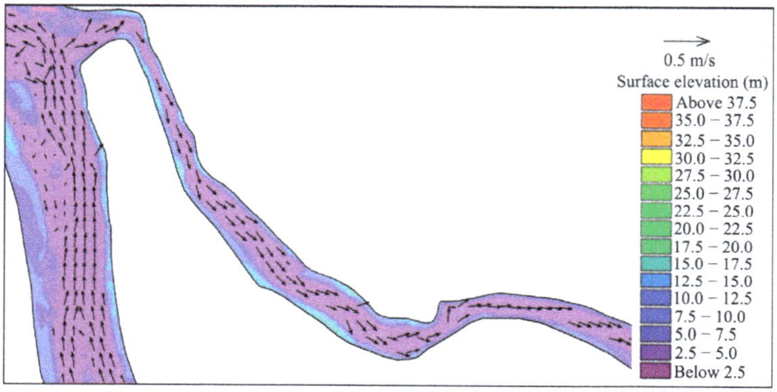

Figure 5. Flow field distribution near Huaiankou at 0:00 on 26 May.

Moments when the WDR becomes negative typically occur during the transition between flooding and ebbing tides at the South and North Ports. This is due to the combined effects of channel morphology, water level changes, and tidal propagation. The flow direction near Kegong in the South Port segment lags behind that in the North Port segment. During this time, the spatial distribution of flow velocity and direction in the South and North Ports differs significantly, resulting in an unstable flow pattern. The flow rates at the Kegong and Wenshanli sections are positive and negative, respectively. As a result, the WDR calculated using Equation (4) becomes negative. Therefore, Equation (5) is used to recalculate the simulation results in order to eliminate the directional effects on the WDR calculation, and these moments are marked with red borders, as shown in Figure 6a. After the revision, the previously negative water diversion ratios at 00:00, 12:00, and 21:00 were adjusted to 26%, 22%, and 9%, respectively. For subsequent data analysis, Equation (5) is used to calculate the WDR for the South and North Ports.

The hourly variations in the WDR of the South and North Ports and the tide level at Kegong during spring and neap tides are shown in Figure 6. During neap tide days (i.e., on 2 June), the WDR of the North Port is relatively low at 02:00 and 06:00, at 3% and 6%, respectively, reaching higher values of 33% and 69% at 07:00 and 19:00. At other times, the WDR fluctuates between 12% and 26%. On spring tide days, the WDR of the North Port gradually decreases during the initial ebb phase as the tide level falls. During the second flood phase, it exhibits a pattern of initially decreasing, subsequently increasing, and then decreasing again. This is likely due to the complex flow dynamics during the transition between flood and ebb tides as described above. In other phases of spring tide days, it generally shows a trend of first increasing and then decreasing.

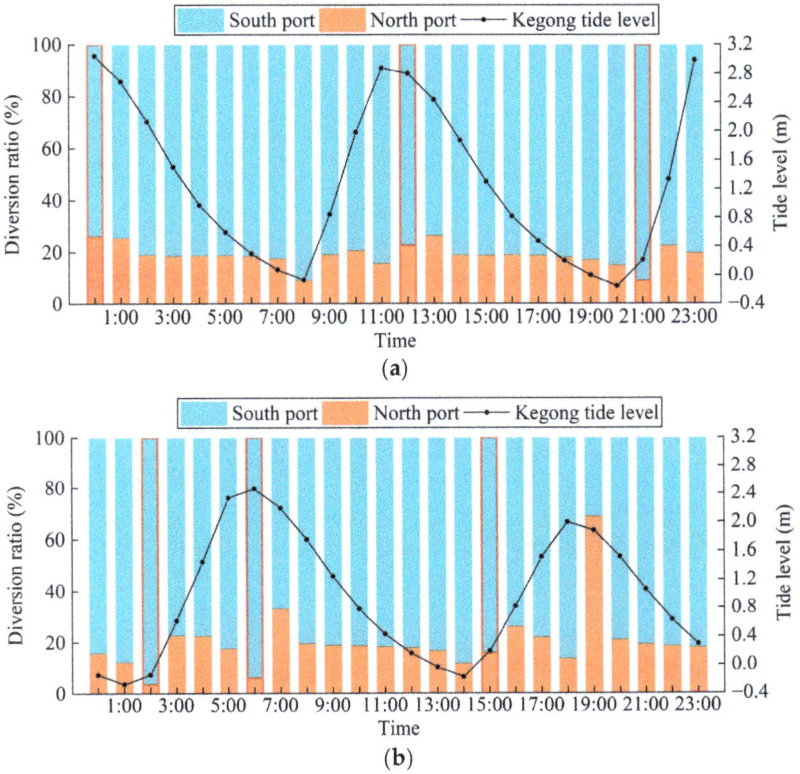

Figure 6. Hourly WDR of the South and North Ports during spring and neap tides in the wet season calculated using Equation (5). (**a**) Spring tide. (**b**) Neap tide.

On neap tide days, the WDR of the North Port exhibits a similar pattern during both the flood and ebb phases, initially increasing and then decreasing. However, during the first flood phase, it also shows the same variation pattern as observed during the second flood phase on spring tide days. In summary, with the variation in the tide level, the WDR of the North Port generally follows a pattern of first increasing and then decreasing, except for certain special moments.

Table 3 presents the average WDR of the North Port during different periods of spring and neap tides in the wet season. The average WDR of the North Port is 19.13% on spring tide days and 20.17% on neap tide days, indicating it is higher during neap tides than spring tides. However, both values are close to 20%, indicating strong agreement between simulation and observed data. This result aligns with previous studies based on hydrological station data and simulation results using the Delft3D model, which indicate that the annual average WDR between the South and North Ports during the wet season in recent years is approximately 4:1 [26,27]. The calculation results further reveal that during the ebb tide phase on neap tide days, the average WDR of the North Port is greater than that on spring tide days. Conversely, during the flooding phase, the WDR is higher on spring tide days than on neap tide days. Regardless of the tidal type, the WDR of the North Port during the ebb tide phase consistently exceeds that during the flooding phase. These patterns are consistent with findings from previous research [27]. Furthermore, the WDR during flood and ebb tidal surges is generally higher than during slack tide moments.

Table 3. Average WDR of the North Port during different periods of spring and neap tides in the wet season.

Spring Tide		Neap Tide	
Over the entire day	19.13%	Over the entire day	20.17%
Ebb tide stage	19.32%	Ebb tide stage	20.76%
Flood tide stage	16.63%	Flood tide stage	16.00%
Ebbing tide (4:00, 15:00)	19.00%	Ebbing tide (10:00, 21:00)	19.50%
Low slack tide (8:00, 20:00)	12.50%	Low slack tide (1:00, 14:00)	12.50%
Flooding tide (9:00, 22:00)	21.00%	Flooding tide (4:00, 16:00)	24.50%
High slack tide (0:00, 11:00)	21.00%	High slack tide (6:00, 18:00)	10.00%

3.2. *The Impact of a Diversion Dyke on the WDR*

The scenario design in this study is based on the current configuration of the diversion dyke in the lower reach of the Minjiang River. The study simulates the effects of the dyke's length and rotational angle on the WDR between the South and North Ports. A schematic diagram of the diversion dyke location is shown in Figure 7. Due to space constraints, this analysis focuses on WDR changes over half a tidal cycle during the spring and neap tide in the wet season. The analysis period for the spring tide is from 03:00 to 15:00 on 26 May, while for the neap tide, it covers the same time period on 2 June.

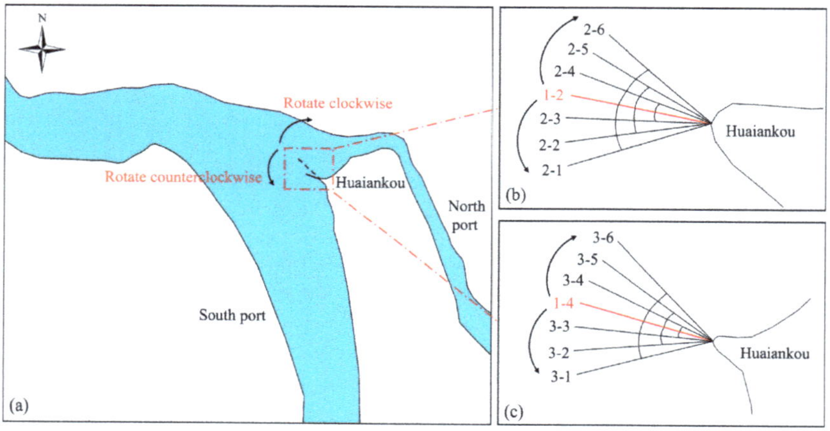

Figure 7. Schematic diagram of diversion dyke operational conditions. (**a**) Schematic diagram of diversion dyke locations. (**b**) Rotation angle scenarios for the 110 m-long diversion dyke. (**c**) Rotation angle scenarios for the 450 m-long diversion dyke.

3.2.1. The Variation in Diversion Dyke Length

The length of the diversion dyke at Huaiankou was determined using satellite imagery, revealing an existing length of approximately 110 m. Based on this, six different diversion dyke length conditions were designed (Table 4).

Table 4. Operating conditions for diversion dyke length.

Operational Condition Number	1-1	1-2	1-3	1-4	1-5	1-6	1-7
Diversion dyke length/m	50	110	300	450	500	550	600

The average WDR of the North Port under different operational conditions is provided in Table 5 and variation is illustrated in Figure 8. From the results, it can be observed that the length of the diversion dyke has a noticeable effect on the WDR of the South and North Ports, with the WDR fluctuating between 18% and 21%. When the dyke length ranges from 50 m to 300 m, the average WDR of the North Port remains relatively stable. However, the WDR starts to increase once the dyke length exceeds 300 m, with a more significant rise observed when the dyke length exceeds 450 m.

Table 5. Average WDR of the North Port for different dyke lengths.

Simulation Time	Diversion Dyke Length/m						
	50	110	300	450	500	550	600
Spring tide	19.10%	19.10%	19.16%	19.67%	19.68%	20.72%	20.76%
Neap tide	18.74%	18.74%	18.71%	18.99%	19.01%	19.40%	19.41%

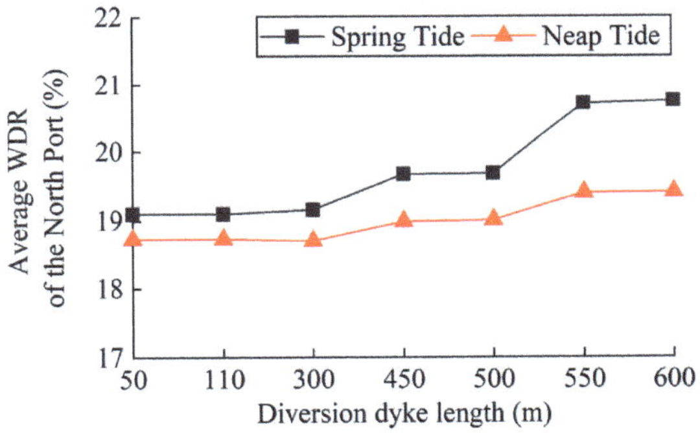

Figure 8. Variation in the average WDR of the North Port with different dyke lengths.

To further analyze the effect of dyke length on the WDR, the range of 450 m to 600 m was subdivided. Simulations were conducted for dyke lengths of 500 m and 550 m. The results indicate that the average WDR of the North Port differs by less than 0.02% when comparing dyke lengths of 450 m and 500 m. Similarly, the difference is within 0.04% between dyke lengths of 550 m and 600 m. Despite the small differences overall, the WDR increases most noticeably as the dyke length increases from 500 m to 550 m.

Based on the analysis above, the WDR of the North Port increases as the length of the diversion dyke extends. This finding aligns with the conclusions of Dai et al. [24]. When the dyke length is less than 300 m, its impact on regulating the WDR between the South and North Ports is minimal. However, once the dyke length exceeds 300 m, its effect on flow regulation becomes progressively noticeable. Within the range of 450 m to 600 m, dyke lengths of 500 m and 550 m are critical thresholds that most significantly affect the WDR variation between the South and North Ports.

3.2.2. The Variation in Diversion Dyke Angle

To further investigate the impact of the diversion dyke on the WDR between the South and North Ports, this study simulates scenarios with varying dyke angles (Table 6). Considering that the dyke length has changed over time due to water flow erosion, both the current dyke length of 110 m and the original dyke length of 450 m are included in the analysis. The scenarios assume that the dyke rotates either counterclockwise or clockwise at specific angles from its current position. When the dyke rotates counterclockwise, it gradually becomes perpendicular to the South Port flow, as illustrated in Figure 7.

Table 6. Operating conditions for diversion dyke angles (negative sign indicates counterclockwise rotation, while positive sign indicates clockwise rotation).

Operating Condition Number for the Dyke Length of 110 m	Operating Condition Number for the Dyke Length of 450 m	Operating Condition Description
2-1	3-1	−30°
2-2	3-2	−20°
2-3	3-3	−10°
1-2	1-4	0°
2-4	3-4	+10°
2-5	3-5	+20°
2-6	3-6	+30°

Based on the simulations of the operating conditions in Table 6, Table 7 presents the calculated average WDR values for the North Port under different diversion dyke angles. The data show that for a dyke length of 110 m, the WDR values for the North Port remain nearly identical across all conditions. Any differences are minimal, within a range of 0.01%, and can be considered negligible. The average WDR of the North Port does not vary with the rotation angle of the diversion dyke. Combined with the simulation results from Section 3.2.1, which demonstrated that WDR values for dyke lengths shorter than 300 m were nearly the same, it can be concluded that the existing diversion dyke at Huaiankou has little to no effect on regulating water flow between the South and North Ports.

Table 7. Average WDR of the North Port for different dyke angles.

Diversion Dyke Length/m	Simulation Time	Rotation Angle of the Diversion Dyke						
		−30°	−20°	−10°	0°	+10°	+20°	+30°
110	Spring tide	19.10%	19.09%	19.09%	19.10%	19.09%	19.09%	19.09%
	Neap tide	18.74%	18.74%	18.74%	18.74%	18.74%	18.74%	18.74%
450	Spring tide	24.98%	21.52%	20.42%	19.67%	19.25%	17.55%	14.18%
	Neap tide	23.60%	19.58%	19.26%	18.99%	18.76%	17.35%	13.98%

As shown in Figure 9, when the dyke length is 450 m, the WDR of the North Port responds more significantly to changes in the rotation angle compared to the scenario with a dyke length of 110 m. Larger rotation angles lead to more pronounced changes in the WDR, consistent with the findings of Yu et al. [28]. When the diversion dyke rotates counterclockwise, its contact with the South Port water flow gradually increases. This intercepts more water from the South Port, resulting in a higher WDR for the North Port. The changes in the WDR for counterclockwise rotations of 10° and 20° are similar. However, at a 30° counterclockwise rotation, the dyke becomes nearly perpendicular to the South Port flow direction, causing a sharp increase in the WDR for the North Port, which exceeds 23%.

Conversely, when the dyke rotates clockwise, its contact with the North Port water flow increases, leading to a continuous decrease in the WDR of the North Port. A 10° clockwise rotation causes only a slight decrease in the North Port's WDR. As the rotation angle increases to 20°, the reduction becomes more pronounced, reaching a maximum decrease at a 30° clockwise rotation, where the North Port's WDR drops below 15%.

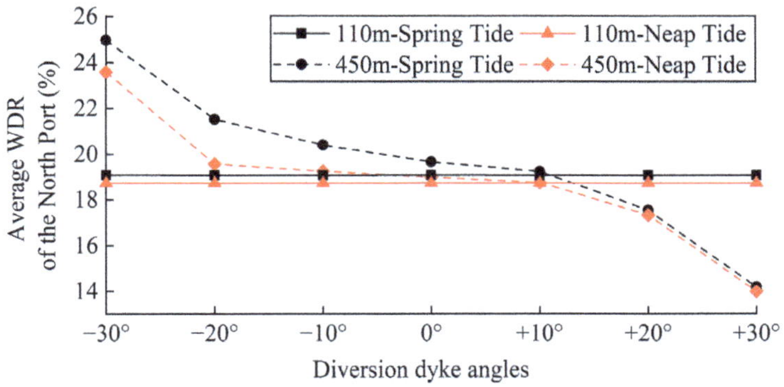

Figure 9. Variation in the average WDR of the North Port with different rotation angles.

4. Conclusions

In this study, a two-dimensional hydrodynamic mathematical model for the lower Minjiang River was developed based on measured data and the MIKE 21 hydrodynamic module. It is noteworthy that the current model focuses on hydrodynamic processes at the tidal cycle scale and does not incorporate sediment transport dynamics, which are critical for long-term morphological adjustments and bifurcation ratio evolution. Future research will address this limitation by coupling sediment transport modules to comprehensively analyze the interplay between hydrodynamics, sediment interactions, and channel morphology. The model was utilized to analyze the dynamic variation patterns of the WDR for the South and North Ports during the spring and neap tides of the wet season and to quantitatively evaluate the impacts of diversion dyke configurations on changes in the WDR for the South and North Ports. The main conclusions are as follows:

(1) During the wet season, the WDR of the North Port is approximately 20%. The WDR is higher during the ebb phase of the neap tide, but lower during its flood phase compared to the spring tide. Regardless of tidal conditions, the WDR of the North Port is consistently higher during the ebb phase than during the flood phase. Furthermore, the WDR during the flooding and ebbing surge periods is higher than during the slack tide moments. During the flood and ebb phases, the WDR of the North Port generally shows a trend of first increasing and then decreasing as the tide level rises and falls.
(2) The WDR of the North Port increases as the length of the existing dyke increases. Once the dyke length exceeds approximately 450 m, the WDR begins to increase significantly. Within the range of 450 m to 600 m, the dyke lengths of 500 m and 550 m are the critical values influencing the variation in the WDR for both the South and North Ports.
(3) The existing diversion dyke with a length of 110 m has little effect on regulating the water flow between the South and North Port, and changes in the angle of the dyke do not significantly affect the WDR. However, the original diversion dyke with a length of 450 m is more sensitive to angle changes. When the rotation angle reaches 30°, the WDR shows a significant change. The larger the counterclockwise rotation angle of the dyke, the more pronounced the increase in the WDR of the North Port.

Author Contributions: Conceptualization, C.C.; Methodology, C.C. and W.Y.; Software, W.Y.; Validation, H.C.; Formal analysis, H.C.; Investigation, W.Y.; Resources, Z.W. and K.H.; Data curation, K.H.; Writing— original draft, W.Y.; Writing—review & editing, C.C. and F.C.; Visualization, Y.C.; Supervision, C.C., Z.W. and F.C.; Funding acquisition, Z.W. and K.H. All authors have read and agreed to the published version of the manuscript.

Funding: This research is supported by the Transportation Science and Technology Project of Fujian Province (JC202316) and the National Natural Science Foundation of China (U22A20585).

Data Availability Statement: Data will be made available on request.

Conflicts of Interest: Authors Zhihui Wang and Kailong Huangfu were employed by the company Fujian Port & Waterway Investigation and Design Institute Limited. The remaining authors declare that the research was conducted in the absence of any commercial or financial relationships that could be construed as a potential conflict of interest.

References

1. Chen, Q.; Zhu, J.R.; Lyu, H.H.; Pan, S.Q.; Chen, S.L. Impacts of topography change on saltwater intrusion over the past decade in the Changjiang. *Estuar. Coast. Shelf Sci.* **2018**, *231*, 106469. [CrossRef]
2. Hu, P.; Lei, Y.L.; Deng, S.Y.; Cao, Z.X.; Liu, H.H.; He, Z.G. Role of bar-channel interactions in a dominant branch shift: The Taipingkou waterway, Yangtze River, China. *River Res. Appl.* **2021**, *37*, 494–508. [CrossRef]
3. Prandle, D.; Lane, A. Sensitivity of estuaries to sea level rise: Vulnerability indices. *Estuar. Coast. Shelf Sci.* **2015**, *160*, 60–68. [CrossRef]
4. Ankita, P.; Mery, B. Changes in river bed terrain and its impact on flood propagation—A case study of River Jayanti, West Bengal, India. *Geomat. Nat. Hazards Risk* **2019**, *10*, 1928–1947. [CrossRef]
5. Kleinhans, M.G. Sorting out river channel patterns. *Prog. Phys. Geogr.-Earth Environ.* **2010**, *34*, 287–326. [CrossRef]
6. Carling, P.; Jansen, J.; Meshkova, L. Multichannel rivers: Their definition and classification. *Earth Surf. Process. Landf.* **2014**, *39*, 26–37. [CrossRef]
7. Best, J. Anthropogenic stresses on the world's big rivers. *Nat. Geosci.* **2019**, *12*, 7–21. [CrossRef]
8. Luchi, R.; Zolezzi, G.; Tubino, M. Modelling mid-channel bars in meandering channels. *Earth Surf. Process. Landf.* **2010**, *35*, 902–917. [CrossRef]
9. Wang, D.W.; Ma, Y.X.; Liu, X.F.; Huang, H.Q.; Huang, L.; Deng, C.Y. Meandering-anabranching river channel change in response to flow-sediment regulation: Data analysis and model validation. *J. Hydrol.* **2019**, *579*, 124209. [CrossRef]
10. Lyu, B.; Li, J.; Hu, P.; Cao, Z.; Liu, H. High-resolution hydro-sediment-morphodynamic modelling of a meandering river reach with mid-channel bars on multiyear timescales: A case study of Shashi Reach in Middle Yangtze River. *J. Hydrol.* **2024**, *635*, 131167. [CrossRef]
11. Song, W.W.; Pang, Y. Research on the influence of diversion flow ratio on the water environment in Nanjing. *Water Supply* **2019**, *19*, 2367–2378. [CrossRef]
12. Alomari, N.K.; Yusuf, B.; Mohammad, T.A.; Ghazali, A.H. Experimental investigation of scour at a channel junctions of different diversion angles and bed width ratios. *Catena* **2018**, *166*, 10–20. [CrossRef]
13. Abdulhafedh, A.Y.; Alomari, N.K.; Al-Janabi, A.M.S. Scour hole reduction at a diversion channel junction using different entrance edge shapes. *Int. J. Sediment Res.* **2023**, *38*, 811–820. [CrossRef]
14. Xu, X.S.; Zhou, C.; Tan, Y.S.; Chen, D.B.; Fu, J.; Chai, C.; Liang, L.F. Analyzing the diversion effect of debris flow in cross channels utilizing two-phase flow theory and the principle of energy conservation. *Water* **2024**, *16*, 2134. [CrossRef]
15. Zhang, J.H.; Yu, M.H. Experimental study of flood diversion in the middle and lower Han River, China. *Can. J. Civ. Eng.* **2014**, *41*, 381–388. [CrossRef]
16. Wang, C.L.; Zhao, Y.; Ning, L.; Bi, J. Permeability Evolution of Coal Subjected to Triaxial Compression Based on In-Situ Nuclear Magnetic Resonance. *Int. J. Rock Mech. Min. Sci.* **2022**, *158*, 105213. [CrossRef]
17. Wang, C.L.; Pan, L.H.; Zhao, Y.; Zhang, Y.F.; Shen, W.K. Analysis of the Pressure-Pulse Propagation in Rock: A New Approach to Simultaneously Determine Permeability, Porosity, and Adsorption Capacity. *Rock Mech. Rock Eng.* **2019**, *52*, 4301–4317. [CrossRef]
18. Gao, Y.Y.; Yang, H.J.; Wang, L.Z.; Zhao, M. Three-dimensional numerical investigation on flow behaviors around a diversion dike. *Phys. Fluids* **2022**, *34*, 125119. [CrossRef]
19. Ranjbar-Zahedani, M.; Keshavarzi, A.; Khabbaz, H.; Ball, J.E. Optimizing flow diversion structure as an effective pier-scour countermeasure. *J. Hydraul. Res.* **2021**, *59*, 963–976. [CrossRef]
20. Iqbal, S.; Tanaka, N. An experimental investigation on Dike stabilization against floods. *Geosciences* **2023**, *13*, 307. [CrossRef]

21. Nandhini, D.; Murali, K.; Harish, S.; Schüttrumpf, H.; Heins, K.; Gries, T. A state-of-the-art review of normal and extreme flow interaction with spur dikes and its failure mechanism. *Phys. Fluids* **2024**, *36*, 051301. [CrossRef]
22. Xie, R.R.; Rao, P.Y.; Pang, Y.; Shi, C.C.; Li, J.B.; Shen, D.D. Salt intrusion alters nitrogen cycling in tidal reaches as determined in field and laboratory investigations. *Sci. Total Environ.* **2020**, *729*, 138803. [CrossRef] [PubMed]
23. Xie, R.R.; Pang, Y.; Luo, B.R.; Li, J.B.; Wu, C.S.; Zheng, Y.Y.; Sun, Q.Y.; Zhang, P.; Wang, F.F. Spatiotemporal variability in salinity and hydraulic relationship with salt intrusion in the tidal reaches of the Minjiang River, Fujian province, China. *Environ. Sci. Pollut. Res.* **2017**, *24*, 11847–11855. [CrossRef]
24. Dai, W.H.; Wu, S.X.; Zhang, Y.; Tang, H.W. Preliminary studies on improvement measures to the diversion ratio of Nanjing Baguazhou branch. *Hydro-Sci. Eng.* **2013**, 1–7. [CrossRef]
25. Yu, Z.M.; Zou, Q.S.; Chen, Y.Q.; Xu, P.; Wang, L.H. Analysis of the variation characteristics of diversion ratios between the South and North Ports of the Minjiang River during flood and ebb tides. In Proceedings of the 2022 Chinese Hydraulic Engineering Society Academic Conference (Volume 5), Beijing, China, 8 November 2022; Yellow River Water Conservancy Press: Beijing, China, 2022; pp. 224–230.
26. Wu, J. Analysis of the variation characteristics and causes of the diversion ratio of North and South Ports in the Mingjiang River. *Hydraul. Sci. Technol.* **2024**, 16–21.
27. Chen, B.Y. Application of the Delft3D 2D hydrodynamic model in the study of tidal range, water levels, and diversion ratio of North and South Ports in the lower reaches of the Mingjiang River. *Hydraul. Sci. Technol.* **2020**, 28–33.
28. Yu, P.P.; Tang, H.W.; Xiao, Y.; Fu, H.; Chen, J.M.; Liang, D.F. Effects of a diversion dyke on river flow: A case study. *Proc. Inst. Civ. Eng.-Water Manag.* **2015**, *168*, 210–219. [CrossRef]

water

MDPI

Article

Spatial and Temporal Shifts and Driving Mechanisms of Embodied Carbon in Water Transport Trade in BRICS Countries

Shanshan Zheng [1,*], Cheng Chen [2] and Peng Qiu [2]

[1] School of International Economics and Trade, Fujian Business University, Fuzhou 350012, China
[2] College of Civil Engineering, Fuzhou University, Fuzhou 350108, China; chencheng_1117@163.com (C.C.); 13959483235@163.com (P.Q.)
* Correspondence: zssfzdx@163.com

Abstract: From an ecological protection perspective, clarifying the spatial and temporal transfer characteristics of embodied carbon in water transport trade among BRICS countries and its driving mechanisms is of great significance for the precise formulation of emission reduction policies. This study integrates the multi-regional input–output model with the LMDI decomposition method to quantitatively analyze the bi-directional flow of embodied carbon in water transport trade among BRICS countries from 1995 to 2018, along with its spatio-temporal differentiation patterns. The driving mechanisms are decomposed across three dimensions: scale, structure, and intensity. By adopting a dual perspective of time-series and spatial correlation, the study systematically uncovers the cross-regional transfer patterns of embodied carbon emissions in water transport trade and examines the interaction pathways of various effects throughout their dynamic evolution. The study finds that (1) the embodied carbon in water transport trade among BRICS countries shows a trend of transnational transfer, with China being the largest net exporter (35.15 Mt in 2018), India and South Africa as net importers (−32.00 Mt and −1.89 Mt in 2018, respectively), and Brazil and Russia shifting from net importers to net exporters; (2) from a temporal perspective, the scale effect drives the growth of embodied carbon emissions (contribution values: 1.23~119.72 Mt for export trade; 4.88~34.36 Mt for import trade), while the intensity effect has a suppressive role (contribution values: −59.08~−1.48 Mt for export trade; −20.56~−5.31 Mt for import trade), and the structural effect is complex in its impact on emissions (contribution values: −17.72~0.45 Mt for export trade; −6.84~13.93 Mt for import trade). Optimizing the trade structure can help reduce carbon emissions; (3) from a spatial perspective, carbon emissions are higher in Southeast Asia and the Northern Hemisphere, and changes in China's carbon emissions (total effect in 2018: 57.01 Mt in export trade and 7.98 Mt in import trade) significantly affect other BRICS countries. Based on the conclusions of the study, it is suggested that BRICS countries should strengthen cooperation to achieve regional emission reduction targets by optimizing the trade structure of water transport, promoting energy structure reforms, advancing green transport technologies and equipment, and establishing a carbon emission regulatory system.

Keywords: ecological protection perspective; embodied carbon emissions; multi-regional input–output modelling; waterway transport industry; LMDI decomposition

1. Introduction

The global shipping industry handles approximately 90% of international trade, but its immense contribution to economic activities comes at a significant environmental cost—shipping. As a major source of greenhouse gas emissions, shipping increasingly drives climate change, and pollution costs have become progressively apparent [1]. Extensive research has been conducted on the environmental impacts of ships and related industries. For example, Premsamarn et al. [2] analyzed Automatic Identification System (AIS) data from the Gulf of Thailand to evaluate the emission reduction effects of Emission Control Area (ECA) policies on NO_X, SO_X, and particulate matter (PM). They identified tankers as the primary regional pollutant source and recommended measures such as low-sulfur fuels and route optimization to significantly reduce emissions, thereby providing a scientific basis for maritime emission reduction strategies in Southeast Asian countries. Similarly, Sha et al. [3] examined vessel entry and exit data from 189 global ports, systematically analyzing emission characteristics across different vessel types, seasons, and geographic regions. Their study confirmed that tankers and container ships are the main contributors to emissions, and they noted that hubs such as Singapore, Rotterdam, and Antwerp exhibit the highest emission intensities, which supports refined port-level emission management. Additionally, da Silva et al. [4] developed an AIS-based ship emission estimation framework by decoding vessel trajectories, integrating technical parameters with operational characteristics, and incorporating port-terminal vessel identification algorithms, significantly improving emission accounting accuracy. Notably, Yeh et al. [5] analyzed exhaust components from 16 different vessel types and found that naphthalene (Nap) accounts for 77–97% of polycyclic aromatic hydrocarbons (PAHs), with some vessels exceeding safety thresholds, posing direct threats to air quality and public health in port cities.

Due to its global and persistent nature, embodied carbon emissions from water transport have drawn considerable attention. Zheng et al. [6] employed input–output analysis combined with the Logarithmic Mean Divisia Index (LMDI) decomposition model to reveal that, despite ongoing efforts to achieve a low-carbon transition, the expansion of economic scale remains the dominant driver of carbon emissions in China's water transport sector, while improvements in energy efficiency and the adoption of clean technologies have effectively mitigated this trend. In this context, carbon intensity metrics have become key tools in balancing economic growth with emission reduction targets. Jaramillo et al. [7] compared four carbon intensity indicators for port tugboats, finding that those based on energy consumption are more practical due to their low variability and strong correlation with operational characteristics, whereas average load indicators, due to high dispersion, are less applicable. This highlights the importance of aligning indicator design closely with vessel-specific attributes. To achieve deep decarbonization, Raftis et al. [8] systematically evaluated the potential of alternative fuels for inland shipping, noting that while liquefied natural gas (LNG) was initially viewed as a transitional solution, battery and hydrogen technologies are emerging as focal points in Europe and Asia. Furthermore, Billington et al. [9] provided empirical evidence that renewable diesel, when used to replace conventional low-sulfur diesel, can reduce tugboat exhaust CO_2 emissions by 6%, offering a feasible pathway for short-term emission reductions.

Accurately quantifying embodied carbon in trade is fundamental for devising effective emission reduction strategies. Embodied carbon refers to the total greenhouse gas emissions generated throughout the production and supply chain of goods and services [10]. The primary methods for its calculation are as follows: (1) bottom-up Life Cycle Assessment (LCA), which is suitable for countries with comprehensive data or specific product analyses but is challenging to extend to the macro level [11]; (2) the IPCC standardized approach, which uses standard coal equivalence and sectoral emission factors to ensure cross-regional comparability for macro-scale studies [12]; and (3) input–output analysis (IOA), a top-down method that quantifies economy-wide emissions and analyzes inter-industry linkages, widely applied in energy and environmental research [13]. These methods provide a scientific basis for identifying high-carbon segments and optimizing supply chain management [14,15]. Furthermore, Al-Abdulqader et al. [16] conducted a cross-country empirical study demonstrating that carbon trading systems are more effective than carbon tax policies in driving emission reductions, emphasizing the need for tailored carbon pricing mechanisms that account for national economic structures and political contexts.

Despite significant advances, a systematic analysis of the spatiotemporal transfer and driving mechanisms of embodied carbon in water transport trade among BRICS countries (Brazil, Russia, India, China, and South Africa) [17] is still lacking. This study utilizes OECD Multi-Regional Input–Output (MRIO) tables and the LMDI decomposition method to elucidate the spatiotemporal patterns and driving factors of carbon emissions in the water transport sector among BRICS nations, thereby providing theoretical support and practical guidance for the formulation of differentiated emission reduction policies and regional low-carbon cooperation.

2. Methodology

2.1. Basic Principles of Input–Output Analysis

The input–output method visualizes the interrelations among various sectors in the national economy by constructing an input–output table. This method uses direct consumption coefficients to establish a functional relationship between production and consumption. According to traditional economic theory, total demand equals total supply; combined with the concept of intermediate products, the input–output balance is expressed as follows:

$$X = AX + Y \quad (1)$$

where X represents the total output column vector, Y is the final demand matrix (comprising domestic final demand Y_D and export demand Y_E), and A is the direct consumption coefficient matrix, indicating the direct inputs required from various sectors to produce one unit of the final product.

Through derivation, total output can be further expressed as follows:

$$X = (I - A)^{-1}Y = CY \quad (2)$$

Here, $(I - A)^{-1}$ is known as the Leontief inverse or complete demand coefficient matrix C, whose elements represent the total inputs required from each sector to obtain one unit of a sector's final product.

The use of complete demand coefficients allows for a more comprehensive reflection of resource consumption during production, which is critical for precise assessments of carbon emissions and resource consumption.

2.2. Calculation Methods for Trade-Embodied Carbon Emissions

2.2.1. Fundamental Formula for Measuring Trade-Embodied Carbon Emissions

Let e^d denote the direct carbon emission per unit output (i.e., the direct carbon emission coefficient) for each industry sector, where the elements of the row vector $e^d = \left[e_1^d, e_2^d, \cdots, e_j^d\right]$ represent the direct carbon emission coefficients for the respective sectors. Let F^d represent the total carbon emissions for all sectors in a country, i.e.,

$$F^d = e^d X \quad (3)$$

where X is the total output column vector. Substituting Equation (3) into Equation (2), we obtain

$$F^d = e^d X = e^d (I - A)^{-1} Y = E^d Y \quad (4)$$

where E^d represents the complete carbon emission coefficient, i.e., $(I - A)^{-1}$, and Y is the final demand vector.

Consequently, the calculation formula for trade-embodied carbon emissions, TC, is given by

$$TC = e^d (I - A)^{-1} TV \quad (5)$$

where TV denotes the trade volume (export volume, import volume, or net export volume).

2.2.2. Classification and Calculation Formulas for Trade-Embodied Carbon Emissions

Drawing upon previous studies [18], the embodied carbon emissions involved in the production process are further subdivided according to the production and consumption countries into four categories:

Category *I*: embodied carbon emissions from domestic production with domestic consumption:

$$X^I = e^d \left(I - A^d\right)^{-1} Y_D = E^d Y_D \quad (6)$$

where e^d represents the domestic direct carbon emission coefficient, $\left(I - A^d\right)^{-1}$ is the domestic full-demand matrix, and Y_D is the final domestic use vector.

Category *II*: embodied carbon emissions from domestic production but exported:

$$X^{II} = e^d \left(I - A^d\right)^{-1} Y_E \quad (7)$$

where Y_E represents the export vector.

Category *III*: embodied carbon emissions from direct consumption of imports:

$$X^{III} = E^d A^m \left(I - A^d\right)^{-1} Y_D + E^{im} Y_m \quad (8)$$

where $E^d A^m \left(I - A^d\right)^{-1} Y_D$ represents the portion of imported products used domestically after processing, and $E^{im} Y_m$ represents the portion of imported products directly used for domestic consumption. $E^d A^m \left(I - A^d\right)^{-1}$ denotes the domestic full carbon emission coefficient for imports, E^{im} denotes the full carbon emission coefficient of the importing country, and Y_m is the final use vector for imported products.

Category *IV*: embodied carbon emissions from re-export of imported intermediate inputs:

$$X^{IV} = E^d A^m \left(I - A^d\right)^{-1} Y_E \quad (9)$$

Based on this, the aggregates are defined as follows:

Total embodied carbon on the production side (EEP):

$$EEP = X^I + X^{II} = e^d \left(I - A^d\right)^{-1} (Y_D + Y_E) = e^d \left(I - A^d\right)^{-1} Y \quad (10)$$

Total embodied carbon on the consumption side (EEC):

$$EEC = X^I + X^{III} = E^d Y_D + E^d A^m \left(I - A^d\right)^{-1} Y_D + E^{im} Y_m \quad (11)$$

Export embodied carbon (EEE):

$$EEE = X^{II} + X^{IV} = E^{d}Y_{E} + E^{d}A^{m}\left(I - A^{d}\right)^{-1}Y_{E} \tag{12}$$

Imported embodied carbon (EEI):

$$EEI = X^{III} + X^{IV} = E^{d}A^{m}\left(I - A^{d}\right)^{-1}Y + E^{im}Y_{m} \tag{13}$$

Net embodied carbon transfer (EETB):

$$EETB = EEE - EEI = EEP - EEC \tag{14}$$

2.3. LMDI Decomposition Model in Temporal Dimension

Grossman and Krueger [19] examined the mechanisms by which trade liberalization impacts environmental pollution, proposing that international trade influences environmental quality through scale, structure, and intensity effects. The scale effect refers to the increase in energy consumption and carbon emissions resulting from trade expansion, while the structure effect reflects changes in trade composition impacting pollutant emissions. The intensity effect indicates that technological innovation can reduce carbon emissions, thereby improving overall environmental quality. Based on an extended Kaya–LMDI framework, this study develops a model for analyzing the driving factors behind embodied carbon emissions in water transport trade:

$$C_{w} = T \times \frac{T_{w}}{T} \times \frac{C_{w}}{T_{w}} \tag{15}$$

where C_w is the total trade-embodied carbon emissions of the country's water transport sector (in million tons); T is the total foreign trade of the country (in million USD); and T_w is the foreign trade volume of the country's waterway transport sector (in million USD).

The equation is further transformed, as shown below:

$$C_{w} = Q \times S_{w} \times R_{w} \tag{16}$$

where $Q = T$ represents the trade scale (i.e., total trade volume), reflecting the scale effect; S_w denotes the trade structure of water transport, i.e., the proportion of water transport in overall trade, which reflects the structural characteristics of the transportation mode and embodies the structure effect; and R_w represents the embodied carbon emission intensity of water transport (embodied carbon emission per unit trade volume), reflecting the intensity effect.

Using the LMDI additive decomposition model, the change in trade-embodied carbon emissions of the country's water transport sector from year 0 to year t (total effect ΔC_w) is determined by the scale effect ΔC_Q, structural effect ΔC_S, and intensity effect ΔC_R, with the following relationship between the effects:

The total effect is decomposed as follows:

$$\Delta C_{w} = C_{w}^{t} - C_{w}^{0} = Q^{t}S_{w}^{t}R_{w}^{t} - Q^{0}S_{w}^{0}R_{w}^{0} = \Delta C_{Q} + \Delta C_{S} + \Delta C_{R} \tag{17}$$

The formulas for calculating each effect are as follows:

Scale effect:

$$\Delta C_{Q} = \frac{C_{w}^{t} - C_{w}^{0}}{lnC_{w}^{t} - lnC_{w}^{0}} \ln\left(\frac{C_{Q}^{t}}{C_{Q}^{0}}\right) \tag{18}$$

Structural effect:

$$\Delta C_s = \frac{C_w^t - C_w^0}{lnC_w^t - lnC_w^0}\ln\left(\frac{C_S^t}{C_S^0}\right) \tag{19}$$

Intensity effect:

$$\Delta C_R = \frac{C_w^t - C_w^0}{lnC_w^t - lnC_w^0}\ln\left(\frac{C_R^t}{C_R^0}\right) \tag{20}$$

2.4. LMDI Decomposition Model in Spatial Dimension

To quantify the spatial differences in the influencing factors of trade-embodied carbon in the water transport industry across different regions (BRICS countries), and to facilitate direct comparisons of spatial differences, the average of the water transport trade-embodied carbon of BRICS countries is used as a benchmark. A spatial LMDI decomposition model is constructed to analyze the spatial differences in the influencing factors of trade-embodied carbon in 1995, 2003, 2011, and 2018. The model is built as follows:

The change in water transport trade-embodied carbon between region i and benchmark u is expressed as follows:

$$\Delta C_w^{(i-u)} = C_w^i - C_w^u = \Delta C_Q^{(i-u)} + \Delta C_S^{(i-u)} + \Delta C_R^{(i-u)} \tag{21}$$

The formulas for calculating each effect are as follows:

Scale effect:

$$\Delta C_Q^{(i-u)} = \frac{C_w^i - C_w^u}{lnC_w^i - lnC_w^u}\ln\left(\frac{C_Q^i}{C_Q^u}\right) \tag{22}$$

Structural effect:

$$\Delta C_S^{(i-u)} = \frac{C_w^i - C_w^u}{lnC_w^i - lnC_w^u}\ln\left(\frac{C_s^i}{C_s^u}\right) \tag{23}$$

Intensity effect:

$$\Delta C_R^{(i-u)} = \frac{C_w^i - C_w^u}{lnC_w^i - lnC_w^u}\ln\left(\frac{C_R^i}{C_R^u}\right) \tag{24}$$

where C_w^i is the water transport trade-embodied carbon emissions of region i during the study period, and C_w^u is the average trade-embodied carbon emissions of the water transport industry across all regions during the study period, i.e., those of benchmark region u; C_Q^u, C_s^u, and C_R^u are the benchmark values for scale effect, structural effect, and intensity effect, respectively; $\Delta C_Q^{(i-u)}$, $\Delta C_S^{(i-u)}$, and $\Delta C_R^{(i-u)}$ are the differences in the contributions of scale, structure, and intensity effects to trade-embodied carbon emissions between region i and the benchmark region u.

2.5. Data Sources and Processing

2.5.1. Data Sources

This study uses the Inter-Country Input–Output (ICIO) Tables from the Organization for Economic Co-operation and Development (OECD) as its primary database. Compared to other mainstream international input–output databases (e.g., EORA, EXIOBASE, WIOD), the OECD ICIO database offers the following advantages: (1) extensive coverage—it comprehensively covers time-series data from 1995 to 2018 for the BRICS countries (Brazil, Russia, India, China, South Africa), meeting the needs of long-term dynamic analysis for emerging economies; (2) standardized industry classification—it is constructed based on the United Nations' International Standard Industrial Classification (ISIC Rev.4), which organizes industries into 45 sectors, ensuring data comparability across countries; (3) traceability of carbon emission data—the accompanying Trade in Embodied CO_2 Database provides industry-specific direct emission factors and detailed energy consumption data, supporting precise calculation of embodied carbon emission.

Therefore, this study is based on the OECD database, selecting 1995–2018 as the primary research period, with the research subject being the water transport sector. The input–output data and import–export trade data are sourced from the OECD ICIO, and the carbon emission coefficient data are obtained from the OECD Trade in Embodied CO_2 Database.

2.5.2. Industry Scope Definition

To ensure data availability and consistency in statistical definitions, this study employs the non-competitive input–output tables from the OECD ICIO, whose industry classification strictly follows ISIC Rev.4. ISIC Rev.4, developed by the United Nations Statistics Division under the Economic and Social Council. It is the internationally recognized standard for classifying economic activities, ensuring consistency and comparability of cross-country data. In the ICIO framework, the classification of economic sectors directly maps to the four-digit codes of ISIC Rev.4.

This study focuses on the embodied carbon emission in water transport trade, with data extracted from the OECD ICIO. The research subject is the D50 water transport sector, which corresponds to ISIC Rev.4's Class 501 (sea and coastal water transport) and Class 502 (inland water transport). This classification covers all water transport activities for both goods and passengers (see Table 1). Through the mapping between ISIC Rev.4 and the ICIO, this study ensures data consistency with international standards and provides a clear industry scope for the analysis.

Table 1. Water transport classification code mapping.

Industry Classification	ICIO Industry Name and Code	ISIC Rev.4 Industry Code
Water transport	D50 Water transport	50 Water transport 501 Sea and coastal water transport 502 Inland water transport

3. Results and Discussion

3.1. *Evolution of the Pattern of Net Trade-Embodied Carbon Flows*

The embodied carbon in export trade was calculated and is shown in Figure 1. China's waterway transport industry continues to lead in total embodied carbon for export trade, increasing from 22.81 Mt in 1995 to 65.74 Mt in 2018, with an average annual growth rate of 4.71 percent. Notably, its growth rate shows phased fluctuations: it grew at an average annual rate of 6.24 percent from 2003 to 2011 due to trade expansion, and then slowed down to −9.35 percent from 2015 to 2018 due to the impact of emission reduction policies. Among the other four countries, Brazil's embodied carbon in export trade in the waterway transport industry showed a "U-shaped" fluctuation: it dropped by 44.3 percent from 1995 to 2001, and then rebounded to 6.11 Mt (an increase of 120 percent) from 2001 to 2007. India's embodied carbon in export trade in the waterway transport industry has been growing steadily, from 0.7 Mt in 1995 to 3.27 Mt in 2018, with the growth rate of the waterway transport industry in the past two years continuing to rise. South Africa's total embodied carbon in export trade in the waterway transport sector was the lowest, fluctuating gently, increasing from 1.01 Mt in 1995 to 1.21 Mt in 2018. Notably, during the global financial crisis (2007–2010), embodied carbon in export trade in China's waterway transport sector declined by 11.39 percent, while Russia led the trend of growth, and India maintained positive growth, reflecting the resilience of its trade.

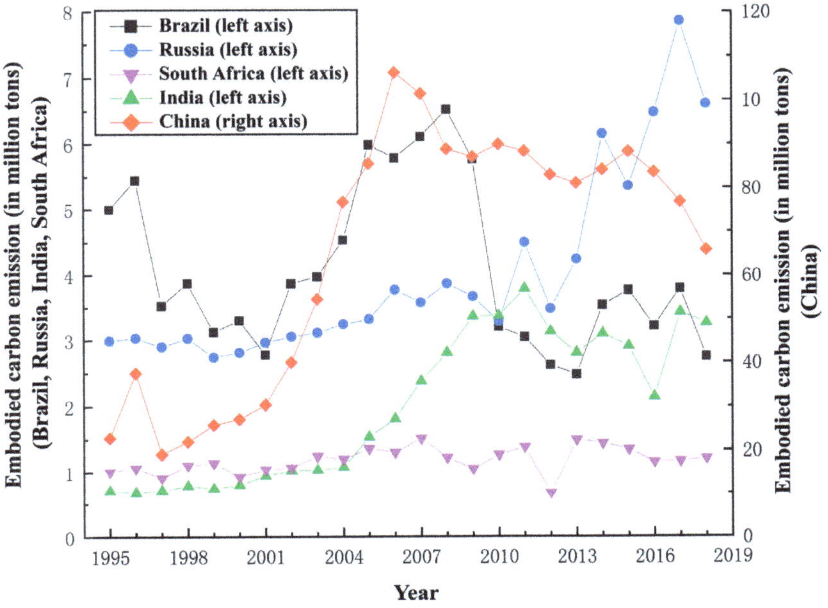

Figure 1. Trends in embodied carbon in BRICS export trade in the waterway transport sector.

The embodied carbon in import trade was calculated and is shown in Figure 2. India's embodied carbon emissions in water transport import trade have experienced the highest growth rate, surging from 2.15 Mt in 1995 to 35.27 Mt in 2018, with an average annual growth rate of 12.93%. Notably, this growth accelerated after 2010, with an average annual increase of 3.77% from 2010 to 2018. China ranks second in terms of growth rate, with the highest total embodied carbon emissions in water transport import trade, reaching 30.59 Mt in 2018. However, its growth rate is slower than India's, with an average annual increase of 10.85%. In contrast, Brazil, Russia, and South Africa showed a downward trend. Brazil's embodied carbon emissions in water transport import trade decreased from 5.94 Mt in 1995 to 1.96 Mt in 2018, reflecting an average annual decline of 4.68%. Russia's emissions rose from 2.89 Mt to 5.08 Mt, though this was influenced by a peak of 10.67 Mt in 2017. South Africa saw its emissions drop from 10.05 Mt to 3.10 Mt, with an average annual decrease of 4.98%, indicating a significant reduction.

Looking at the data in phases, the global financial crisis of 2008 triggered a pullback in import carbon emissions across several countries. For instance, China's import carbon emissions fell from 26.51 Mt in 2008 to 25.43 Mt in 2010, while Brazil's emissions dropped from 3.59 Mt to 2.22 Mt in the same period. In contrast, India's emissions increased from 15.61 Mt to 26.24 Mt, reflecting a rebound of 68%. After 2015, South Africa experienced a sharp decline in import carbon emissions, from 9.90 Mt to 3.10 Mt, aligning with its low-carbon port renovation policies.

The embodied carbon in net export trade was calculated and is shown in Figure 3. China has consistently been a net exporter, with its net transfer increasing from 19.95 Mt in 1995 to a peak of 79.91 Mt in 2006, before declining to 35.15 Mt in 2018, a decrease of 56.01%. This decline supports the assertion that "emission reduction policies have begun to show results." India's net import gap has continued to expand, from −1.45 Mt in 1995 to −32.00 Mt in 2018, with an average annual growth rate of 11.3%. Since 2000, Brazil has shifted from a net importer (−0.19 Mt) to a net exporter (0.79 Mt in 2018), though with significant fluctuations, peaking at 2.93 Mt in 2008. Russia, after a role reversal in 2009, remained a net importer from 2010 to 2017, but in 2018 became a net exporter of 1.53 Mt, reflecting its shift in energy export structure. South Africa's net import gap has gradually narrowed from −9.04 Mt in 1995 to −1.89 Mt in 2018, due to reductions in its import scale and improvements in energy efficiency.

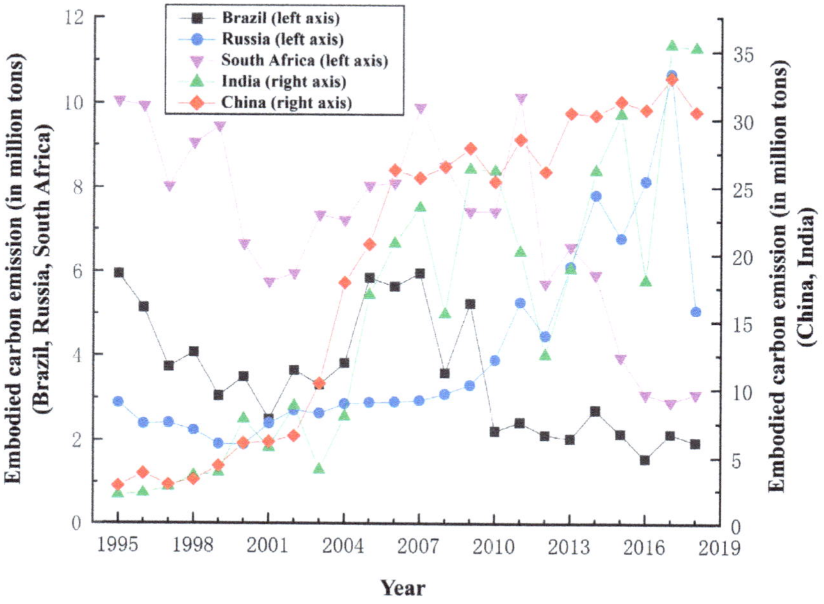

Figure 2. Trends in embodied carbon in BRICS import trade in the waterway transport sector.

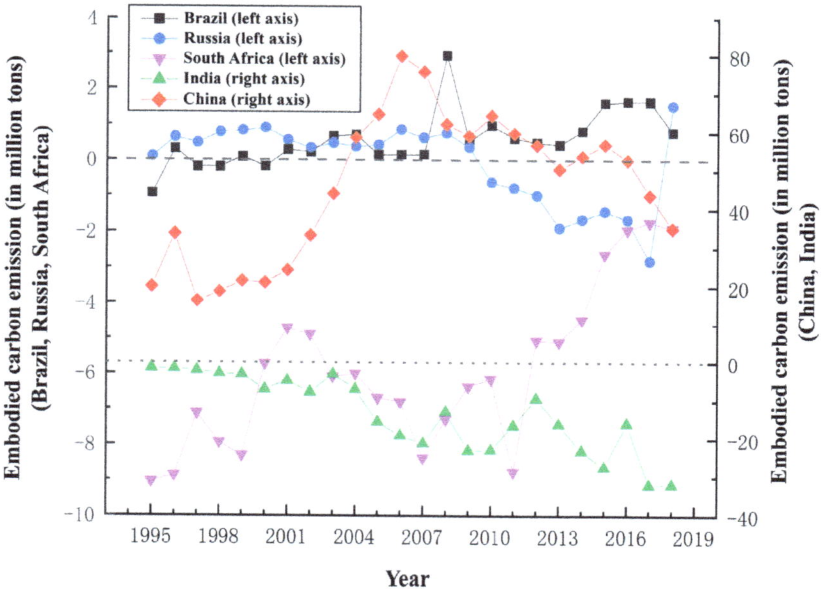

Figure 3. Trends in embodied carbon in BRICS net export trade in the waterway transport sector.

3.2. Decomposition of Driving Effects in Temporal Dimension

Leveraging the temporal LMDI model, this study decomposes the driving factors of embodied carbon emissions within the export and import trade of the water transport industry across BRICS nations from 1995 to 2018, with a particular focus on the temporal heterogeneity characteristics of scale, structure, and intensity effects. The findings reveal significant national disparities and stage-wise fluctuations in both export and import trade effects (Table 2: Export trade, Table 3: Import trade).

Table 2. Decomposition of driving factors of embodied carbon in export trade of waterway transportation industry.

Country	Effect	Contribution Value (Mt)			
		1995~2003	2003~2011	2011~2018	1995~2018
Brazil	Scale effect	1.90	4.46	−0.33	5.98
	Structural effect	−4.23	−0.90	0.02	−4.51
	Intensity effect	1.30	−4.49	0.02	−3.72
	Total effect	−1.03	−0.92	−0.30	−2.25
Russia	Scale effect	1.60	5.05	−0.68	7.95
	Structural effect	−0.06	−0.78	1.65	0.33
	Intensity effect	−1.42	−2.89	1.14	−4.67
	Total effect	0.12	1.38	2.11	3.62
India	Scale effect	0.67	3.40	0.76	4.33
	Structural effect	0.18	0.28	−0.82	0.19
	Intensity effect	−0.52	−0.92	−0.46	−1.95
	Total effect	0.33	2.75	−0.52	2.56
China	Scale effect	43.60	100.75	23.93	119.72
	Structural effect	−4.51	12.11	−37.14	−17.72
	Intensity effect	−7.50	−78.92	−9.39	−59.08
	Total effect	31.59	33.94	−22.60	42.93
South Africa	Scale effect	0.35	1.25	−0.19	1.23
	Structural effect	0.40	0.01	0.07	0.45
	Intensity effect	−0.51	−1.11	−0.04	−1.48
	Total effect	0.24	0.13	−0.17	0.20

Table 3. Decomposition of driving factors of embodied carbon in import trade of waterway transportation industry.

Country	Effect	Contribution Value (Mt)			
		1995~2003	2003~2011	2011~2018	1995~2018
Brazil	Scale effect	0.26	4.28	−0.44	4.88
	Structural effect	−1.24	0.82	0.16	0.29
	Intensity effect	−1.66	−5.97	−0.19	−9.15
	Total effect	−2.63	−0.87	−0.48	−3.98
Russia	Scale effect	0.62	5.33	−0.94	5.60
	Structural effect	0.70	0.66	0.33	1.90
	Intensity effect	−1.57	−3.36	0.42	−5.31
	Total effect	−0.25	2.63	−0.19	2.19
India	Scale effect	2.20	18.42	4.28	32.06
	Structural effect	1.77	2.71	8.25	13.73
	Intensity effect	−2.02	−5.01	2.52	−12.67
	Total effect	1.95	16.11	15.05	33.12

Table 3. *Cont.*

Country	Effect	Contribution Value (Mt)			
		1995~2003	2003~2011	2011~2018	1995~2018
China	Scale effect	6.70	25.65	10.80	34.36
	Structural effect	1.27	16.82	1.13	13.93
	Intensity effect	−0.41	−24.34	−9.89	−20.56
	Total effect	7.56	18.12	2.04	27.73
South Africa	Scale effect	2.08	9.10	−0.95	6.68
	Structural effect	−0.10	−0.57	−6.42	−6.84
	Intensity effect	−4.70	−5.71	0.32	−6.80
	Total effect	−2.72	2.81	−7.05	−6.95

The scale effect is the core driver of carbon emission growth, but the paths differ significantly across countries. Specifically, in terms of exports, China's export scale effect contribution is the highest, reaching 119.72 Mt from 1995 to 2018, with a total contribution rate of 278.91%. Notably, from 2003 to 2011, the contribution rate surged to 296.87%, primarily due to the sharp increase in port throughput driving bulk commodity exports. India's export scale effect contribution is relatively low, at just 4.33 Mt, but from 2011 to 2018, its contribution rate turned negative, dropping by −144.86%, reflecting stagnation in water transport exports, likely due to bottlenecks in port infrastructure. South Africa's scale effect showed significant volatility, with a contribution rate of 929.12% from 2003 to 2011, mainly driven by a short-term surge in coal exports. However, from 2011 onwards, the contribution turned negative, possibly due to fluctuations in international energy prices leading to a contraction in exports. Russia's export scale effect remained overall positive, contributing 7.95 Mt, but from 2011 to 2018, it turned negative (−0.68 Mt). Brazil's export scale effect was relatively stable, with a contribution of 5.98 Mt, experiencing a slight decline only from 2011 to 2018.

In terms of imports, scale expansion is particularly evident in developing countries. India's import scale effect contributed the most, with 32.06 Mt from 1995 to 2018, accounting for 96.79% of its total embodied carbon emissions. However, after 2010, the growth rate slowed to 28.46%, reflecting the gradual substitution of imports by domestic manufacturing. China's import scale effect exhibited a phased characteristic: from 2003 to 2011, the contribution rate was 141.52%, corresponding to the deep integration of global supply chains; after 2011, it surged to 528.28%, indicating the carbon leakage risks associated with domestic consumption upgrades.

Structural effects exhibit regional heterogeneity. China's structural effect contribution remains positive, indicating its impact on embodied carbon emissions is above the average, promoting carbon emissions. Brazil's contribution value sharply dropped to −0.18 Mt in 2011, but it remained positive in other years with a decreasing trend, suggesting that while its impact was above the average, it was gradually converging towards the mean. The structural effect contributions of the other three countries were all negative, indicating their structural effects were below the average, thus inhibiting export-related embodied carbon emissions. Overall, before 2011, the structural effect differences between countries expanded, but in recent years, due to industrial adjustments and optimization, these contributions have decreased, narrowing the spatial distribution gap between countries.

The import structural optimization effect is significantly influenced by policy tools. South Africa's import structural effect contribution rate was 91.01% (contribution value of −6.42 Mt) from 2011 to 2018, closely linked to its implementation of a green tariff policy. Brazil's import structural effect was unusually negative from 2003 to 2011 (−93.62%), primarily due to the substitution of iron ore imports with local mining. India's import structural effect contributed 54.79% (contribution value of 8.25 Mt) from 2011 to 2018, reflecting its expanding industry's increasing dependence on high-carbon intermediate goods.

Intensity effects generally play a role in emission reduction. In terms of exports, intensity effects reflect the differences in technological advancements, with China showing the most notable improvement. From 1995 to 2018, China's intensity effect contributed −59.08 Mt, with a contribution rate of −137.62%. Particularly from 2003 to 2011, the contribution rate reached −232.55%, confirming that China's energy efficiency revolution in the water transport sector had begun to show results. India's intensity effect was overall negative, with a cumulative contribution rate of −76.19%, but from 2011 to 2018, it turned positive, with a contribution rate of 87.23%, indicating that the technological upgrades in India's water transport industry lagged behind its economic growth. Russia's intensity effect exhibited fluctuations, being positive from 2011 to 2018, with a contribution rate of 54.09%, and negative in other years.

For imports, the intensity effect reveals the asymmetry of technological spillovers. China's import intensity effect from 2011 to 2018 had a contribution rate of −483.46%, the highest among the five countries. This could be attributed to the adoption of ultra-large container ships (ULCVs), which reduced carbon emissions per unit of freight, reflecting China's strong ability to absorb low-carbon technologies. South Africa, in contrast, had a contribution rate of −4.51% during the same period, reflecting limited improvements in technology. Brazil's import intensity effect was unusually positive in 2003–2011 (685.32%), which may be linked to the rebound effect due to the failure of biofuel technology promotion.

In terms of total effects, from 1995 to 2018, China's export total effect was 42.93 Mt, ranking the highest among the five countries, with the scale effect contributing 278.91%. India had the fastest growth, with a total effect of 2.56 Mt, and the structural effect contributed 41.45%. Russia's total effect increased from 0.12 Mt from 1995 to 2003 to 2.11 Mt from 2011 to 2018, driven by energy strategy adjustments.

For import total effects, from 1995 to 2018, India had the largest total effect, contributing 33.12 Mt, predominantly driven by the scale effect (96.79%). China's import total effect was 27.73 Mt, far exceeding other countries, with technological improvements (intensity effect) offsetting some of the growth (−74.15%). South Africa's import total effect worsened from −2.72 Mt (1995–2003) to −6.95 Mt (2011–2018), primarily driven by structural effects (contribution rate of 98.36%).

3.3. Characteristics of the Distribution in Spatial Dimensions

This study employs the spatial LMDI model, using the average embodied carbon emissions in the import/export trade of the water transport industry in the BRICS countries from 1995 to 2018 as a benchmark. The model reveals the spatial heterogeneity characteristics of the scale, structure, intensity, and total effects. By comparing the differences between each influencing factor and the average level, the model facilitates direct comparisons between regions. A positive decomposition result indicates that the factor's impact on carbon emissions is above the average, contributing to an increase in emissions. Conversely, a negative result suggests that the factor's impact is below the average, leading to a suppression of carbon emissions.

3.3.1. Spatial Dimension Analysis of Embodied Carbon in Export Trade

The spatial differentiation characteristics of export-related embodied carbon in the water transport trade of BRICS countries exhibit distinct country-specific differences (see Figures 4 and 5).

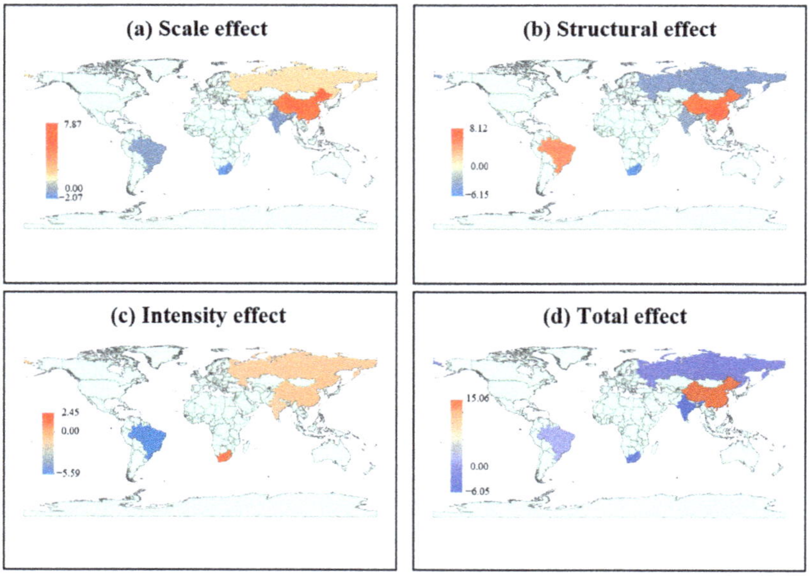

Figure 4. Spatial LMDI decomposition of the embodied carbon in the export trade of the water transport industry of the BRICS countries in 1995.

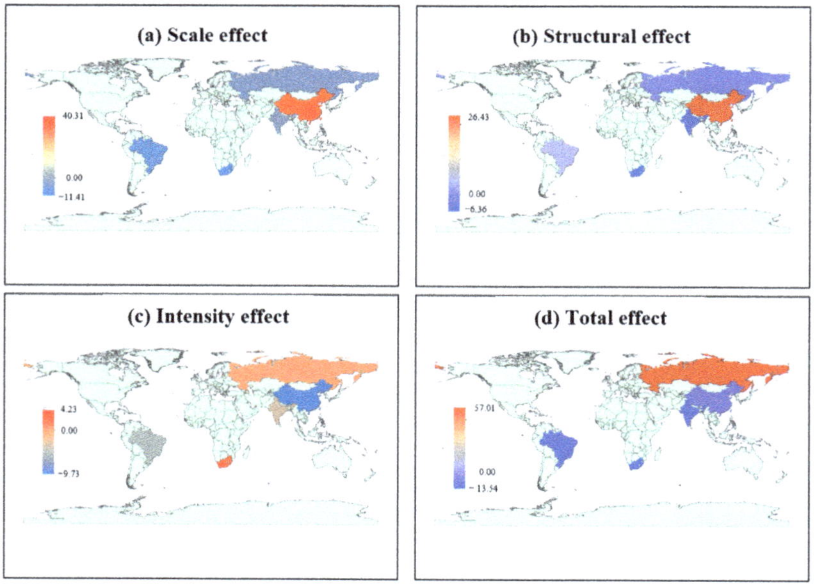

Figure 5. Spatial LMDI decomposition of the embodied carbon in the export trade of the water transport industry of the BRICS countries in 2018.

In terms of scale effects, the export scale effects of the BRICS countries significantly influence the spatial differences in embodied carbon emissions. China's export scale effect contribution is the highest, increasing from 7.87 Mt in 1995 to 40.31 Mt in 2018, and peaking at 46.92 Mt in 2011, far exceeding the average level of the five countries. This indicates that the expansion of China's trade volume plays a dominant role in the growth of embodied carbon emissions. The contributions of the other four countries are negative, indicating that their export scale effects are below the average level and have a suppressive effect on carbon emissions. Since 2003, due to international trade friction, China's export growth has slowed, while the total trade volume of the other four countries has increased, reducing the spatial differences in scale effects.

Structural effects show obvious regional heterogeneity. China's structural effect contribution remains positive, indicating that its structural effect has a greater impact than the average level, promoting carbon emissions. Brazil's contribution sharply dropped to −0.18 Mt in 2011, but remained positive in other years, showing a decreasing trend. This suggests that Brazil's structural effect, though above the average level, is gradually converging to the mean. The contributions of the other three countries are all negative, indicating that their structural effects are below the average, thus suppressing carbon emissions. Overall, before 2011, the structural effect differences between the countries expanded. However, in recent years, due to industrial adjustments and optimization, the contribution rates have declined, narrowing the spatial distribution gap between the countries.

The intensity effects reflect technological differences. South Africa's spatial intensity effect contribution has always been positive, with a contribution of 4.23 Mt in 2018, indicating that its energy efficiency level is above the average of the five countries. The intensity effect contributions of Brazil, Russia, and India are all negative, suggesting that their energy efficiency improvements and environmental policies are less effective than the regional average. Before 2003, China's intensity effect was below the average, but its contribution value of 1.93 Mt in 2003 exceeded the average level, reflecting that during this period, China's water transport trade expanded at the cost of the environment. After 2011, China's intensity effect turned negative, indicating that technological progress in China's water transport industry effectively suppressed spatial embodied carbon emissions.

In terms of total effects, from 1995 to 2018, China's export total effect was always positive and far exceeded the average of the five countries, with a contribution of 42.93 Mt. Russia and India had contribution values close to the mean, indicating the "high in the East, low in the West" spatial distribution of embodied carbon emissions in BRICS exports. From the perspective of changes, the export embodied carbon is gradually shifting towards Southeast Asia, and countries are narrowing the spatial distribution gap with average levels.

3.3.2. The Spatial Dimension Analysis of Embodied Carbon in Import Trade

The spatial differentiation characteristics of embodied carbon in the import trade of the BRICS countries exhibit significant asymmetry compared to export trade, with the driving mechanisms of the scale, structure, and intensity effects showing distinct country-specific heterogeneity (see Figures 6 and 7).

In terms of import scale effects, the spatial distribution of the countries is similar to that of export trade. China's import scale effect contribution steadily increased from 2.09 Mt in 1995 to 25.05 Mt in 2018; although its growth rate was slower than exports, it still shows significant growth. The import scale effect contributions of the other four countries remained negative, suppressing the spatial emissions of embodied carbon in water transport trade. Over time, the gap between each country's scale effect and the average level has widened, indicating that the spatial distribution differences in the embodied carbon emissions from import trade in water transport have increased.

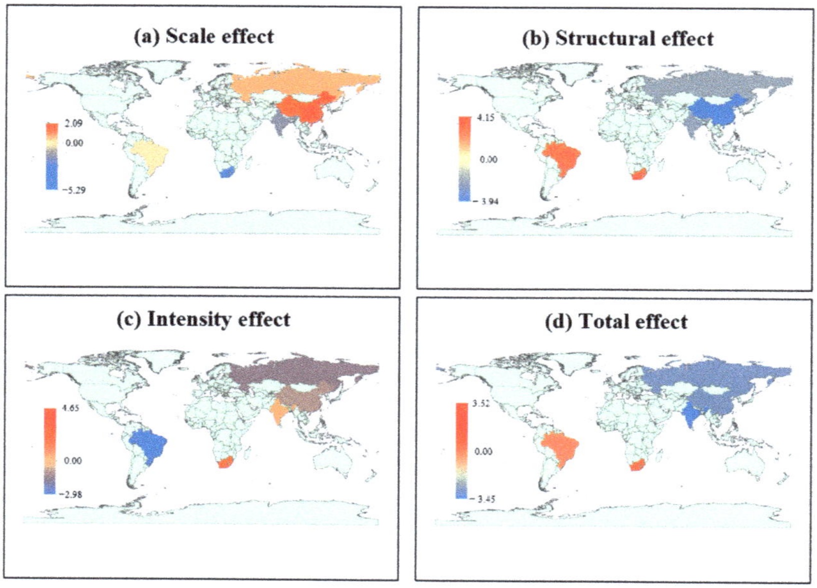

Figure 6. Spatial LMDI decomposition of the embodied carbon in import trade of the water transport industry of BRICS countries in 1995.

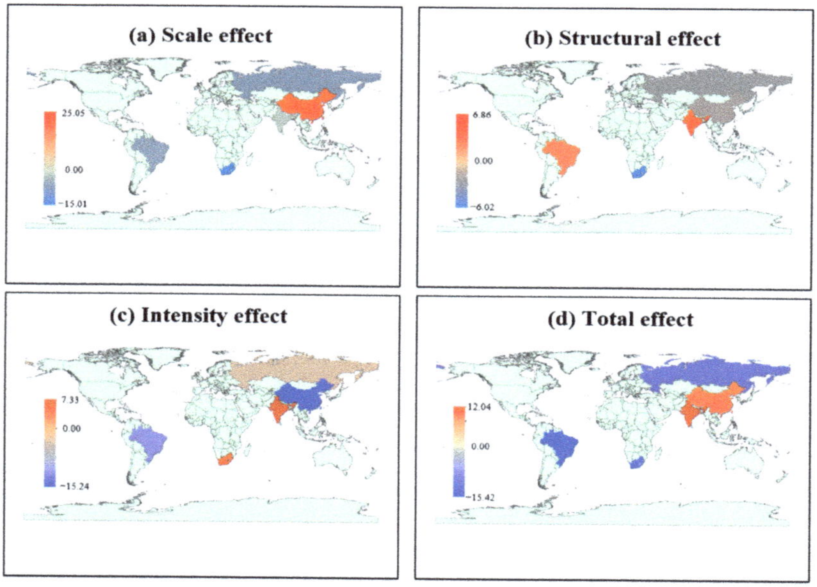

Figure 7. Spatial LMDI decomposition of the embodied carbon in import trade of the water transport industry of BRICS countries in 2018.

Import structural effects exhibited significant fluctuations. Brazil's import structural effect contribution decreased from 2.97 Mt in 1995 to 2.70 Mt in 2018, but it remained above the average level, indicating that Brazil's water transport industry's structural effect promoted carbon emissions. South Africa's contribution turned negative in 2018 (−6.02 Mt), and this value was much lower than the average level of the five countries, further widening the gap with the mean. India, in contrast, had a positive contribution in 2018 (6.86 Mt), indicating that its import structural effect promoted carbon emissions. China and Russia's import structural effects remained negative, and their gap with the average level gradually shrank. Meanwhile, Brazil, India, and South Africa showed increasing gaps, demonstrating that different import trade structures either promoted or suppressed the embodied carbon emissions in water transport trade.

In terms of import intensity effects, South Africa's contributions were high and consistently positive. Brazil and Russia's contributions were negative, suggesting their impact was lower than the average level. Both India and China experienced sudden changes in their intensity effect contribution rates in 2003. However, the trends were opposite: China shifted from promoting carbon emissions growth to suppressing it, indicating that technological advancements in China's import trade and increased environmental awareness effectively suppressed spatial embodied carbon emissions.

Throughout the study period, the total effects of each country showed a "dual-core drive" pattern. Brazil, Russia, and South Africa had total effects below the average level of the five countries, while China and India had total effects above the average level, promoting carbon emission growth. India's growth continued to rise, while China's growth showed a slight decline. In terms of spatial distribution changes, the embodied carbon from import trade is gradually shifting towards Southeast Asia, with all countries, except India, striving to reduce embodied carbon emissions from water transport export trade.

3.4. Discussion on How to Manage Carbon Emissions

This study employs a multi-regional input–output framework combined with the LMDI decomposition method to comprehensively analyze embodied carbon emissions in water transport trade among BRICS countries from 1995 to 2018. Our findings reveal significant spatiotemporal heterogeneity in the driving factors across the five BRICS nations. In export trade, China consistently acts as the primary net exporter, with its embodied carbon emissions peaking prior to the implementation of effective emission reduction policies; in contrast, India's import embodied carbon has surged rapidly, reflecting an increasing reliance on high-carbon intermediate products. Meanwhile, Brazil, Russia, and South Africa exhibit substantial fluctuations, indicating diverse stages of industrial restructuring and policy intervention. These results suggest that, particularly for China and India, scale effects continue to drive carbon emission growth, while improvements in energy efficiency and the adoption of clean technologies significantly mitigate these emissions.

From a practical standpoint, precise quantification of embodied carbon emissions is essential for designing effective emission reduction policies. With a clear understanding of each country's carbon contribution, targeted measures—such as adopting low-carbon fuels, upgrading shipping technology, and implementing regional carbon pricing—should be pursued. As Agarwala [20] notes, hydrogen propulsion systems may offer a promising decarbonization pathway for maritime shipping. Overall, our study provides a solid empirical and theoretical basis for managing global maritime carbon emissions and formulating coordinated reduction strategies.

4. Summaries and Conclusions

This study employs a multi-regional input–output model combined with the LMDI decomposition method to quantify embodied carbon emissions in water transport trade among BRICS countries from 1995 to 2018. By decomposing the effects into scale, structure, and intensity components, our analysis reveals distinct spatiotemporal dynamics. China consistently leads in export trade, with its net export embodied carbon peaking in 2006 before declining markedly due to effective emission reduction policies. In contrast, India's import trade experiences a rapid scale effect that intensifies carbon emissions, while Brazil,

Russia, and South Africa display variable transitions between net import and export statuses. Temporal analysis shows that scale effects are the primary drivers of carbon growth, although technological progress significantly offsets this in China. Spatially, embodied carbon emissions follow a "high in the East, low in the West" pattern. Overall, these findings underscore that achieving a sustainable, low-carbon transformation in the water transport sector is imperative. We assert that targeted, region-specific policies and coordinated decarbonization strategies are essential for mitigating carbon leakage and fostering global climate action.

Author Contributions: Methodology, S.Z.; Data curation, P.Q.; Writing—original draft, S.Z.; Writing—review & editing, C.C.; Supervision, S.Z. All authors have read and agreed to the published version of the manuscript.

Funding: This research was funded by China's Fujian Provincial Department of Education Social Sciences Research Project for Young and Middle-Aged Faculty (General Program): Research on the Evaluation of Green Development Level and Improvement Path in Fujian Province (Grant number: JAS24101).

Data Availability Statement: The original contributions presented in this study are included in the article. Further inquiries can be directed to the corresponding author.

Conflicts of Interest: The authors declare no conflict of interest.

References

1. Kondratenko, A.A.; Zhang, M.Y.; Tavakoli, S.; Altarriba, E.; Hirdaris, S. Existing technologies and scientific advancements to decarbonize shipping by retrofitting. *Renew. Sustain. Energy Rev.* **2025**, *212*, 115430. [CrossRef]
2. Premsamarn, P.; Win, T.K.; Watanabe, D. Assessing the impact of emission control areas policy on ship emissions in the Gulf of Thailand using AIS data. *Marit. Technol. Res.* **2025**, *7*, 275311. [CrossRef]
3. Sha, M.; Zhai, J.Y.; Wan, Z.; Wu, Z.D.; Jin, Y.; Zhu, E.Y. Insights into the global characteristics of shipping exhaust emissions at berth. *J. Mar. Sci. Eng.* **2024**, *12*, 1527. [CrossRef]
4. da Silva, J.N.R.; Santos, T.A.; Teixeira, A.P. Methodology for predicting maritime traffic ship emissions using automatic identification system data. *J. Mar. Sci. Eng.* **2024**, *12*, 320. [CrossRef]
5. Yeh, C.K.; Tzu, F.M.; Chen, P.Y.; Shen, H.C.; Yuan, C.S.; Lin, C.; Pu, H.P.; Ngo, H.H.; Bui, X.T. Emission characteristics of naphthalene from ship exhausts under global sulfur cap. *Sci. Total Environ.* **2023**, *902*, 166172. [CrossRef] [PubMed]
6. Zheng, S.S.; Chen, C.; Xie, S.K. An LMDI-Based analysis of carbon emission changes in China's waterway transportation sector. *Sustainability* **2025**, *17*, 325. [CrossRef]
7. Jaramillo, J.; Gutierrez, J.; Macia, Y.M. Comparative analysis of carbon intensity indicators applicable to harbor tugboats. *Sustainability* **2025**, *17*, 1706. [CrossRef]
8. Raftis, C.C.; Vanelslander, T.; van Hassel, E. A global analysis of emissions, decarbonization, and alternative fuels in inland navigation—A systematic literature review. *Sustainability* **2023**, *15*, 14173. [CrossRef]
9. Billington, J.A.; Yang, B.; Sobieralski, W.; Fehrenbacher, C.; Hoch, G.M.; Damiano, A.L.; Nadafianshahamabadi, R.; Mitchell, A.R.; Quiros, D.C. Climate and emissions benefits of renewable diesel: In-use testing of a tier 2 commercial harbor craft. *Emiss. Control Sci. Technol.* **2025**, *11*, 2. [CrossRef]
10. Hu, Y.J.; Wang, B.L.; Dong, X.Y. A burden-sharing model shaping the embodied carbon emission and considering regions' efforts to reduce emissions in China's power sector. *J. Environ. Manag.* **2025**, *373*, 123440. [CrossRef] [PubMed]
11. Hosseinzadeh-Bandbafha, H.; Tabatabaei, M.; Aghbashlo, M.; Khanali, M.; Khalife, E.; Shojaei, T.R.; Mohammadi, P. Data supporting consolidating emission indices of a diesel engine powered by carbon nanoparticle-doped diesel/biodiesel emulsion fuels using life cycle assessment framework. *Data Brief* **2020**, *30*, 105428. [CrossRef] [PubMed]
12. Huang, T.Y.; Ding, L.P.; Yu, Y.Q.; Huang, L.; Yang, L.Y. From AR5 to AR6: Exploring research advancement in climate change based on scientific evidence from IPCC WGI reports. *Scientometrics* **2023**, *128*, 5227–5245. [CrossRef]
13. Yu, Y.; Li, S.; Sun, H.; Taghizadeh-Hesary, F. Energy carbon emission reduction of China's transportation sector: An input–output approach. *Econ. Anal. Policy* **2021**, *69*, 378–393. [CrossRef]
14. Lee, G.; Kim, H.; Min, K.; Hwang, T.; Kim, E.; Lee, J.; Kang, D. Carbon emission reduction strategies in urban water sectors: A case study in Incheon Metropolitan city, South Korea. *Sustainability* **2025**, *17*, 1887. [CrossRef]

15. Al-Ayouty, I. Towards environmental sustainability: An input–output analysis to measure industry-level carbon dioxide emissions in Egypt. *Sustainability* **2025**, *17*, 1035. [CrossRef]
16. Al-Abdulqader, K.S.; Ibrahim, A.J.; Ong, J.; Khalifa, A.A. Does carbon pricing matter? Evidence from a global sample. *Energies* **2025**, *18*, 1030. [CrossRef]
17. Asif, M.; Li, J.Q.; Zia, M.A.; Hashim, M.; Bhatti, U.A.; Bhatti, M.A.; Hasnain, A. Environmental sustainability in BRICS economies: The nexus of technology innovation, economic growth, financial development, and renewable energy consumption. *Sustainability* **2024**, *16*, 6934. [CrossRef]
18. Lin, B.; Sun, C. Evaluating carbon dioxide emissions in international trade of China. *Energy Policy* **2010**, *38*, 613–621. [CrossRef]
19. Grossman, G.M.; Krueger, A.B. Economic growth and the environment. *Q. J. Econ.* **1995**, *110*, 353–377. [CrossRef]
20. Agarwala, N. Is hydrogen a decarbonizing fuel for maritime shipping? *Marit. Technol. Res.* **2024**, *6*, 271244. [CrossRef]

MDPI AG
Grosspeteranlage 5
4052 Basel
Switzerland
Tel.: +41 61 683 77 34

Water Editorial Office
E-mail: water@mdpi.com
www.mdpi.com/journal/water

www.ingramcontent.com/pod-product-compliance
Lightning Source LLC
LaVergne TN
LVHW070451061126
851858LV00011B/210

* 9 7 8 3 7 2 5 8 8 2 9 5 3 *